Applied Introductory Circuit Analysis for Electrical and Computer Engineers

Michael Reed
University of Virginia

Ron Rohrer
Intersouth Partners

Prentice Hall
Upper Saddle River, New Jersey 07458

Library of Congress Cataloging-in-Publication Data

Reed, Michael
    Applied introductory circuit analysis for electrical and
computer engineers / Michael Reed, Ron Rohrer.
        p.   cm.
    Includes bibliographical references and index.
    ISBN 0-13-787631-9
    1. Electric circuit analysis. I. Rohrer, Ron II. Title.
    TK454.R44   1999
    621.319dc21   98-40881
                                                CIP

Publisher: *Tom Robbins*
Production editor: *Edward DeFelippis*
Editor-in-chief: *Marcia Horton*
Managing editor: *Eileen Clark*
Assistant vice president of production and manufacturing: *David W. Riccardi*
Art director: *Jayne Conte*
Cover designer: *Bruce Kenselaar*
Manufacturing buyer: *Pat Brown*
Editorial assistant: *Dan DePasquale*

The author and publisher of this book have used their best efforts in preparing this book. These efforts include the development, research, and testing of the theories and programs to determine their effectiveness. The author and publisher make no warranty of any kind, expressed or implied, with regard to these programs or the documentation contained in this book. The author and publisher shall not be liable in any event for incidental or consequential damages in connection with, or arising out of, the furnishing, performance, or use of these programs.

Printed in the United States of America

10 9 8 7 6 5 4 3 2 1

ISBN 0-13-787631-9

Prentice-Hall International (UK) Limited, *London*
Prentice-Hall of Australia Pty. Limited, *Sydney*
Prentice-Hall Canada Inc., *Toronto*
Prentice-Hall Hispanoamericana, S.A., *Mexico*
Prentice-Hall of India Private Limited, *New Delhi*
Prentice-Hall of Japan, Inc., *Tokyo*
Simon & Schuster Asia Pte. Ltd., *Singapore*
Editora Prentice-Hall do Brasil, Ltda., *Rio de Janeiro*

Contents

PREFACE   ***ix***

    Acknowledgments   x

INTRODUCTION   ***1***

    Microelectronics   1

    About this Book   2

1. FUNDAMENTAL ELECTRICAL CONCEPTS   ***3***

    1.1   Introduction   3

    1.2   Conventions   3

        1.2.1   Understood Units, 3
        1.2.2   Engineering Notation, 4
        1.2.3   Functions of Time, 5

    1.3   Charge, Current, and Voltage   5

        1.3.1   Charge, 5
        1.3.2   Current, 6
        1.3.3   Voltage, 8
        1.3.4   Hydraulic Analogy, 9

    1.4   Power   10

        1.4.1   Associated Reference Directions, 10
        1.4.2   Unassociated Reference Directions, 11

    1.5   Circuits, Nodes, and Branches   12

    1.6   Branch and Node Voltages   14

    1.7   Kirchhoff's Voltage and Current Laws   16

        1.7.1   Kirchhoff's Voltage Law, 16
        1.7.2   Kirchhoff's Current Law, 19

    1.8   Circuit Elements   19

        1.8.1   Voltage Source, 19
        1.8.2   Current Source, 20
        1.8.3   Resistance, 21
        1.8.4   Capacitor, 22

        1.8.5   *Inductance, 25*
        1.8.6   *Switch, 27*

1.9   Combining Circuit Elements   28

        1.9.1   *Elements in Series, 29*
        1.9.2   *Elements in Parallel, 33*
        1.9.3   *Forbidden Combinations, 38*

1.10   Voltage- and Current-Divider Circuits   40

1.11   Resistive-Circuit Examples   42

1.12   Power and Energy Relationships   45

        1.12.1   *Power/Energy Relationships for Resistors, 45*
        1.12.2   *Power/Energy Relationships for Capacitors, 46*
        1.12.3   *Continuity of Capacitor Voltage, 46*
        1.12.4   *Power/Energy Relationships for Inductors, 48*
        1.12.5   *Continuity of Inductor Current, 48*

1.13   Summary   49

2.   GATE DELAY AND RC CIRCUITS   **61**

2.1   Introduction: Delays in Logic Circuits   61

2.2   Transition Times in CMOS   63

2.3   Inside the CMOS Inverter   64

        2.3.1   *Interconnected CMOS Inverters, 68*

2.4   Solving First-Order *RC* Circuits   72

        2.4.1   *The General Case, 72*
        2.4.2   *Solving the Differential Equation, 74*

2.5   *RC* Delays in Integrated Circuits   76

        2.5.1   *Inverter Pull-Down, 76*
        2.5.2   *Inverter Pull-Up, 78*

2.6   Significance of the Time Constant τ 80

2.7   Maximum Inverter-Pair Switching Speed   83

        2.7.1   *Case I: Inverter Pair Initially at Rest, 84*
        2.7.2   *Case II: Inverter Pair Switching Continuously, 87*

2.8   Algebraic Analysis of Inverter-Pair Switching Speed   91

        2.8.1   *Increasing Switching Speed, 93*

2.9   Energy and Power Dissipation in Digital Systems   94

        2.9.1   *Energy Dissipation, 94*
        2.9.2   *Power Dissipation by the Computer, 98*

2.10   Other First-Order *RC* Circuits   98

2.11   Summary   101

3.   INTERCONNECTS AND RC LADDER CIRCUITS   **115**

3.1   Introduction   115

3.2   Resistance and Capacitance of Interconnects   116

3.3   Interconnect Models   119

3.4   Single-*RC*-Lump Approximation of an Interconnect   122

    3.4.1   Maximum Interconnect Length, 127
    3.4.2   Interconnect Scaling, 129

3.5   Two-*RC*-Lump Interconnect Approximation   130

3.6   Analysis of the Two-Section *RC* Ladder Circuit   133

    3.6.1   Derivation of the Differential Equations, 134
    3.6.2   Determining the dc Steady-State Response, 136
    3.6.3   The Initial Conditions, 137
    3.6.4   Determining the Transient Solution, 138

3.7   Natural Frequencies and Higher-Order Circuits   153

3.8   Timing Delays Using the Two-Lump Model   156

3.9   Timing Delays Using Higher-Order Interconnect Models   159

3.10 Summary   159

4.   FANOUT AND CAPACITIVE COUPLING   **171**

4.1   Introduction   171

4.2   Fanout   171

4.3   Fanout and Interconnects   175

    4.3.1   One-Lump Model, 175
    4.3.2   Two-Lump model, 175
    4.3.3   Multiple Interconnects, 175

4.4   Capacitive Coupling and Crosstalk   179

    4.4.1   The Pi-Connection Model, 180
    4.4.2   Adjacent Interconnect Termination, 182

4.5   Capacitive Coupling to a Grounded Adjacent Line   182

4.6   Capacitive Coupling to a Floating Adjacent Line   185

4.7   Capacitive Coupling to an Adjacent Active Line   189

4.8   The Capacitance Matrix   196

4.9   Summary   199

5. PACKAGE INDUCTANCE AND RLC CIRCUIT ANALYSIS   **213**

5.1   Introduction   213

5.2   Modelling the Effects of Package Inductance   214

5.3   First-Order *RL* Circuits   215

5.4   *RLC* Circuit Model of Coupled Inverter Gates   220

5.5   dc Steady-State Response of *RLC* Circuits   221

5.6   Series *RLC* Circuit Differential Equations   222

5.7   Natural Frequencies of the Series *RLC* Circuit   224

5.8   Series *RLC* Circuit Responses   226

      5.8.1   *The Overdamped Response, 226*
      5.8.2   *The Critically Damped Response, 231*
      5.8.3   *The Underdamped Response, 239*
      5.8.4   *The Undamped Response, 245*

5.9   Application to Digital-System Switching Speed   251

5.10   Gate Conductance and *RLGC* Circuits   255

      5.10.1   *Gate Conductance, 256*
      5.10.2   *Analysis of the RLGC Circuit, 257*

5.11   Neglecting Unimportant Components in Circuit Analysis   260

      5.11.1   *Current Discontinuity in an RC Circuit, 260*
      5.11.2   *Natural Frequencies in a Highly Overdamped RLC, 261*

5.12   Summary   266

6. PRINTED-CIRCUIT INTERCONNECTS AND LOSSLESS TRANSMISSION LINES 279

6.1   Introduction   279

6.2   *n*-lump Transmission-Line Model   279

6.3   Analysis of the Lossless Transmission-Line Model   281

      6.3.1   *Derivation of the Telegrapher's Equations, 281*
      6.3.2   *Verification of the Wave Solutions, 285*

6.4   Interpretation of the Wave Solutions   287

6.5   Reflections on Transmission Lines   294

6.6   Transmission-Line Models   297

    6.6.1   dc Model, 297
    6.6.2   Source End Model, 298
    6.6.3   Load-End Model, 299

6.7   Solving Circuits with Transmission Lines   300

    6.7.1   Transmission Line Matched at Both Ends, 300
    6.7.2   Transmission Line with Matched Load Only, 303
    6.7.3   Transmission Line Unmatched at Both Ends, 307

6.8   Capacitively Loaded Transmission Line   311

6.9   Summary   318

7.   CLOCK SKEW AND SIGNAL REPRESENTATION   327

7.1   Introduction   327

7.2   Clock Skew in Synchronous Logic Circuits   327

7.3   Periodic Signals   330

7.4   Time-Domain Analysis of Clock Skew   331

    7.4.1   Circuit Model, 331
    7.4.2   Qualitative Behavior, 333
    7.4.3   Quantitative Analysis, 334

7.5   The Big Picture   337

7.6   Fourier Series Signal Representation   339

    7.6.1   Fourier Coefficients, 341
    7.6.2   Square-Wave Clock Signal, 344
    7.6.3   Shifted Square Wave, 346
    7.6.4   Periodic Exponential Sawtooth, 351

7.7   Superposition   357

    7.7.1   Sinusoidal Steady-State Response of an RC Circuit, 357
    7.7.2   Multiple Sinusoidal Sources, 360
    7.7.3   Applying Superposition, 362

7.8   Complex Form of the Fourier Series   366

7.9   Summary   371

8.   CLOCK DEGRADATION AND AC STEADY-STATE ANALYSIS   383

8.1   Introduction   383

8.2   Sinusoids and Phasors   383

    8.2.1   Sinusoidal Signal Representation, 383
    8.2.2   Phasors, 385

8.3   Impedance and Admittance   390

    8.3.1   Impedance of Basic Circuit Elements, 390
    8.3.2   Impedances in Series, 394
    8.3.3   Impedances in Parallel: Admittance, 396
    8.3.4   Components of Impedance and Admittance, 400
    8.3.5   Asymptotic Behavior of Reactance, 402

8.4   Clock-Signal Analysis Using ac Steady-State Theory   403

    8.4.1   First-Order RC Interconnect, 403
    8.4.2   Two-Section RC Ladder Interconnect, 406
    8.4.3   Interconnect Model Including Package Inductance, 413

8.5   Transfer Functions   414

    8.5.1   Input-Output Relationship, 414
    8.5.2   Bode Plots, 416

8.6   *RLC* Circuits in the Frequency Domain   419

    8.6.1   Series RLC Circuit, 419
    8.6.2   Mechanical Analogy, 427
    8.6.3   Quality Factor and Bandwidth, 428
    8.6.4   Clock-Signal Degradation, 433

8.7   ac Steady-State Analysis Using Complex Fourier Series   436

8.8   Summary   440

9.   OTHER CIRCUIT-ANALYSIS CONCEPTS   **453**

9.1   Introduction   453

9.2   Thevenin and Norton Equivalent Models   453

9.3   Controlled Sources   459

9.4   Root-Mean-Square Values   460

9.5   Nodal Analysis   463

    9.5.1   Circuits with Grounded Voltage Sources, 464
    9.5.2   Augmented Nodal Analysis, 469
    9.5.3   Modified Nodal Analysis, 471
    9.5.4   Modified Nodal Analysis with Inductance, 474

9.6   Summary   479

SELECTED ANSWERS   **493**

INDEX   **497**

Preface

There are plenty of excellent introductory circuits textbooks to choose from, and even more that are not so good. So why have we written yet another such book given that the odds are high it will be one of the latter? Our primary motivation has come from a simple observation: our students today are not the same as we were. It's true; students today grew up with computers and cable TV, not soldering irons and shortwave radios. By and large they are not content merely to ingest facts that we promise will be useful, someday. They demand to know, "What is this stuff good for... **right now**?"

In the days of yore, when chemistry, physics, and calculus were competing for the attention of a beginning electrical engineer, it was relatively easy to capture his or her attention with the much more interesting introductory circuit theory. But in modern curricula digital systems design may be taught concurrently with, or even before, introductory circuits. Now those courses are both easy and exciting, and our students say, "we can even design real things with it!" In contrast, circuit theory doesn't seem so interesting anymore.

It is axiomatic that circuit theory is fundamental to the subsequent understanding of both electrical and computer engineering. But our modern students don't always agree, having already designed digital systems without knowing Kirchhoff's Voltage Law from English common law. There might have been a time when the statement of some of our students, "I'm going to be a *computer engineer*, so I don't need to know this boring circuits stuff!" contained a kernel of truth. But now we even hear from our *electrical engineering* students, that in a world rapidly becoming all digital, "Who needs this analog stuff?" In this book we show that both computer *and* electrical engineers need to know elementary circuit analysis in order to appreciate and attempt to overcome fundamental limitations even on digital systems behavior.

Our goal is to get across that "analog stuff" in such a way that our students will see exactly what it is good for, and maybe even learn something about it. Our recurring motivational theme is, how does the passive interconnect limit the switching speed and signal integrity of digital systems? In this way we give what are interesting practical applications to the most elementary circuit analysis concepts.

This approach, moreover, reflects actual industrial design practice. Digital systems are designed, even automatically synthesized, to be functionally correct, but they must be electrically analyzed to determine whether they will perform to specified speed or may suffer signal integrity problems such as crosstalk or excessive signal ringing. The effects of passive interconnect are becoming increasingly dominant. With integrated circuit technology projected to have minimum feature sizes less than 0.1 μm early in the twenty-first century the application theme threaded through this book will become increasingly relevant. Using the fundamental tools provided in this book, beginning electrical and computer engineering students can appreciate and understand these effects.

We came to this theme after attempting several others. It is real and relevant. It encompasses almost all of elementary circuit theory. It is appealing to the broadest audience of electrical and computer engineers.

To be true to our belief that students must be motivated to learn, we've tried to stray as little as possible from our digital system performance limitations theme. We have not included alternative solution schemes when one would suffice. We have introduced modeling elements as needed and have not included those that are not relevant to our theme. We have tried to provide the necessary theory on a *just-in-time* basis so that students wouldn't be bored and lose interest.

This book then is not the usual elementary circuits grab bag of solution techniques and abstract modeling elements. You won't find mutual inductance, delta-wye conversions, or interpretations of Parseval's theorem; nor will you learn the birthplace of Charles Ohm or what a neper is. And it does not include the SPICE circuit simulation program. Once students learn SPICE, we've observed, they wonder why they've been subjected to "all of that difficult and useless analysis!" (i.e., the *boring analog stuff*). Rather we attempt to show them that simple symbolic analytical models can lead to great design insight. Development of this insight, we think, is something that should come before involved modeling of circuits using SPICE.

Over the past six years we have been reasonably successful using this book in a one semester course for sophomores. We presume a rudimentary knowledge of calculus and linear algebra, but no background in differential equations or physics.

As a result of our efforts, our students no longer ask, "What is this good for?" Rather they say, "Is this stuff good for anything else?" which we interpret as a smashing success: they are still full of the natural curiosity which attracted them to engineering in the first place. Some of our students don't learn every bit of this material, but they all appreciate that it is useful.

You may not agree with our philosophy, or you might just not like this book. But, if you think that maybe you would like to try something different, you will agree that this book is certainly just that. Most of your students will like it, and many will even learn something from it.

Acknowledgments

A number of students contributed to early versions of course notes which were incorporated into this book: Brian Zimmerman, Chris Koszarsky, Rodney Phelps, John Muza, Berend Ozceri, Kerry Hagan, Pascal Renucci, and Srini Gopalaswamy. In addition, we would like to thank the generations of students which suffered through preliminary editions, especially those who pointed out many typographical and other errors. Finally, we wish to thank the many anonymous referees whose insightful comments have helped improve this text.

Special thanks go to Mark Miller for a critical reading of the manuscript, and to our editors at Prentice Hall, Tom Robbins and Eric Svendsen, who periodically cracked the whip.

Parts of this book were written while one of the authors (MLR) was on sabbatical at the Swiss Federal Institute of Technology, Zurich. The generous hospitality of Prof. Dr. Henry Baltes and the staff of the Physical Electronics Laboratory are gratefully acknowledged.

Michael L. Reed   *Ron A. Rohrer*
University of Virginia   *Intersouth Partners, Inc.*

For Mary Rose and Casey

Introduction

Microelectronics

The nineteenth century was an era of electricity. The basic laws that govern electrical behavior were consolidated by James Maxwell in 1873. By the turn of the century much of the industrialized world was heavily reliant on electricity.

In 1907 Lee De Forest invented the triode vacuum tube, ushering in the electronics era that dominated the first half of the twentieth century. The vacuum tube enabled amplification, the process of turning a small electrical signal input into a much larger replica of itself. Amplification is important in radio and television broadcasting and reception, as well as to provide excessively loud music on a local basis. Being inherently nonlinear, vacuum tubes also support switching: a set of small electrical input signals turn on or off to cause a larger set of electric output signals to be turned on or off (according to some logical specification), which in turn can control subsequent switches. Unfortunately, vacuum tubes were not without limitations. In addition to being fragile and bulky, they generated enormous amounts of heat, necessitating elaborate and expensive cooling mechanisms. Consequently, complicated digital systems employed electromechanical relays, metal switches open and closed by electromagnets, until after the middle of the twentieth century.

The invention of the transistor by John Bardeen, Walter Brattain and William Shockley at AT&T Bell Laboratories in 1948, initiated the microelectronics revolution. As with vacuum tubes, transistors can provide both amplification and controlled switching, but they are much smaller and produce relatively little heat, and therefore are more reliable and practical. Soon transistors, being entirely electronic, came to replace the much slower electromechanical relays as high speed switching elements, prefacing a new era of electronic information processing. The first all transistor computer was the IBM 360 series introduced in 1964.

The microelectronics revolution accelerated in 1958 with the simultaneous independent invention of the integrated circuit (IC) by Jack Kilby of Texas Instruments and Robert Noyce of Fairchild Semiconductor Company. All the transistors of an integrated circuit are fabricated simultaneously along with their interconnecting wires. Integrated circuit technology permits inexpensive mass production, associated economies of scale, and complicated circuit functions, such as microprocessors and gigabit memories. Much of the circuitry of an FM stereo receiver, a compact disc player or even a high definition television set can be integrated with very few external components necessary.

This era of miniaturized electronics did not start as such. The earliest integrated circuits had only a few transistors. But the complexity of integrated circuits has grown exponentially since the late 1950s and experts predict that such growth will continue well into the twenty-first century. Over time, IC's have become larger, with more and smaller transistors per square centimeter, operating at successively higher clock speeds. By the year

2000, industry forecasts predict a device density approaching 60 million transistors per square centimeter (Table I.1).

Table I.1 Trends of IC die size and transistor density through 2000.

	1988	1991	1994	2000
Number of transistors (millions)	4	16	64	512
Die size (cm^2)	0.3	0.8	2.0	8.8
Device density (millions of transistors/cm^2)	13	20	32	58

About this Book

Clearly, an integrated circuit with millions of transistors is a very complicated system. In some sense, it is difficult to comprehend how anyone can understand how such a system operates and behaves. Yet, by zeroing in on a small set of limiting components, we can accurately predict some of the large-scale behaviors of the overall system.

In this book, we will use the example of a digital integrated circuit—the building block of nearly every commercial electronic device—as a vehicle for introducing circuit analysis. In particular, we will use simple but effective models of logic gates to predict, with increasing sophistication, the speed of integrated circuits. We wish to answer a simple question: what limits the clock speed of a system like a personal computer? In answering this question, we introduce the fundamental ideas of circuit analysis, and show how these ideas can be applied to digital systems.

While we will focus on a single problem, interconnect delay in digital systems, the ideas and techniques used can of course be applied to all electronic circuits. Moreover, it has been our experience that students have no difficulty in applying these techniques to other problems, even though they were learned using this approach.

Some other features of this book include the following:

- Nearly every step in every derivation and example is shown explicitly. It looks like a lot of math, but it is no more than other approaches.

- For the most part, example problems use values corresponding to real systems, picofarads and kilohms. We believe it is useful for students to think in terms of actual engineering quantities as early as possible.

- System equations are represented in terms of state variables to take advantage of the emphasis on linear algebra in modern engineering curricula.

Our overriding philosophy is that our students should learn a core set of topics really well. Our idea of what constitutes a core set might disagree with yours, so some traditional material (i.e., Thevenin/Norton equivalents, controlled sources, rms values) which isn't essential for understanding interconnect delay in digital systems is relegated to Chapter 9. These sections are self-contained, so they can be covered whenever you feel the urge.

Fundamental Electrical Concepts

1

1.1 Introduction

This chapter introduces definitions of fundamental electrical and circuit concepts essential for developing an understanding of the material in the remainder of the text. First, we establish some conventions regarding how we write and use symbols which stand for various quantities. Next, we look at the fundamental quantities encountered in circuit analysis: charge, current, voltage, and power. We define and explore circuit topologies and solution methods and look in detail at various electronic elements used in digital systems.

1.2 Conventions

1.2.1 Understood Units

When writing a variable in an expression, the dimension associated with that variable is understood to be expressed in a standard unit. Unless otherwise specified, the standard units used in this text correspond to the familiar units of the SI system: potential is measured in volts, current in amperes, time in seconds, resistance in ohms, capacitance in farads. (The one recurring exception to this rule is the unit for length. Physical lengths are understood to be expressed in centimeters, instead of the usual SI unit of meters. This follows standard practice in the IC industry, where lengths are generally given in centimeters or micrometers.[1]) The standard unit for each variable will be stated when the concept of each is first introduced.

As an illustration, consider the following expression, which describes the voltage in a circuit containing a resistor and capacitor:

$$v_C(t) = -5e^{-\frac{t}{2 \times 10^{-6}}} + 5$$

In this expression, the value of $v_C(t)$ is *understood* to be expressed in volts (the standard unit of voltage), and t is *understood* to be expressed in seconds (the standard unit of time).

---

[1] "Micron" is engineering lingo for "micrometer." $1 \ \mu m = 10^{-6} \ m = 10^{-4} \ cm$.

Suppose we wish to evaluate this expression for $t = 4$ μs. We first need to express t in the standard unit:

$$t = 4 \times 10^{-6} \qquad (1.1)$$

Equation (1.1) is understood to be in seconds; therefore we need not write a unit after the value. Next, we substitute this value in the voltage expression:

$$v_C(4 \times 10^{-6}) = -5e^{\frac{4 \times 10^{-6}}{2 \times 10^{-6}}} + 5$$

$$= 4.32$$

where it is *understood* that $v_C(t) = 4.32$ volts. By following this convention, we avoid cluttering the algebraic derivations with repetitious symbols designating the variable units. (For completeness, the unit is generally included at the conclusion of the problem, in what would be considered the "final answer.")

Example 1.1

The inductor current in a critically damped RLC circuit is given by

$$i_L(t) = 1.2 \times 10^{-6}e^{-50t} - 60 \times 10^{-6}te^{-50t}$$

What are the units of $i_L(t)$, t, and each numerical value in this expression?

Currents are understood to be in amperes, and time is understood to be in seconds. Using this information, and the fact that the argument of an exponential function is dimensionless, we can deduce the following:

$i_L(t)$, 1.2×10^{-6} are in amperes (A)

t is in seconds (s)

50 is in inverse seconds (s^{-1})

60×10^{-6} is in A/s

1.2.2 Engineering Notation

In many instances a standard unit is either too small or too large a measure for convenience. For example, the digital signals in computers are electrical currents, whose standard unit is the ampere, but the magnitudes of the actual currents are much smaller. It is standard practice to speak of a quantity in multiples of 10^{3k} units, where k is an integer (positive or negative). While it is completely correct to refer to a current whose value is 8.5×10^{-5} amperes, it is conventional to express this value as 85 microamperes (85 μA). Figure 1.1 summarizes these standard denominations of units. This system of standard prefixes is known as *engineering notation*. We will always express the final numerical answer to a problem in these multiples.

peta–	P–	10^{15}	milli–	m–	10^{-3}
tera–	T–	10^{12}	micro–	μ–	10^{-6}
giga–	G–	10^{9}	nano–	n–	10^{-9}
mega–	M–	10^{6}	pico–	p–	10^{-12}
kilo–	k–	10^{3}	femto–	f–	10^{-15}
			atto–	a–	10^{-18}

Figure 1.1 Standard prefixes, abbreviations, and their values, arranged in descending order of magnitude.

1.2.3 Functions of Time

In most systems, the quantities of interest are functions of time. In some situations, however, a quantity is expressly designed to be a constant value. For example, the voltage available at a common U.S. electrical utility socket can be expressed as

$$v(t) = 165\cos(2\pi 60t)$$

which is a function of time, whereas the voltage available across the terminals of a battery is given by

$$V = 1.5$$

The point of these two examples is that we will always express time-dependent quantities with *lower-case* symbols: v, i, q. If we write v, it is understood to be $v(t)$, even without the explicit functional dependence on time shown. Quantities that do not vary with time will always be expressed with *upper-case* symbols: V, I, Q. (Of course, time-dependent quantities can, as a special case, be constant over some range.)

1.3 Charge, Current, and Voltage

We now introduce and define three basic electrical quantities: charge, current, and voltage.

1.3.1 Charge

All matter is made up of fundamental building blocks known as atoms, which are in turn composed of myriad particles. From the perspective of an electrical or computer engineer dealing with electronic circuits, the important property of these particles is their **charge**. Charge is related to the electrical forces among various particles, such as electrons and protons. Following a convention introduced by Benjamin Franklin, the electron carries a negative charge, the proton a positive charge of equal magnitude.

The standard unit of charge is the **coulomb** (abbreviation: C). The coulomb is defined as follows: two small, identically charged particles which are separated by one meter in a vacuum and repel each other with a force of $10^{-7}c^2$ newtons, each possess a charge of (plus or minus) one coulomb (c is the velocity of light: 2.997925×10^{10} cm/s). In terms of this unit the charge of an electron is approximately -1.602×10^{-19} C; thus a charge of negative one coulomb (-1 C) consists of the accumulated charge of about 6.24×10^{18} electrons. The symbol associated with static charge, or charge that does not vary with time, is Q; the symbol given to charge that may vary with time is the lower-case q.

In a computer, the digital bits of information stored in semiconductor memories take the form of charge. For example, a digital "1" is stored in a memory chip as the presence of a certain quantity of charge; the absence of this charge is interpreted as a digital "0."

1.3.2 Current

Current is the name we give to the net motion of charge from one place to another. Current is of crucial importance in electronic circuits, as their usefulness to us depends on this process of charge transfer. Most useful circuits have an important characteristic: they can vary the rate at which charge is transferred, as a function of time. For example, communication systems depend completely on the controlled variation in the rate of charge transfer, which is eventually perceived by us as audio or video information. In a computer, the time-varying process of charge transfer is the physical realization of computation, the rearrangement of bits of digital information.

Current flows along a path. The current present in a path has both a magnitude and a direction associated with it. We must unambiguously specify both the direction along the path and the sign of the numerical magnitude of the current. This concept is illustrated in Figure 1.2.

Figure 1.2 Equivalent ways of specifying flow of electrical current along a conductor. The arrow specifying the reference direction can be placed next to, or on top of, the line signifying the current path; it makes no difference. In every case shown here, there is 2×10^{-3} C of positive charge flowing past any point on the wire each second, from right to left.

When presented with a circuit problem to solve, we will assign a symbol to the current along the path. The objective is to find a mathematical expression or numerical value which corresponds to the current flowing in the actual circuit. When specifying this current, there are four interrelated details which together provide an unambiguous description: the reference direction (specified by the arrow); the variable associated with the current (such as $i(t)$); the magnitude of the quantity associated with $i(t)$ (i.e., 2 mA); and the sign of the quantity associated with $i(t)$.

The significance of the arrow in the designation of current is this: if the value of the current $i(t)$ next to the arrow is greater than zero, then the arrow represents the direction of movement of **positive charge,** Thus, if we write I next to a right-pointing arrow, and further specify that $I > 0$, then positive charges are moving to the right along the path. From a circuit standpoint, this is equivalent to the flow of negative charges moving to the left.

A circuit diagram without notations for the various currents confronts us with a problem: we must choose and define current variables and their associated directional arrows. This task sometimes seems overwhelming to the neophyte because the choices are completely arbitrary. We can see that they are arbitrary by reflecting on the obvious result that the function of the circuit cannot depend on any labels we paste on it. In the same way that we cannot make a radio fly under its own power simply by painting "Boeing" on the front, it cannot possibly make any difference whether we specify a current inside the radio as $i_1(t)$ to the right or $i_2(t)$ to the left; the circuit will function the same regardless of any labels. What *will* change with different notation are the values of the numbers associated with the variables, which is our way of observing the behavior of the circuit. The important thing to remember is that once a direction has been assigned, we must be consistent in using it during circuit analysis. (While the choice of directions is arbitrary, it is often true that an intelligent choice makes the process of solving the circuit easier. Methods of choosing appropriate directions are illustrated by example throughout this text.)

Mathematically, current is the rate of charge transfer. It can be visualized as the instantaneous rate at which charge moves past a reference in a specified direction, as illustrated in Figure 1.3. Current is conventionally assigned the variable i, and defined as

$$i = \frac{dq}{dt} \tag{1.2}$$

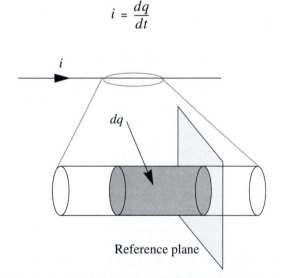

Figure 1.3 Current flow in a conductor. The charge dq flows past a surface (defined by a reference plane perpendicular to the direction of current) in time dt. The rate of charge transfer, dq/dt, is the current i.

The standard unit for current is the **ampere** (abbreviation: A), which is equal to 1 C/s (coulomb per second).

In practical terms, an ampere is a huge amount of current, appropriate for characterizing devices capable of converting bread into toast. For electronic circuits like computers and their peripherals, typical circuit-board-level currents are much smaller, in the range of milliamperes (mA) and microamperes (μA). Inside integrated circuits (ICs) values of current may be measured in nanoamperes (nA), picoamperes (pA), and femtoamperes (fA).

If the current is constant in time, then we refer to it as **direct current**, or dc for short. The upper-case symbol I is used for dc; upper-case symbols for other constant circuit variables are sometimes called dc variables. (In fact, constant currents are often referred to as "dc currents," even though this usage might appear redundant.)

1.3.3 Voltage

Consider a general circuit element with two terminals. It does not matter at this point what the element is—it could be a resistor, light bulb, or blender. We will symbolize this general element with the blob shape shown in Figure 1.4. The two terminals a and b are electrical connections through which current can flow into and out of the element.

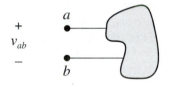

Figure 1.4 General two-terminal circuit element. The **terminals** a and b are electrical connections through which current can flow into and out of the element. The voltage v_{ab} between the two terminals is a measure of the energy required to move charge through the element.

Suppose we connect this element into an electrical circuit such that a dc current flows through it. We then make measurements on the element and determine that energy is being expended to move the current from terminal a, through the element and out of terminal b. We characterize this energy by an electrical **voltage** (also called a **potential difference**) which exists *across* the element.

The voltage across a terminal pair is a measure of the work required to move charge through the element. Formally, voltage is defined as work per unit charge:

$$v_{ab} = \frac{w_{ab}}{q} \tag{1.3}$$

where w_{ab} is the work required to move charge q from point b to point a through a potential difference v_{ab}.

The unit of voltage is the **volt** (abbreviation: V). One volt is the potential difference between two points, if it takes one joule (abbreviation: J) of energy to move one coulomb of charge from one of those points to the other. Note that potential difference can exist regardless of whether current is flowing. If a conducting path exists between two points with a potential difference, then charge will flow, resulting in an electric current. On the other hand, if there is no such path, the potential difference will continue to exist even in the absence of current flow. (For example, a voltage exists across the terminals of a house socket with or without a lamp plugged into it.)

As was the case with current, there is the issue of determining the direction and sign of the voltage difference between two points in a circuit. Reversing the polarity (sign) of the voltage difference amounts to a reversal of the direction of energy flow when a current is present. Again, we must unambiguously specify several details: which two points; a reference direction (denoted with + and − signs); the variable associated with the voltage (i.e., $v(t)$); and the magnitude and sign of the quantity associated with $v(t)$, for example, $v(t) = 4.0$ V or $v(t) = -1.2$ V.

Since voltage is **always** measured between two terminals, we must specify the two terminals in some way. In accordance with common practice, the two terminals (and also the reference direction) are specified by assigning + and − signs, as shown in Figure 1.4. Another way is to label the two points with letters, and subscript the voltage variable with the two letters corresponding to the two points in the circuit. Figure 1.4 illustrates the correspondence between the subscript order and the signs: the + sign is placed next to the point corresponding to the first subscript, the − sign next to the point corresponding to the second subscript. If the subscripts are omitted, then the reference direction must be shown with + and − signs.

As is the case with the current arrow, it is important to understand that the pair of plus–minus signs does not indicate the "actual" polarity of the voltage; it is merely part of a convention that enables us to talk unambiguously about the voltage across any terminal pair. The sign of the voltage follows from the definition of voltage and can be expressed as follows: given a positive current which enters through terminal a and leaves through terminal b, such that an external source must expend energy to maintain this current, then terminal a is **positive** with respect to terminal b. Equivalently, terminal b is **negative** with respect to terminal a.

In all electrical circuits, we speak of current *through* an element and voltage *across* the element. Current is a measure of how much charge is moving past a reference point, so it is meaningful to speak of the value of current at a particular point. However, voltage relates the work that can be done on a unit charge in moving the charge *between* two points. Clearly, this depends highly on which two points we are talking about. Thus it is NOT meaningful to speak of the voltage *at* a point; it is only meaningful to refer to the voltage *between* two points. (Later, we will speak of voltage at a point (node voltage), but always with the understanding that it is referenced to a zero ground potential, which is the second point.)

1.3.4 Hydraulic Analogy

Flow of electrical charge is analogous to flow of any other kind of particles. It is helpful to draw an analogy between electrical current and water flow. If we think of positive charges as water, then current (measured in amperes, coulombs per second) is analogous to volumetric flow rate (measured in liters per second); potential difference (a measure of work per unit charge) is analogous to pressure head or height if the water is driven by gravity.

Using this analogy, many of the concepts of electrical circuits become easier to interpret. For example, it is easy to see that a difference in height can produce a flow of water, but the existence of the height difference does not mean that any water is present; thus, a voltage difference can exist without a current. Also, water can flow in a completely level channel; this corresponds to current flow in an ideal wire where there is no potential difference across the ends. Energy must be expended to raise water from a low to a high level against the influence of gravity; thus it is natural to speak of a higher position in a hydraulic system as analogous to a higher potential in an electrical circuit.

1.4 Power

Power is the rate at which work is done. Its unit is the **watt** (abbreviation: W), equal to one joule per second (J/s). The amount of power delivered to or from a circuit element is equal to the work per unit charge associated with the element, times the number of charges per second flowing through it. Therefore, the power is equal to the product of the voltage across the element and the current through it:

$$p = vi \qquad (1.4)$$

To determine whether an element is drawing power from the circuit, or returning power to the circuit, we must consider the voltage and current reference directions.

1.4.1 Associated Reference Directions

The reference directions are **associated** when the current direction arrow enters the terminal of a circuit element with the + sign and leaves through the terminal with the – sign, as illustrated in Figure 1.5.

Figure 1.5 Circuit elements defined with **associated** reference directions for voltage and current. The current reference arrow is drawn such that current enters the element terminal assigned the + sign for the voltage reference. This is equivalent to having the current arrow leaving the terminal. This convention should be followed whenever practical.

If we use Eq. (1.4) and assign associated reference directions to an element as illustrated in Figure 1.5, then power is being delivered **to** the element when the product of the voltage and current values is **positive**. On the other hand, power is being delivered **from** the element if the product of the voltage and current values is **negative**. It is generally a good practice to use the associated reference directions whenever possible. However, reference directions that are not associated (i.e., unassociated) may be specified in some circuits. In such cases, caution must be exercised, as the terminal characteristics of the element (i.e., how the current through the element depends on the voltage across the element) will differ from the standard definition, which always assumes associated reference directions.

1.4.2 Unassociated Reference Directions

The reference directions are **unassociated** when the current reference arrow points into the terminal assigned with a – sign, as illustrated in Figure 1.6.

Figure 1.6 Circuit elements defined with **unassociated** reference directions for voltage and current. The current reference arrow is drawn such that current enters the element terminal assigned the – sign for the voltage reference. Although this assignment of directions is not incorrect, it should be avoided.

If we use Eq. (1.4) and assign unassociated reference directions to the voltage and current variables as illustrated in Figure 1.6, then power is being delivered to the element when the product of the voltage and current values is negative, and power is released by the element to the remainder of the circuit if the product of the voltage and current values is positive. Assignment of unassociated reference directions should be avoided.

A summary of these power relationships is given in Figure 1.7. Throughout this book we will, for the most part, use associated reference directions in order to develop a consistent methodology for analyzing circuits.

Reference Direction	vi Product	Condition
Associated	$p > 0$	Absorbing power
Associated	$p < 0$	Releasing power
Unassociated	$p > 0$	Releasing power
Unassociated	$p < 0$	Absorbing power

Figure 1.7 Summary of power relationships and their dependence on reference directions.

Example 1.2

Figure 1.8 illustrates a flashlight circuit with a 3-V battery and a light bulb which draws 20 mA of current (I = 0.02 A). Determine the power associated with each element, and whether the element is absorbing or releasing power.

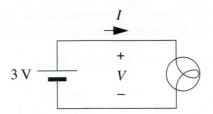

Figure 1.8 Flashlight circuit with 3-V battery and light bulb which draws 20 mA of current.

We can draw the elements separately with their voltage and current reference symbols, using either the associated or unassociated reference conventions. Using the associated reference convention, we have the assignment as shown in Figure 1.9.

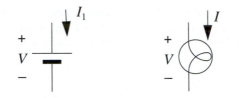

Figure 1.9 Assignment of associated voltage and current reference directions to the battery and light bulb of Figure 1.8.

With these assignments, we have for the battery:

$$p_{battery} = I_1 V = (-0.02)(3) = -0.06$$

The reference arrow for current I_1 is pointing *into* the + reference sign of the voltage V, so I_1 and V are associated. I_1 is –20 mA because it flows opposite to I, which was already specified to be +20 mA. Putting these together, we have $p_{battery}$ = –60 mW, which we interpret as releasing power.

For the light bulb,

$$p_{bulb} = IV = (0.02)(3) = 0.06$$

The reference arrow for current I is pointing *into* the + reference sign of the voltage V, so I (= 20 mA) and V are associated. The power associated with the bulb is p_{bulb} = 60 mW. Since this is a positive number, and we have used associated reference directions, the light bulb is absorbing power.

1.5 Circuits, Nodes, and Branches

A **circuit** is an interconnection of electrical elements. All circuits are composed of **elements**, such as transistors, capacitors, switches, and other components, that have distinct electrical characteristics. These elements can be connected in a variety of ways depending on the task the circuit is designed to accomplish. Abstract circuit elements are shown as the shaded rectangular regions in Figure 1.10 and following figures.[2]

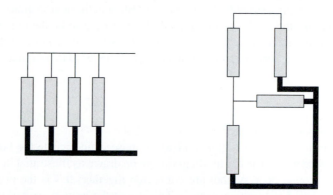

Figure 1.10 Examples of nodes in electronic circuits. The circuit elements are represented by gray rectangles, while a representative node in each circuit is designated by a bold line.

The connections of wires among the elements are called **nodes**. A **node** is any set of unbroken line segments in a circuit diagram that joins two or more circuit elements. All points of a node are at the same potential. Examples of nodes are the bold lines in Figure 1.10. What is significant is not the manner in which a node is drawn, but the node's topology (what it connects). As an example consider the two circuits of Figure 1.11. Both circuits are topologically equivalent, because the upper and lower nodes form the same set of connections among the elements.

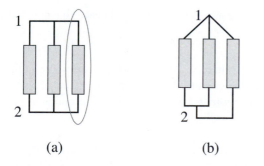

(a)   (b)

Figure 1.11 Examples of circuits that are topologically equivalent. The connections among the circuit elements in (a) and (b) are equivalent, even though they look different.

It is convenient in analyzing circuits to choose one node of the circuit to serve as a reference point for all voltage values. The proper choice of a reference node can simplify considerably the analysis process (but cannot, of course, change in any way the actual oper-

---

[2] Circuit elements must have at least two terminals, but may have more. For example, bipolar junction transistors are three-terminal devices; field-effect transistors used in CMOS circuits have four terminals; certain types of transformers and power semiconductor devices may have five or more electrical connections. In this text we will deal primarily with two-terminal elements and a few three-terminal devices.

ation of the circuit, since it is just a label). The reference node is called the **ground node** or **common.** The significance of the ground node is that, by convention, its potential is taken to be zero; therefore, the voltage between another node and the ground node can be expressed using a simplified notation (described in the next section). If a ground node is not already specified, it is usually a good idea to take it to be the one connecting the greatest number of circuit elements, or the one at the bottom of the circuit diagram. In many cases these two situations coincide. Various symbols are used to designate the ground node:

In this book we will use the leftmost symbol.

In the circuit of Figure 1.11(a), suppose we designate the bottom node (node 2) as the ground node. The circuit diagram can be drawn as illustrated in Figure 1.12. All nodes with a ground node symbol are connected together; this is the origin of the term "common" (short for "common node"). This use of the ground symbol is helpful in preventing clutter in a circuit diagram, since widely separated elements electrically connected through the ground node need not have a line drawn between them.[3]

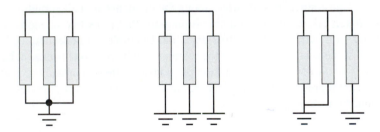

Figure 1.12 Equivalent ways of representing a circuit using a ground node. All points connected to any ground node are connected together.

Another useful notion is that of a circuit **branch**. A branch is a single path in a circuit composed of one two-terminal element and the nodes at either end of that element. For example, the circled element in Figure 1.11(a), along with the nodes at either end, comprises a branch.

1.6 Branch and Node Voltages

Two different notations are used to specify voltages in a circuit. Often, both are used in a single circuit. We call these notations branch voltages and node voltages. Keep in mind, however, that voltage is always defined as the difference in potential between two points.

[3.] Note that the crossed wires at the bottom of this circuit are connected; this is indicated by the dot where the two wires meet: ⬩. Crossings without a dot ⊥ mean that the two conductors are not connected to each other.

Branch voltage is simply the voltage across a branch. To specify a branch voltage, it is necessary either to use + and – signs on the two nodes or to designate each node with labels which then show up as subscripts on the voltage variable. For example, in Figure 1.5 the voltages v_1 and v_2 are branch voltages, since they represent the potential difference between two terminals of an element.

Node voltage is defined as the potential at a particular node with reference to the ground node. A node voltage is just like a branch voltage with the + sign on the node of interest and the – sign on the ground node.

Figure 1.13 illustrates both notations. The symbols v_a and v_b represent node voltages, while the v_1 and v_2 symbols specify branch voltages. Branch voltages can always be thought of as the difference between two node voltages. This leads to the double-subscript notation for branch voltages: if v_a and v_b are two node voltages, then we designate the branch voltage (with the + sign at v_a and the – sign at v_b as v_{ab}. Mathematically, $v_{ab} \equiv v_a - v_b$.

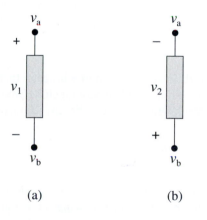

(a)   (b)

Figure 1.13 Branch voltages (v_1, v_2) and node voltages (v_a, v_b). (a) The branch voltage v_1 is equal to the difference $v_a - v_b$. The branch voltage can be equivalently specified as v_{ab}. (b) The branch voltage v_2 is equal to the difference $v_b - v_a$. This branch voltage can be equivalently specified as v_{ba}.

Example 1.3

In the circuit shown in Figure 1.14 express the branch voltages v_1 through v_6 in terms of the three node voltages v_a, v_b, and v_c.

Applying the conventions for specifying branch and node voltages, we have:

$$v_1 = v_a - 0 = v_a \qquad\qquad v_2 = v_b - 0 = v_b$$

$$v_3 = 0 - v_c = -v_c \qquad\qquad v_4 = v_a - v_b$$

$$v_5 = v_c - v_b \qquad\qquad\qquad v_6 = v_a - v_c$$

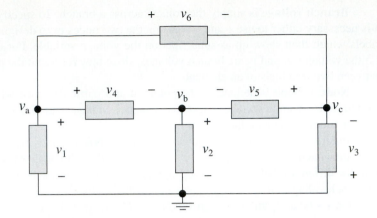

Figure 1.14 Circuit for Example 1.3. v_a, v_b and v_c are node voltages; v_1 through v_6 are branch voltages.

1.7 Kirchhoff's Voltage and Current Laws

When dealing with circuits and analyzing how they work, Kirchhoff's laws are fundamental. Using Kirchhoff's two laws, and knowledge of element behavior (such as Ohm's law), we can construct equations which describe the operation of any circuit, no matter how complex.

1.7.1 Kirchhoff's Voltage Law

Refer to Figure 1.15(a). Applying the relationship between branch and node voltages, described in Figure 1.13, to each of the branches results in the following set of equations:

$$v_1 = v_a - v_b$$

$$v_2 = v_b - v_c$$

$$v_3 = v_a - v_d$$

$$v_4 = v_b - v_d$$

$$v_5 = v_c - v_d$$

(These equations are true regardless of which of the four nodes is chosen as the ground node.) The interconnection of circuit elements like that shown in Figure 1.15 imposes con-

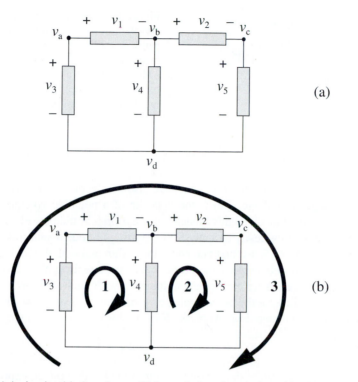

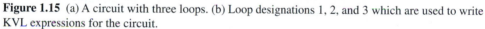

Figure 1.15 (a) A circuit with three loops. (b) Loop designations 1, 2, and 3 which are used to write KVL expressions for the circuit.

straints on the branch and node voltages of the circuit. These constraints are described by **Kirchhoff's voltage law**, or KVL for short:

> *KVL: The algebraic sum of the branch voltage drops around any closed loop is zero.*

To interpret this statement, we must first understand the meaning of "loop" and "voltage drop."

A **loop** is defined as any closed path in a circuit, beginning and ending on the same node. Looking at Figure 1.15(a), we see that this circuit has three possible loops, which are illustrated in Figure 1.15(b). For example, loop 1 follows a path through nodes a-b-d-a. To use KVL to write an equation, we choose a particular loop in the circuit and a direction of travel around the loop.

A **voltage drop** is the difference in potential between two nodes along the travel of a loop. The salient point regarding voltage drops is that, as we travel along the path of the loop, we must encounter the + sign reference of the branch voltage before the − sign; otherwise the potential difference is called a voltage rise. Clearly, the only difference between a voltage drop and a voltage rise is the polarity; a voltage drop of 3 V is equivalent to a voltage rise of −3 V, and a voltage drop of −50 mV is equivalent to a voltage rise of 50 mV. Figure 1.16 illustrates this idea.

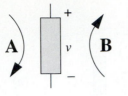

Figure 1.16 Voltage drops and voltage rises. In this figure, the arrows represent portions of two possible loop paths, **A** and **B**. v is the branch voltage across the element, which is always the same regardless of the KVL path. In path **A**, we first encounter the + sign, therefore v is the voltage drop, and $-v$ is the voltage rise. In traversing path **B**, we first encounter the − sign, so in this case we call v the voltage rise, and $-v$ the voltage drop.

We see from Figure 1.16 that an element's branch voltage is dependent on how the voltage is defined across it, and which node is encountered first (the + signed or − signed node) when going around the loop. We also see from Figure 1.16 that a positive voltage drop (negative voltage rise) is the opposite of a negative voltage drop (positive voltage rise). That is, if an element's branch voltage *drop* is v, then the element's voltage *rise* is $-v$.

We are now ready to apply KVL. Starting at node a, and following path 1 in the circuit of Figure 1.15, we first encounter the + sign associated with branch voltage v_1. Since we entered this element via the + sign, v_1 is a voltage drop. We next encounter branch voltage v_4, again entering the element at the node with the + sign, so v_4 is also a voltage drop. Finally, we traverse the branch voltage v_3, but this time we enter the branch via the − sign, so $-v_3$ is the voltage drop. Applying KVL to this path results in the following equation:

$$v_1 + v_4 - v_3 = 0 \qquad (1.5)$$

If we substitute the previously determined node voltage expressions for each of the branch-voltage terms in Eq. (1.5), we obtain:

$$(v_a - v_b) + (v_b - v_d) - (v_a - v_d) = 0 \qquad (1.6)$$

which is obviously satisfied.

Applying KVL to paths 2 and 3 in Figure 1.15 results in

$$v_2 + v_5 - v_4 = 0 \qquad (1.7)$$

$$v_1 + v_2 + v_5 - v_3 = 0 \qquad (1.8)$$

KVL is essentially a statement of the conservation of energy. Recall that voltage is defined as the amount of work per unit charge necessary to move a charge between two points. In traversing a loop in a circuit, we always end up where we started, on the same node. Since all points along this node are at the same potential, there is no change in potential in going around the loop.

Another way to visualize KVL is to think of voltage as gravitational potential. In walking from one point to another, you may encounter drops as you go downhill, and rises as the path goes uphill. However, no matter where you start, or what path you take, you will end up at the same height (i.e., potential) in a path that closes upon itself, since a closed path returns you to the same place.

1.7.2 Kirchhoff's Current Law

Just as KVL expresses a relationship among voltages in a circuit, **Kirchhoff's current law**, or KCL for short, expresses a constraint on the currents in a circuit.

> *KCL: The sum of the currents entering a node is equal to the sum of the currents leaving the node.*

A current is considered to be entering a node if the reference arrow points from the circuit element into the node; it is leaving the node if the reference arrow points from the node into the circuit element. (Again, the actual value of the currents, positive or negative, is irrelevant at this point; we are concerned only with the direction of the reference arrows in order to apply KCL.)

KCL is illustrated in Figure 1.17, which shows a node consisting of the intersection of three wires. We see that if 5 μA of current enters the node, then according to KCL, a total of 5 μA must leave the node.

KCL is essentially a statement about charge conservation. An electric circuit performs useful functions by moving charges around, but does not actually create or destroy any charged particles. (The analogy to hydraulics is that in a closed system of plumbing, we do not create or leak away any water, only redistribute it.)

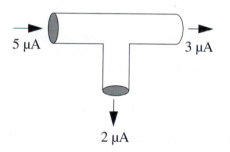

5 μA

3 μA

2 μA

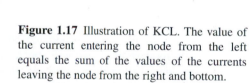

Figure 1.17 Illustration of KCL. The value of the current entering the node from the left equals the sum of the values of the currents leaving the node from the right and bottom.

1.8 Circuit Elements

The two fundamental circuit laws, KVL and KCL, tell us something about voltages and currents in all circuits. KVL relates voltages to voltages, and KCL relates currents to currents. Our objective is to solve for all circuit voltages and currents as a function of time. To meet this objective, we need to have relationships between branch voltages and branch currents. These relationships are determined by the circuit elements connected between nodes.

1.8.1 Voltage Source

An **ideal independent voltage source** is a circuit element that provides a known voltage across its two terminals, independent of the current flowing through it. It does not matter what is connected to the voltage source; the voltage across the two terminals is always the specified value. The symbol for an ideal independent voltage source is shown in Figure 1.18.

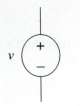

Figure 1.18 Symbol for an ideal independent voltage source. The voltage across the two terminals is v, which may be a function of time but is independent of the current through the source. The current through the source may take on any value.

A special case is the ideal independent dc voltage source, where the designation "dc" means that the voltage is constant (i.e., independent of time). It is designated by the symbol shown in Figure 1.19(a), which is the same as before, except for the upper-case variable which designates a time-independent quantity. In many circuits the minus-sign-referenced node of a voltage source is connected to the ground node. Figure 1.19(b) shows an equivalent representation of a voltage source which is used in such circuits to reduce clutter.

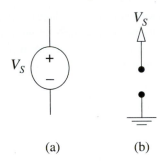

(a)   (b)

Figure 1.19 Symbols for an ideal independent dc voltage source. In (b), the negative terminal of the voltage source is connected to the ground node.

Although "dc" originated as an abbreviation for "direct current," in accordance with modern usage we will speak of dc voltage sources as those whose **voltage** is independent of time. As we will see, this does NOT mean that the *current* through a dc voltage source is constant, direct, or unidirectional. Also, we will generally refer to ideal independent voltage sources simply as "voltage sources," with the "ideal independent" part being understood.

1.8.2 Current Source

An **ideal independent current source** is a circuit element that provides a known current i independent of the voltage across it. The circuit symbol used to depict an ideal independent current source is shown in Figure 1.20. The arrow shows the reference direction for the current. It is the direction in which positive charges are moving if the value associated with i is positive. As with voltage sources, we will usually drop the "ideal independent" modifiers and simply say "current source."

It is a common error when analyzing circuits to set the voltage across a current source equal to zero volts. Just as the current through an ideal voltage source can be any value, the voltage across an ideal current source can take on any value, seldom zero!

Figure 1.20 Circuit symbol for an independent ideal current source. The voltage across the source may take on any value.

1.8.3 Resistance

The **resistance** of an element relates the voltage across the element to the current through the element. Resistance is measured in **ohms** (abbreviation: Ω) and depends on several factors: the physical size and shape of the element, the properties of the material it's made of, and temperature. The latter two components are characterized by an intrinsic property known as **resistivity**, which is given the symbol ρ and has units of Ω-cm. (For example, the conductors interconnecting gates on an IC are generally made of an aluminum alloy with a typical resistivity of $\rho = 2.8\ \mu\Omega$-cm.) The resistance of a uniform element of material is given by

$$R = \rho\frac{l}{A_R} \tag{1.9}$$

where A_R is the cross sectional area perpendicular to the current flow, and l is the length of the element. Figure 1.21 illustrates the geometry of a rectangular segment of material. In this case the resistance is

$$R = \rho\frac{l}{wh}. \tag{1.10}$$

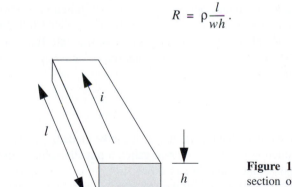

$A_R = h \times w$

Figure 1.21 The resistance of a uniform section of a material is determined by its resistivity, length, and cross-sectional area (denoted by the shaded region) perpendicular to the direction of current flow.

Note that if the current were traveling in a different direction (say, perpendicular to the direction indicated in Figure 1.21), A_R and l would be different. Therefore the *resistance* of the element would be different (but the *resistivity* of the material does not change).

If we increase the length of the element by a factor α, its resistance scales by the same factor (i.e., its resistance is α times larger), assuming all other parameters are kept constant. If either the width w or height h of an interconnect is decreased, this results in a decrease in the cross-sectional area and a resulting increase in the resistance. As IC's become more complex, all of these factors come into play, resulting in significant increases in interconnect resistance. In the next chapter we will see how the interconnect resistance affects computer performance.

A circuit element with a resistance R is known as a resistor.[4] The symbol for a resistor is shown in Figure 1.22.

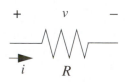

Figure 1.22 The symbol for a resistor of value R (ohms). Note the use of associated reference directions for voltage and current.

For many materials, the relationship between the voltage and the current of a resistor is given approximately by **Ohm's law**, which states that the voltage across a resistor is proportional to the current through the resistor. The proportionality constant is the resistance R. In symbols,

$$v = Ri \qquad (1.11)$$

We observe from this expression that the units of resistance, ohms, have dimensions of voltage/current; in standard units, $1\,\Omega = 1$ V/A.

(Note that the value of voltage across a resistor, as a function of current, assumes that associated reference directions have been used. If unassociated reference directions are assigned to a resistance, then we must insert a negative sign into Eq. (1.11). This is the case for other elements too, and underscores the importance of using associated reference directions whenever practical.)

1.8.4 Capacitor

A **capacitor** is a circuit element that stores charge. It can be made by sandwiching an insulator between two conducting plates, a structure called a **parallel-plate capacitor**. A perspective view of a parallel-plate capacitor is shown in Figure 1.23. A_C is the area of the conducting plates, and d is the thickness of the dielectric material between the two plates.

The electrical properties of a capacitor can be understood by considering what happens when current flows into one terminal. As positive charge flows into the capacitor, it begins to accumulate on the plate connected to that terminal, since it cannot continue to flow through the insulating layer. This accumulation of positive charge attracts negative

---

[4.] If a circuit designer wants a certain value of resistance in a particular place in a circuit, we call that resistance a *resistor*; a resistor is an example of an *electronic component*. The circuit designer would prefer zero-resistance interconnects, but this is not possible in conventional ICs. The (unwanted) resistance associated with the interconnect is called a *parasitic* resistance.

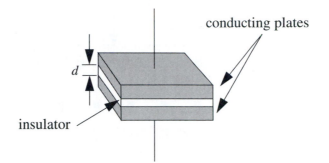

Figure 1.23 Structure of a parallel-plate capacitor. Two conducting plates of area A_C are separated by a dielectric layer of thickness d.

charge to the opposite plate. This negative charge has to come from somewhere; it can be supplied only by the wire connected to the opposite plate. This flow of negative charge constitutes an electrical current oriented out of the opposing plate, which is equal in magnitude to the current flowing into the capacitor. From a circuit point of view, it appears that a current is flowing directly through the capacitor; on the microscopic level, charge continues to accumulate on both plates. That is, for every positive charge deposited on the first plate there is a corresponding negative charge attracted to the opposing plate.

Figure 1.24 illustrates what happens within a capacitor as current flows into it. The accumulation of charge results in an electric field $\mathscr{E}$ pointing toward the plate with the negative charge.[5] The magnitude of the electric field in a parallel-plate capacitor is given approximately by:

$$|\mathscr{E}| = \frac{q}{\varepsilon A_C} \tag{1.12}$$

where q is the magnitude of the charge on each plate, and ε is a constant, known as the **permittivity**, which depends on the dielectric material separating the plates. ε is usually expressed as

$$\varepsilon = \kappa \varepsilon_0 \tag{1.13}$$

where κ is a dimensionless number called the **dielectric constant** of the material, and $\varepsilon_0 = 8.854 \times 10^{-14}$ F/cm is the permittivity of free space.

In integrated circuits, both the gate capacitors and the parasitic capacitance associated with interconnect often have as the dielectric a material known as silicon dioxide, SiO_2, which has a dielectric constant of approximately 3.8. The permittivity of SiO_2 is given by

$$\varepsilon_{SiO_2} = 3.8\varepsilon_0 = 3.365 \times 10^{-13} \text{ F/cm}$$

[5] Recall that the electric field describes the force per unit charge: the force on a charge q is given by $q\mathscr{E}$. The direction of the electric field is defined as the direction in which a hypothetical positive charge would move as a result of the field.

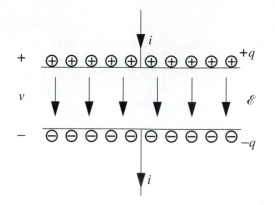

Figure 1.24 The charging of a capacitor. The positive charge accumulated on the top plate is equal to the magnitude of negative charge on the bottom plate. The accumulation of charges results in an electric field, $\mathscr{E}$, directed from the plate having the positive charges toward the plate having the negative charges. Note that the current is defined by the direction in which the positive charges flow.

In analyzing circuits, we need to know the voltage across the element rather than the value of the electric field. We can calculate the voltage drop across the capacitor by integrating the electric field from one plate to the other. The result is simply

$$v = |\mathscr{E}|d \qquad (1.14)$$

where v is the voltage across the capacitor. Substituting Eq. (1.12), the expression for the electric-field magnitude, into Eq. (1.14) results in

$$v = \frac{qd}{\varepsilon A_C} \qquad (1.15)$$

and solving for q yields

$$q = \varepsilon \frac{A_C}{d} v \qquad (1.16)$$

Equation (1.16) relates the charge $+q$ and $-q$ stored on each of the capacitor plates to the voltage v across the capacitor. The relationship between these two quantities is linear; i.e., if the charge on each plate is doubled, the voltage is also doubled. We call the constant of proportionality, which depends on the geometry and construction of the capacitor, the **capacitance**. Capacitance is given the symbol C. For the parallel-plate capacitor we have been considering, the capacitance is given by

$$C = \varepsilon \frac{A_C}{d}$$

Capacitors need not have the parallel-plate geometry; they can be of all shapes and configurations. Any device capable of storing charge acts as a capacitance. We have, for all kinds of capacitors, the general relationship

$$q = Cv \tag{1.17}$$

Since q has units of coulombs, and v is measured in volts, the fundamental unit of capacitance is a coulomb per volt (C/V) which is called a **farad** (abbreviation: F). Note that because it takes energy to store charge in a capacitor, a capacitance is also an energy-storage device.

We can derive the terminal characteristics of the device by using the definition of current, Eq. (1.2):

$$i = \frac{dq}{dt}$$

Differentiating Eq. (1.17) results in

$$\frac{dq}{dt} = \frac{dCv}{dt} = C\frac{dv}{dt}$$

Finally, we find that the current-voltage relationship for a capacitor is given by

$$i = C\frac{dv}{dt} \tag{1.18}$$

In this equation, i is the current flowing through the capacitor, C is the capacitance, and v is voltage across the capacitor. This expression tells us that the **current through a capacitor is proportional to the time derivative of the voltage across it**; as long as the voltage is changing, there is current flowing through the capacitor. It follows that there is zero current through a capacitor if the voltage across it is constant, and vice versa.

The symbol for a capacitor is shown in Figure 1.25.

Figure 1.25 Circuit symbol for a capacitor, along with associated reference directions for current and voltage.

1.8.5 Inductance

Another circuit element which stores energy is the **inductor**. A common way of making an inductor is to wind a wire into a helix or coil, as shown in Figure 1.26.

Passing a current through a conductor results in a magnetic field encircling the wire. For the solenoid shown in Figure 1.26 the resulting magnetic field **B** is illustrated in Figure 1.27.

We will illustrate the phenomenon of **inductance** using the solenoid, but it is important to remember that inductance is an intrinsic property of all conductors, regardless of their shape. Inductance arises as a consequence of the interaction between the current and the magnetic field set up by the current.

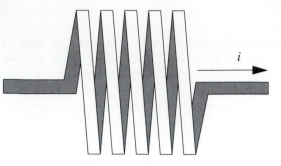

Figure 1.26 An inductor formed from a coil of wire. This geometry of inductor is known as a *solenoid*.

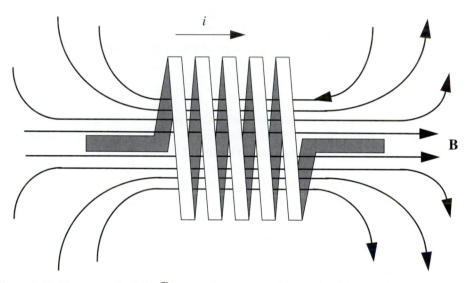

Figure 1.27 The magnetic field, **B**, set up by a current i in a solenoid. A change in the current results in a change in the magnetic field, which in turn induces a voltage which tends to oppose the original change in current.

If the current through the solenoid is constant, so is the magnetic field. In this case, there is no relative movement between the conductor and the magnetic lines of force. However, if the current is time varying, the magnetic field also varies, in step with the current. This causes the magnetic lines of force to cut across the conductor. The movement of the **B** field relative to the conductor generates a voltage which tends to oppose the change in current. The value of this voltage is given by Faraday's law, which for a solenoid is

$$v = NA\frac{d}{dt}|\boldsymbol{B}| \tag{1.19}$$

In this equation v is the voltage, N is the number of turns of wire, and A is the cross-sectional area of the inductor. We can relate the magnitude of the magnetic field to the current in the wire using Ampere's law; the result for a solenoid is

$$|\boldsymbol{B}| = \frac{N\mu i}{l} \qquad (1.20)$$

where l is the axial length of the solenoid and μ is a constant known as the **permeability**. Except for inductors which have a core of magnetic material with a high relative permeability, we will assume that $\mu \approx \mu_0$, the permeability of free space, which has a value of 1.257×10^{-8} Ω–s/cm.

Combining Eqs. (1.19) and (1.20), we have

$$v = \left[\frac{N^2 A\mu}{l}\right]\frac{di}{dt} \qquad (1.21)$$

Equation (1.21) shows that the voltage across an inductor is proportional to the time rate of change of the current through the inductor. The constant of proportionality is known as the **inductance** and is given the symbol L. The inductance of a solenoid is equal to the term in square brackets in Eq. (1.21). The general terminal relationship for any kind of inductor is

$$v = L\frac{di}{dt} \qquad (1.22)$$

where v is the voltage drop across the inductor. A consequence of Eq. (1.22) is that a constant current flowing through an inductor corresponds to a zero voltage drop across it. Conversely, if there is zero volts across the inductor, the current through it is constant in time.

The circuit symbol for an inductor is shown in Figure 1.28, where associated reference directions consistent with Eq. (1.22) have been used. The unit of inductance is the **henry** (abbreviation: H), which is equivalent to an ohm-second (Ω–s).

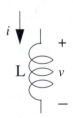

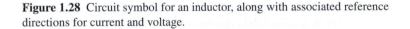

Figure 1.28 Circuit symbol for an inductor, along with associated reference directions for current and voltage.

1.8.6 Switch

The operation of an electrical switch is familiar to all; it passes current when closed, and does not pass current when open. There is a power switch on all computers; the keys of the keyboard are switches; switches are on our lamps, radios, toasters, cars, and coffee makers. What else is there to know about switches?

The apparent simplicity of a switch belies its usefulness as a circuit element and as a model for other elements. For example, the detailed explanation of transistor operation is quite involved, yet at the most basic level it acts as a switch. We will find in subsequent chapters that modeling a transistor as a switch is a most useful concept, allowing us to make reasonable and accurate estimates of the maximum operating speed of a computer.

The symbol for an ideal switch is shown in Figure 1.29. An ideal switch has two states, open and closed. The "switching event" takes place at the time indicated; by convention the switch is drawn in the position it assumes before the switching event. The switch in Figure 1.29 has a switching event at $t = t_0$ and is shown in the open position; thus, by convention, the switch is open for $t < t_0$, and closed for $t > t_0$. The origin of the time coordinate is often taken such that $t = 0$.

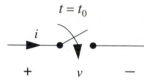

Figure 1.29 Symbol for a ideal switch which closes at $t = t_0$, along with associated reference directions for current and voltage.

"Open" and "closed" are defined as follows: an open switch has $i = 0$, with no restriction on the voltage across it; a closed switch has $v = 0$, with no restriction on the current through it. Upon closer inspection, we see that an open switch can be replaced with a current source of value zero, and a closed switch is identical to a zero-valued voltage source. These limiting cases are important enough to merit their own terminology and symbols, illustrated in Figure 1.30.

Ideal switches neither draw nor supply power. This can be seen from the power relationship $p = vi$; either the voltage or current associated with the switch is always zero, therefore the power p is likewise zero, regardless of the position of the switch.

Suppose we have an ideal switch which closes at $t = 0$. This means that $i = 0$ for $t < 0$ and $v = 0$ for $t > 0$. This definition does not strictly define the state of the switch exactly at $t = 0$. We will deal with this ambiguity by superscripting the zero in $t = 0$ with − and + signs, which signify an infinitesimally small increment of time before and after a switch closure. For example, we will specify an inductor voltage at the instant just before the switch closure with the notation $v_L(0^-)$; the inductor voltage just after a switch closure will be indicated by $v_L(0^+)$.

It is also possible to have switches which open at the switching event. The symbol for such a switch is shown in Figure 1.31.

1.9 Combining Circuit Elements

Using our knowledge of KVL, KCL, and the laws governing the current-voltage behavior of the various circuit elements, we are ready to solve any circuit. These fundamental rela-

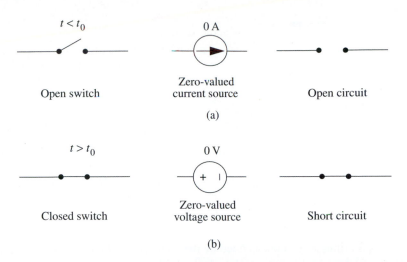

Open switch $t < t_0$

Zero-valued current source 0 A

Open circuit

(a)

Closed switch $t > t_0$

Zero-valued voltage source 0 V

Short circuit

(b)

Figure 1.30 (a) An open switch has a current-voltage relation identical to a current source with $i = 0$; this is called an **open circuit** and is often represented as two unconnected dots, as shown on the right. (b) A closed switch is indistinguishable from a voltage source with $v = 0$. We call this a **short circuit** and represent it with the symbol on the right.

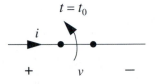

Figure 1.31 Symbol for a ideal switch which opens at $t = t_0$, along with associated reference directions for current and voltage.

tions are tools which allow us to write equations describing the circuit response. Once these equations have been solved, the behavior of the circuit is known.

Prior to applying these tools, often we combine certain circuit elements into simpler, equivalent, elements, if possible. This often simplifies the resulting expressions. In this section we show how to reduce particular combinations.

1.9.1 Elements in Series

A common arrangement of circuit elements is the **series** connection. Elements are connected in series if they share the same current.

Consider the circuit shown in Figure 1.32. The three resistors R_1, R_2, and R_3 are all in series with the voltage source. Our objective is to find the current i. This can be accomplished by replacing the series combination of the three resistors with a single equivalent resistance R_{eq}.

First, we note that all of the elements in this circuit have the same current going through them. For example, if we examine the node A, we observe that the current leaving the bottom of R_1 is the same current entering R_2. The same observation can be made at the three other nodes of the circuit, so the current in the loop is the same everywhere.

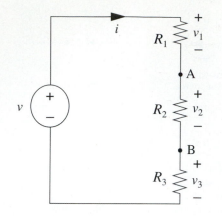

Figure 1.32 Series resistor circuit. The three resistors are in series with the voltage source. All elements in this circuit share the same current.

Starting at the node defined by the connection of the voltage source and R_3, application of KVL clockwise around the loop gives

$$-v + v_1 + v_2 + v_3 = 0$$

or

$$v = v_1 + v_2 + v_3$$

Using Ohm's law [Eq. (1.11)] for each voltage in this expression:

$$v = iR_1 + iR_2 + iR_3$$

Factoring out the i, we get

$$v = (R_1 + R_2 + R_3)i \tag{1.23}$$

If we represent these three resistors by one equivalent resistor of resistance R_{eq} (Figure 1.33), we see that this resistor must satisfy

$$v = R_{eq}i \tag{1.24}$$

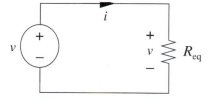

Figure 1.33 Equivalent for the circuit of Figure 1.32. The three series connected resistors have been replaced by an equivalent resistance R_{eq}.

Equations (1.23) and (1.24) are in terms of the same i and v. Dividing Eq. (1.23) by Eq. (1.24) and rearranging terms, we find that

$$R_{eq} = R_1 + R_2 + R_3 \tag{1.25}$$

This shows that we can replace these three resistors with one equivalent resistance R_{eq} equal to the sum of the resistances of the individual resistors. We can generalize Eq. (1.25) and say that for n resistors in series, the equivalent resistance R_{eq} is the sum of the n resistances:

$$R_{eq} = R_1 + R_2 + \cdots + R_n \qquad (1.26)$$

A similar result holds for inductors in series and voltage sources in series. Consider the series connection of two inductors shown in Figure 1.34. We would like to determine an expression for the voltage drop across the two inductors. Using KVL and the inductor current-voltage relationship [Eq. (1.22)], we can write:

$$v = v_1 + v_2$$

$$= L_1 \frac{di}{dt} + L_2 \frac{di}{dt}$$

$$= (L_1 + L_2) \frac{di}{dt}$$

If we let

$$L_{eq} = L_1 + L_2$$

this can be written as

$$= L_{eq} \frac{d}{dt}(i$$

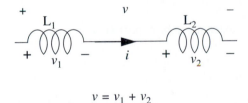

$$v = v_1 + v_2$$

Figure 1.34 Example of a series connection of two inductors, part of a larger circuit.

This shows that we can treat the series connection of two inductors as a single equivalent inductance, whose value is the sum of the inductances of the two individual elements. Generalizing this result, we can replace n inductors in series with an equivalent inductance L_{eq} equal to the sum of the n individual inductances:

$$L_{eq} = L_1 + L_2 + \cdots + L_n$$

Applying KVL to a series connection of n voltage sources, we similarly find that they can be replaced with an equivalent voltage source of value

$$V_{eq} = V_1 + V_2 + \cdots + V_n$$

To treat capacitors in series, we examine the circuit shown in Figure 1.35. Applying KVL clockwise around the loop, we find

$$-v + v_1 + v_2 + v_3 = 0$$

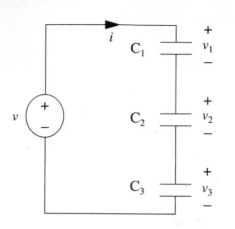

Figure 1.35 Series capacitor circuit. The three capacitors can be replaced with a single equivalent capacitance.

or

$$v = v_1 + v_2 + v_3 \qquad (1.27)$$

Differentiating each term in Eq. (1.27) gives us

$$\frac{dv}{dt} = \frac{dv_1}{dt} + \frac{dv_2}{dt} + \frac{dv_3}{dt} \qquad (1.28)$$

The current-voltage relationship for a capacitor can be expressed as

$$\frac{dv}{dt} = \frac{i}{C} \qquad (1.29)$$

Applying Eq. (1.29) to each of the derivative terms on the right hand side of Eq. (1.28) yields

$$\frac{dv}{dt} = \frac{i}{C_1} + \frac{i}{C_2} + \frac{i}{C_3}$$

Factoring out i yields

$$\frac{dv}{dt} = \left(\frac{1}{C_1} + \frac{1}{C_2} + \frac{1}{C_3} \right) i \qquad (1.30)$$

Figure 1.36 shows the circuit where the three series capacitors have been replaced with an equivalent capacitance C_{eq}. This circuit is governed by

$$\frac{dv}{dt} = \frac{1}{C_{eq}}i \tag{1.31}$$

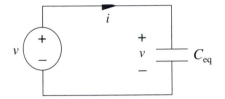

Figure 1.36 Equivalent for the series capacitor circuit of Figure 1.35. The three series-connected capacitors have been replaced by an equivalent capacitance C_{eq}.

Comparing Eqs. (1.31) and (1.30), we see that the equivalent capacitance is computed from this expression:

$$\frac{1}{C_{eq}} = \frac{1}{C_1} + \frac{1}{C_2} + \frac{1}{C_3} \tag{1.32}$$

Equation (1.32) shows that the equivalent reciprocal capacitance C_{eq}^{-1} of the capacitors in series is given by the sum of the individual reciprocal terms C_1^{-1}, C_2^{-1}, and C_3^{-1}. In general, for n capacitors in series, the equivalent capacitance C_{eq} is given by

$$\frac{1}{C_{eq}} = \frac{1}{C_1} + \frac{1}{C_2} + \cdots + \frac{1}{C_n} \tag{1.33}$$

Note that the equivalent capacitance of series-connected capacitors is always less than any of the individual capacitors.

1.9.2 Elements in Parallel

Another oft-used configuration of elements is the **parallel** connection. Elements are in parallel when they are connected between the same two nodes. Parallel elements share the same voltage.

Consider the two-node circuit shown in Figure 1.37. The three capacitors C_1, C_2, and C_3 are all connected in parallel with the voltage source v. Applying KCL at either node,

$$i = i_1 + i_2 + i_3 \tag{1.34}$$

The current-voltage relationship for a capacitor is

$$i = C\frac{dv}{dt}$$

which, when applied to each current in Eq. (1.34), results in

$$i = C_1\frac{d}{dt}(v) + C_2\frac{dv}{dt} + C_3\frac{d}{dt}(v)$$

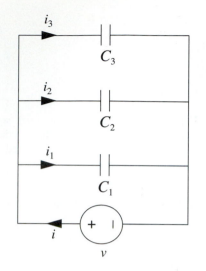

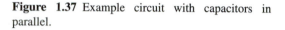

Figure 1.37 Example circuit with capacitors in parallel.

Factoring out the $\dfrac{dv}{dt}$ term,

$$i = (C_1 + C_2 + C_3)\frac{dv}{dt}$$

This can be written in the equivalent form

$$i = C_{eq}\frac{d}{dt}(v) \tag{1.35}$$

where the equivalent capacitance C_{eq} is given by

$$C_{eq} = C_1 + C_2 + C_3 \tag{1.36}$$

Expressed in words, C_{eq} is the sum of the individual parallel capacitances. We can generalize Eq. (1.36) and say that for n capacitors in parallel, the equivalent capacitance C_{eq} is

$$C_{eq} = C_1 + C_2 + \cdots + C_n$$

Example 1.4

"Virtually Parallel" Capacitors

In the next chapter we will encounter a configuration where two capacitors have one node in common, while each of the other nodes is held at a different constant voltage, as shown in Figure 1.38. This arrangement is an approximate representation of the input of a CMOS logic gate and is thus of considerable importance. In Figure 1.38 we see that C_1 has one end tied to V_S and the other tied to node v_C; C_2 also has one end tied to node v_C while its other end is tied to ground.

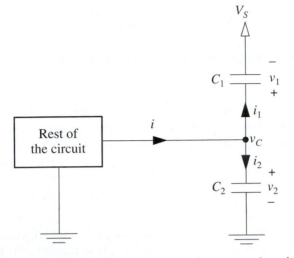

Figure 1.38 An example of a circuit with capacitors that share one node and each have the other node tied to a constant voltage. Note the assignment of associated reference directions for the two branch voltages v_1 and v_2.

We will show that this configuration has a simpler model with a single equivalent capacitance. Applying KCL at node v_C:

$$i = i_1 + i_2 \tag{1.37}$$

Substituting the current-voltage relationship for each capacitor into Eq. (1.37), we find that

$$i = C_1 \frac{dv_1}{dt} + C_2 \frac{dv_2}{dt} \tag{1.38}$$

The two branch voltages v_1 and v_2 can be related to the node voltages using KVL:

$$v_1 = v_C - V_S$$
$$v_2 = v_C$$

Inserting these relationships into Eq. (1.38),

$$i = C_1 \frac{d}{dt}((v_C - V_S)) + C_2 \frac{dv_C}{dt} \tag{1.39}$$

Since V_S is constant and the derivative of a constant is zero,

$$i = C_1 \frac{dv_C}{dt} + C_2 \frac{dv_C}{dt}$$

After combining terms,

$$i = C_{eq}\frac{dv_C}{dt}$$

where the equivalent capacitance C_{eq} is given by

$$C_{eq} = C_1 + C_2$$

Another way of looking at the arrangement of Figure 1.38 is this: it is the series combination of a voltage source (value V_S) and a capacitor (C_1), in parallel with another capacitor (C_2). The combination has an equivalent capacitance the same as if the voltage source were a short circuit; hence the terminology, "virtually parallel." Therefore, for the purposes of calculating the current flowing into the "virtually parallel" combination of capacitors, the equivalent capacitance is the same as if the two capacitors were actually in parallel.

Continuing our discussion of elements in parallel, consider the circuit of Figure 1.39, which shows two current sources in parallel with a resistor. Using KCL, we see that the current i flowing through the resistor is given by

$$i = i_1 + i_2$$

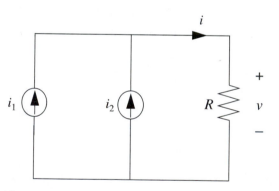

Figure 1.39 Circuit with current sources in parallel.

Clearly, if we were to replace the parallel combination of current sources with a single equivalent current source of value i, the conditions in the circuit would remain the same. Therefore, current sources in parallel can be replaced with an equivalent source equal to the sum of the individual currents.

Resistors in parallel are shown in Figure 1.40. The current i flowing into the parallel combination of resistors is found using KCL:

$$i = i_1 + i_2 + i_3$$

Applying Ohm's law to the current through each resistor,

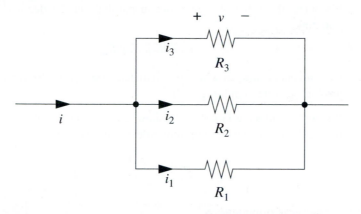

Figure 1.40 Parallel combination of resistors. The procedure for combining resistors in parallel into an equivalent resistance also applies to inductors in parallel.

$$i = \frac{v}{R_1} + \frac{v}{R_2} + \frac{v}{R_3}$$

which is the same as

$$i = \left(\frac{1}{R_1} + \frac{1}{R_2} + \frac{1}{R_3} \right) v$$

The term in parentheses can be replaced with an equivalent reciprocal resistance R_{eq}^{-1} given by

$$\frac{1}{R_{eq}} = \frac{1}{R_1} + \frac{1}{R_2} + \frac{1}{R_3}$$

In general, a parallel combination of n resistors can be replaced with one equivalent resistance R_{eq} given by

$$\frac{1}{R_{eq}} = \frac{1}{R_1} + \frac{1}{R_2} + \frac{1}{R_3} + \cdots + \frac{1}{R_n} \tag{1.40}$$

The combination of two resistors in parallel is quite common. We denote the equivalent value of two parallel resistors with this symbol:

$$R_1 \parallel R_2$$

which is read "R_1 parallel R_2." The equivalent resistance of two parallel resistors is given by the product over the sum:

$$R_1 \parallel R_2 = \frac{R_1 R_2}{R_1 + R_2} \tag{1.41}$$

Likewise, the value of n resistors in parallel can be denoted with the shorthand notation:

$$R_1 \parallel R_2 \parallel \ldots \parallel R_n$$

the equivalent value of which is calculated from Eq. (1.40) (and is NOT equal to the product over the sum for $n > 2$!).

It is a straightforward exercise to show that the equivalent inductance of n inductors in parallel is found using the same rule as for resistors:

$$\frac{1}{L_{eq}} = \frac{1}{L_1} + \frac{1}{L_2} + \frac{1}{L_3} + \cdots + \frac{1}{L_n}$$

1.9.3 Forbidden Combinations

Certain combinations of circuit elements, namely, ideal voltage sources in parallel and ideal current sources in series, are forbidden. To understand why, consider the circuit of Figure 1.41(a) which shows a parallel connection of two voltage sources, v_1 and v_2, where $v_1 \neq v_2$. Using KVL, the voltage between the two nodes v can be expressed as either

$$v = v_1$$

or

$$v = v_2$$

from which it follows that

$$v_1 = v_2$$

which contradicts the original assumption that the voltage sources have unequal values. Similarly, the series combination of unequal current sources in Figure 1.41(a) violates KCL.

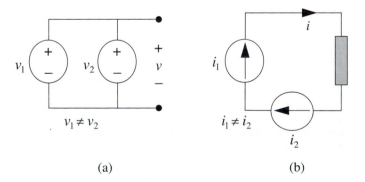

(a)   (b)

Figure 1.41 Forbidden combinations of circuit elements. (a) Parallel combination of unequal voltage sources. (b) Series combination of unequal current sources. Each of these combinations violates one of Kirchhoff's circuit laws.

Recalling that a short circuit is a zero-valued voltage source, and an open circuit is a zero-valued current source, we see that short-circuited ideal voltage sources and open-circuited ideal current sources are likewise forbidden combinations.

It may appear that this rule is rather restrictive; how can one decide that a circuit is somehow "invalid" and cannot be solved? In fact, the problem here is not that the circuits shown in Figure 1.41 cannot be solved, but that they cannot even be built, because doing so would violate physical law. The key concept in both of these examples is that the sources are "ideal," which is but a convenient idealization of reality.

To expand on this concept, consider a 1.5-V battery and a 9-V battery configured in parallel, as shown in Figure 1.42(a). There is nothing "forbidden" about this combination, since anyone can easily acquire two batteries and wire them up as shown. (It's not a good idea, though.) The question is, what value of voltage V would be measured between the two nodes?

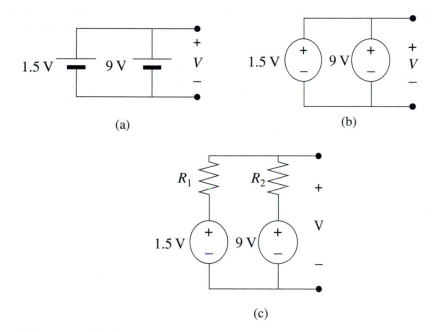

(a)

(b)

(c)

Figure 1.42 Two batteries of unequal voltage connected in parallel. (a) Circuit diagram for this configuration. The short heavy bar is the negative terminal of the battery. (b) Inaccurate first-order model of the circuit. (c) A more accurate model of the circuit.

Treating the batteries as ideal independent dc voltage sources leads to the model shown in Figure 1.42(b). We have already seen that this combination of ideal voltage sources in parallel is forbidden, so this description of the battery combination is inadequate. A better approximation is shown in Figure 1.42(c), where each battery is modeled as the series combination of an (ideal) dc voltage source and a (relatively small) resistor. This circuit does not violate any physical laws, and it predicts a value of V intermediate between 1.5 V and 9 V. (The value depends on the values of resistors R_1 and R_2. See Exercise 1.23.)

The key point is that ideal sources don't really exist. Ideal voltage and current sources cannot actually be built because, by definition, they can supply an infinite amount of power for an infinite amount of time, which is not possible with known technology. Real voltage sources, such as batteries, are more closely approximated with the combination of an ideal source and other components. In solving circuits, it is helpful to use abstractions like ideal sources, perfectly conducting wires, and other idealizations, as models of real circuits, because these approximations most often greatly simplify the analysis and usually lead only to negligible error. This example shows that we must be careful to distinguish between the real component and an idealization of that component.

1.10 Voltage- and Current-Divider Circuits

The circuit shown in Figure 1.43 is known as a **voltage divider**. The individual resistors in the series string each have voltage drop proportional to the their resistance. To illustrate this, we first find the loop current i. Replacing the series combination of resistors with the equivalent resistance given by Eq. (1.25), we find that

$$i = \frac{1}{R_1 + R_2 + R_3}v$$

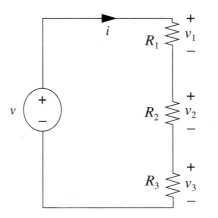

Figure 1.43 Voltage-divider circuit. The voltage across any resistor is proportional to its resistance.

The voltage drop across resistor R_1, v_1, is found using Ohm's law:

$$v_1 = iR_1$$

$$= \frac{1}{R_1 + R_2 + R_3}vR_1$$

$$= \frac{R_1}{R_1 + R_2 + R_3}v$$

Similarly, the voltage drops across the other two resistors are

$$v_2 = \frac{R_2}{R_1 + R_2 + R_3}v$$

$$v_3 = \frac{R_3}{R_1 + R_2 + R_3}v$$

This example illustrates the general voltage divider rule: in a series combination of n resistors, with total voltage drop v, the voltage across resistor k is given by

$$v_k = \frac{R_k}{R_1 + R_2 + \cdots + R_n}v$$

A related circuit is the **current divider**, Figure 1.44, which is a parallel combination of resistors fed with a total current i. It is easily shown that the current in any branch is inversely proportional to the resistance of that branch:

$$i_k = \frac{R_{eq}}{R_k}i$$

where the equivalent resistance of the parallel combination of n resistors is given by Eq. (1.40), repeated here:

$$\frac{1}{R_{eq}} = \frac{1}{R_1} + \frac{1}{R_2} + \frac{1}{R_3} + \cdots + \frac{1}{R_n}$$

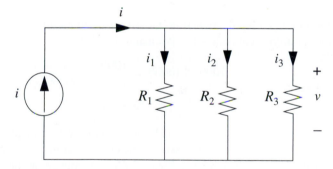

Figure 1.44 Current-divider circuit. The current in any branch is inversely proportional to the resistance of that branch.

The formulae for the currents in a current divider with two branches have a particularly simple form:

$$i_1 = \frac{R_2}{R_1 + R_2}i$$

$$i_2 = \frac{R_1}{R_1 + R_2}i$$

1.11 Resistive-Circuit Examples

To illustrate the basic principles of circuit analysis, we consider a few examples of resistive circuits—i.e., circuits which do not contain inductors or capacitors. (Circuits with inductors and capacitors occupy the remainder of the text.)

Example 1.5

Determine the current I in the circuit of Figure 1.45.

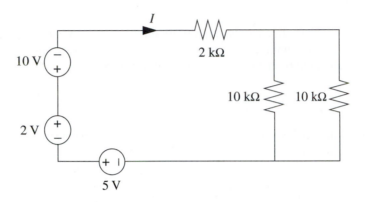

Figure 1.45 Resistive circuit with several voltage sources for Example 1.5.

We first note that the two 10-kΩ resistors are in parallel. Therefore we can combine them into one equivalent resistor using Eq. (1.41):

$$10000 \parallel 10000 = \frac{10000 \times 10000}{10000 + 10000}$$

$$= 5 \times 10^3$$

We next recognize that the 2-kΩ resistor and the 5-kΩ equivalent resistance of the parallel combination are in series. These may subsequently combined into a resistance of 7 kΩ. The circuit in Figure 1.45 now appears in the simpler form shown in Figure 1.46.

At this point we see that the three voltage sources are in series. Applying KVL, they can be combined into a single voltage source of value

$$V_{eq} = 5 + 2 + (-10) = -3$$

After this simplification, the circuit appears as shown in Figure 1.47. We can now easily determine the current using Ohm's law:

$$I = \frac{V}{R} = \frac{-3}{7 \times 10^3} = -429 \ \mu A$$

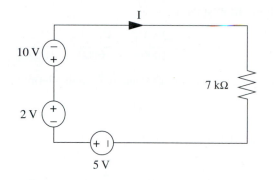

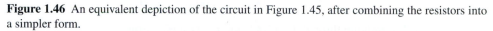

Figure 1.46 An equivalent depiction of the circuit in Figure 1.45, after combining the resistors into a simpler form.

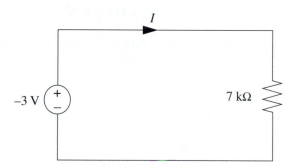

Figure 1.47 Circuit of Figure 1.45, after combining the series-connected voltage sources into a single equivalent source.

Example 1.6

Determine the voltage V_3 in the circuit below.

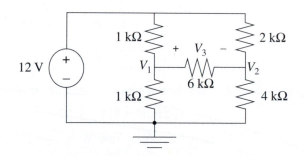

We first note that V_3 can be expressed as

$$V_3 = V_1 - V_2$$

Applying KCL at the V_1 node:

$$\frac{12 - V_1}{1000} = \frac{V_1 - V_2}{6000} + \frac{V_1}{1000}$$

Multiplying through by 6000 and rearranging terms:

$$13V_1 - V_2 = 72 \tag{1.42}$$

Applying KCL at the V_2 node:

$$\frac{12 - V_2}{2000} = \frac{V_2 - V_1}{6000} + \frac{V_2}{4000}$$

Multiplying through by 12000 and rearranging terms,

$$-2V_1 + 11V_2 = 72 \tag{1.43}$$

The simultaneous solution of Eqs. (1.42) and (1.43) will yield the voltage values. One way to proceed is to express these as a system of equations in matrix form:

$$\begin{bmatrix} 13 & -1 \\ -2 & 11 \end{bmatrix} \begin{bmatrix} V_1 \\ V_2 \end{bmatrix} = \begin{bmatrix} 72 \\ 72 \end{bmatrix}$$

Multiplying the top row by 11:

$$\begin{bmatrix} 143 & -11 \\ -2 & 11 \end{bmatrix} \begin{bmatrix} V_1 \\ V_2 \end{bmatrix} = \begin{bmatrix} 792 \\ 72 \end{bmatrix}$$

Replace the top row with the sum of the two rows:

$$\begin{bmatrix} 141 & 0 \\ -2 & 11 \end{bmatrix} \begin{bmatrix} V_1 \\ V_2 \end{bmatrix} = \begin{bmatrix} 864 \\ 72 \end{bmatrix}$$

The top row is now in terms of one variable, V_1. Solving,

$$V_1 = \frac{864}{141} = 6.13 \text{ V}$$

Inserting this value into the equation expressed by the second row,

$$(-2)6.13 + 11V_2 = 72$$

$$V_2 = 7.66 \text{ V}$$

Finally, we find V_3 from

$$V_3 = 6.13 - 7.66 = -1.53 \text{ V}$$

1.12 Power and Energy Relationships

An important aspect of electronic circuits is the rate of energy transfer among the various elements. For example, resistors always dissipate electrical energy (i.e., convert it into heat). In fact, the resistor symbol is often used in circuit diagrams to express any "permanent" removal of energy—e.g., a light bulb. In a computer or other system, this heat must be removed lest it produce an inordinate temperature rise. In this section we develop relationships between circuit variables (such as voltage and current) and the energy associated with circuit elements. (In the discussion that follows, we assume that all voltage and current directions have been assigned using the associated reference convention.)

1.12.1 Power/Energy Relationships for Resistors

Using Ohm's law and the general power formula for a circuit element,

$$p(t) = v(t)i(t)$$

we find that the power associated with a resistor, $p_R(t)$, is

$$p_R(t) = v(t)i(t) = [i(t)R]i(t) = Ri^2(t) \tag{1.44}$$

or, equivalently,

$$p_R(t) = v(t)i(t) = v(t)\left[\frac{v(t)}{R}\right] = \frac{v^2(t)}{R} \tag{1.45}$$

where $v(t)$ is the voltage drop across the resistor and $i(t)$ is the current through the resistor. Note that $p_R(t)$ is **always** a positive quantity. Since we used associated reference directions to define current through and voltage across a resistor, this result means that resistors **always** absorb power from the remainder of the circuit (and subsequently convert it into heat); they can never act as a source of energy.

At times it is useful to calculate the total amount of energy delivered to an element. We can calculate this energy by integrating power over time. For example, the energy absorbed by a resistor from $t = 0$ up to time t is given by the expressions

$$E_R(t) = \int_0^t p_R(t)dt = \int_0^t Ri^2(t)dt = R\int_0^t i^2(t)dt$$

or

$$E_R(t) = \int_0^t p_R(t)dt = \int_0^t \left(\frac{v^2(t)}{R}\right)dt = \frac{1}{R}\int_0^t v^2(t)dt$$

In either case we see that $E_R(t) \geq 0$, confirming that resistors always dissipate energy. In these expressions, the lower limit of $t = 0$ corresponds to the time at which the system was powered up. If we wish to determine the energy absorbed by the resistor over a different interval, then we insert the appropriate limits into the integral.

1.12.2 Power/Energy Relationships for Capacitors

Capacitors are devices which store energy in the form of an electric field, set up by the separation of charges. Unlike resistors, which always convert power supplied from the circuit into heat, capacitors do not dissipate energy; they either take energy from the rest of the circuit and store it or else release energy to the rest of the circuit.

Inserting the current-voltage relationship for a capacitor into the general power relationship $(p(t) = v(t)i(t))$, we can express the instantaneous power being delivered to the capacitor as

$$p(t) = Cv(t)\frac{d}{dt}v(t) \tag{1.46}$$

Note that there is no restriction on the sign of the value Eq. (1.46) can take on; therefore, it is possible for power to flow in or out of the capacitor.[6]

How much energy can a capacitor store? This can be determined by integrating the capacitor power over time:

$$E_C(t) = \int_0^t p(t)dt = \int_0^t Cv(t)\frac{d}{dt}v(t)dt$$

This can be put into a more useful form by recognizing that the integrand can be written as $Cv\ dv$:

$$E_C(t) = C\int_{v(0)}^{v(t)} v\,dv = \frac{1}{2}Cv^2\Big|_{v(0)}^{v(t)}$$

We further assume that the capacitor is initially uncharged (i.e., $v(0) = 0$). The final result is that the energy stored in a capacitor of value C with v volts across it is given by

$$E_C(t) = \frac{1}{2}C[v(t)]^2 \tag{1.47}$$

Note that, although the instantaneous power associated with a capacitor can be either positive or negative, the energy stored in the capacitor is always a nonnegative quantity.

1.12.3 Continuity of Capacitor Voltage

We see from Eq. (1.47) that the energy stored in a capacitor is proportional to the square of the voltage across it. As the voltage changes, charge flows into and out of the device,

---

[6.] If the amount of charge in the capacitor is an increasing function of time, we say that the capacitor is *charging*; if it is decreasing, the capacitor is *discharging*.

changing the energy stored within. An interesting question presents itself: is it possible for the stored energy to change abruptly? Put another way, can $E_C(t)$ be a discontinuous function?

This question is of importance in digital systems because the transitions between logic states are designed to be as fast as possible. Figure 1.48 shows a **step function**, an instantaneous, discontinuous change in logic level which we want to occur when a logic gate switches states. Because logic levels are represented in computers as stored charge in capacitors, the problem of discontinuous changes in $E_C(t)$ is significant.

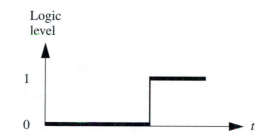

Figure 1.48 Example of a step function. In a computer, information is represented as logic levels which take on discrete values of 0 and 1. Ideally, transitions between levels occur instantaneously as switches in the computer open and close. An instantaneous, abrupt change in value is called a **step function**.

The answer to the question is no. Recall that power is defined as the time derivative of energy. This means that a discontinuity in $E_C(t)$ implies an infinite amount of power is being transferred within the circuit, since the derivative of a step function is unbounded. Although it is possible to draw a circuit schematic which uses only ideal elements that could supply infinite power, it is impossible to construct a **real** circuit capable of doing so. Therefore, we will make use of the real-world result that stored energy must be a continuous function of time.

This result has an important corollary: **capacitor voltages are continuous functions of time**. This is a direct consequence of the continuity of stored energy but can also be seen from Eq. (1.18), the branch relation for capacitance. A discontinuity in capacitor voltage would result in an infinite current through the capacitor. Again, it is possible to draw a circuit diagram which features infinite current, but actually building one would be impossible. In this book we will restrict ourselves to analyzing circuits which we are likely to encounter; thus, the restriction that capacitor voltages must be continuous is not at all limiting. The principle of capacitor-voltage continuity will prove to be extremely useful in solving circuit problems.

This restriction on capacitor voltage in no way prevents discontinuities in capacitor *current*. In fact, we will encounter many examples where the capacitor current can be described using a step function, especially in circuits containing switches.

1.12.4 Power/Energy Relationships for Inductors

Inductors store energy in the form of a magnetic field. Like capacitors, they can accept power from a circuit, or return it to the circuit, but they do not dissipate energy like resistors. In this section we derive an expression for the stored energy in an inductor as a function of current.

Again, the power associated with a two-terminal circuit element is given by

$$p(t) = v(t)i(t)$$

Inserting the current-voltage relationship for an inductor, the instantaneous power being delivered to the element is

$$p(t) = Li(t)\frac{d}{dt}i(t) \tag{1.48}$$

As before, there is no restriction on the sign of the value Eq. (1.48) can take on; therefore, it is possible for power to flow into or out of the inductor.

The energy stored in an inductor is found by integrating the inductor power over time:

$$E_L(t) = \int_0^t p(t)dt = \int_0^t Li(t)\frac{d}{dt}i(t)dt$$

The integrand can be written as $L\,i\,di$:

$$E_L(t) = L \int_{i(0)}^{i(t)} i\,di = \frac{1}{2}Li^2\Big|_{i(0)}^{i(t)}$$

We further assume that the inductor has an initial current of zero; this gives the final result that the energy stored in an inductor of value L, with i amperes of current flowing through it, is

$$E_L(t) = \frac{1}{2}L[i(t)]^2 \tag{1.49}$$

1.12.5 Continuity of Inductor Current

We saw in Section 1.12.3 that the stored energy in a capacitor cannot change instantaneously, otherwise an infinite amount of power (and capacitor current) would flow. An analogous situation exists with inductors. The stored energy associated with an inductance, given by Eq. (1.49), cannot change instantaneously; this implies that **inductor currents are continuous functions of time**. A discontinuity in inductor current would require an infinite amount of voltage (since $v = Ldi/dt$); this is not possible in actual circuits.

Although inductor currents must be continuous, inductor *voltages* can change instantaneously and abruptly without restriction.

1.13 Summary

This chapter has introduced some fundamental electrical concepts which are needed to analyze electronic circuit behavior. These concepts and laws lay the groundwork for linear circuit analysis. The following is a summary of the important points presented in this chapter.

The **voltage** between two points is given by

$$v_{ab} = v_a - v_b = \frac{w_{ab}}{q}$$

where w_{ab} is the work required to move charge q from point b to point a through a potential difference v_{ab}.

The **current** through an element is defined as

$$i = \frac{dq}{dt}$$

Branch voltages are the difference between two **node voltages**. Node voltages are referenced to the **ground** node.

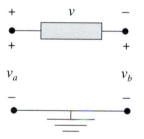

$$v = v_{ab} = v_a - v_b$$

Kirchhoff's voltage law (KVL) states that the algebraic sum of the branch voltage drops around any closed loop is zero. **Kirchhoff's current law** (KCL) states that the sum of the currents entering a node is equal to the sum of the currents leaving a node.

The **power** associated with an element is given by

$$p(t) = v(t)i(t)$$

Assuming associated reference directions are used:

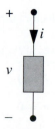

then power is delivered **to** an element when $p > 0$ and is being delivered **from** an element when $p < 0$.

The **resistance** of a uniform material of **resistivity** ρ, with cross-sectional area A and length l is

$$R = \rho \frac{l}{A}$$

The voltage drop across a resistor is given by **Ohm's law**:

$$v = Ri$$

The power dissipated by a resistor is

$$p = \frac{v^2}{R} = Ri^2$$

which is always nonnegative.

A parallel plate capacitor with plate area A, dielectric thickness d, and permittivity ε has **capacitance**:

$$C = \varepsilon \frac{A}{d}$$

The current-voltage relation for any **capacitor** is

$$i = C \frac{dv}{dt}$$

and the energy stored in a capacitor is

$$E_C = \frac{1}{2} C v^2$$

which is a nonnegative quantity. Capacitor voltages are continuous in time, but currents through a capacitor need not be.

The current voltage relation for an **inductor** is

$$v = L \frac{di}{dt}$$

and the energy stored in an inductor is

$$E_L = \frac{1}{2} L i^2$$

which is always nonnegative. Inductor currents are continuous in time; their voltages need not be.

An **ideal switch** which closes at $t = t_0$ has

$$i = 0 \qquad t \le t_0^-$$

$$v = 0 \qquad t \ge t_0^+$$

An open switch is equivalent to a zero-valued current source, also known as an **open circuit**. A closed switch is the same as a zero-valued voltage source, which is called a **short circuit**.

Care must be taken when using combinations of idealized elements such as voltage sources, current sources and switches. It is possible to draw a circuit with nonphysical realizations, which may result from otherwise reasonable attempts to simplify the modeling.

Problems

1.1 The inductor current in an underdamped RLC circuit is given by

$$i_L(t) = e^{-\alpha t}(I_1 \cos \omega_d t + I_2 \sin \omega_d t) + I_L$$

Determine the units of each of the variables in this expression.

1.2 The capacitor voltage in an undamped LC circuit is given by

$$v_C(t) = V_1 \cos \omega_0 t + V_2 \sin \omega_0 t + V_C$$

Determine the units of each of the variables in this expression.

1.3 The *time constant* τ of an *RC* circuit is given by $\tau = RC$. Suppose an *RC* circuit has a resistance R of 1200 Ω and a capacitance C of 4×10^{-8} F. Express the component values and time constant in engineering notation and the appropriate units.

1.4 An *RC* circuit has a time constant of 3.4 ns. If the capacitance is 24.2 pF, determine the value of resistance R, and express in the appropriate engineering units.

1.5 The instantaneous current flowing through a series *RC* circuit with a time constant of 41 µs is 3.4 mA. If the voltage across the resistor at this same time is 2.0 V, determine the value of the capacitor, and express this value in engineering units.

1.6 Given $V_1 = 3$ V, $V_5 = -5$ V, $V_8 = 13$ V, $V_9 = -7$ V, and $I_2 = 1$ mA, $I_3 = -4$ mA, $I_4 = 6$ mA, $I_6 = -2$ mA, $I_7 = 10$ mA, $I_{10} = -15$ mA, find V_2, V_3, V_4, V_6, V_7, V_{10} and I_1, I_5, I_8, I_9. Note that the branch voltages of all elements are defined using the **associated reference** convention; the voltage V_1 is shown explicitly for guidance.

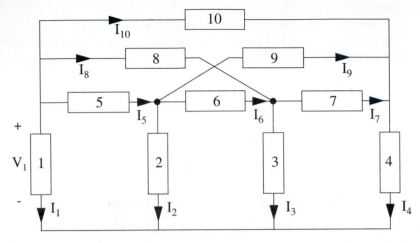

1.7   In the circuit of Exercise 1.6, use KCL to write algebraic expressions relating the currents (in symbolic form) at each node. Express these equations in matrix form. What is the size of the matrix? How many nodes are in this circuit?

1.8   Sketch the current-voltage relations for (a) a 5-V ideal independent voltage source, and (b) an ideal independent current source of value −10 mA. Using these graphs, explain the meaning of the terms "ideal" and "independent."

1.9   The voltage $v(t)$ in the following circuit varies as:

$$v(t) = 3\sin(120\pi t) + \cos(120\pi t)$$

(a) What is the maximum current through the capacitor?

(b) What is the maximum current through the resistor?

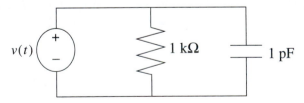

1.10   The voltage $v(t)$ in the circuit below is constant value of 5 V.

(a) What is the maximum current through the capacitor?

(b) What is the maximum current through the resistor?

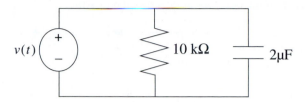

1.11 In the circuit below, determine the values of each of the three voltages across the resistors, and the current i.

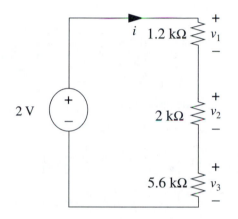

1.12 In the circuit of Exercise 1.11, determine the power dissipated by each of the resistors, and the total power supplied by the 2-V source.

1.13 In the circuit of Exercise 1.11, how much energy is dissipated by the 2-kΩ resistor over the time period the source supplies 15 mJ?

1.14 Is the parallel combination of a voltage source and a current source a "forbidden" combination? Explain why or why not.

1.15 Show that a zero-value voltage source has the same current-voltage relationship as a zero-value resistor.

1.16 Show that a zero-value current source has the same current-voltage relationship as a resistor of infinite value.

1.17 In the circuit below, the value of resistor R_1 is twice that of R_2 and half that of R_3. Determine the values of each of the three voltages across the resistors.

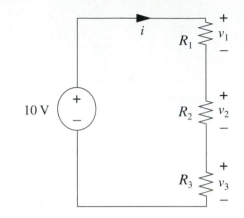

1.18 In the circuit below, determine the values of v_1, v_2 and i.

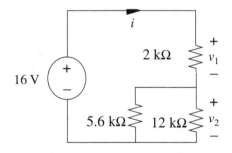

1.19 In the circuit below, determine the values of v_1, v_2 and R, given that i is 11 mA.

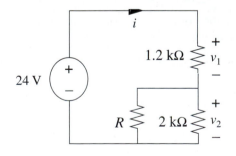

1.20 In the circuit below, the value of resistor R_2 is twice that of R_1 and half that of R_3. Determine the values of v_1 and v_2.

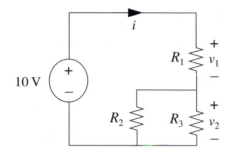

1.21 In the circuit below, the value of resistor R_2 is four times that of R_1. R_1 is half the value of the parallel combination of R_2 and R_3. Determine the values of v_1 and v_2.

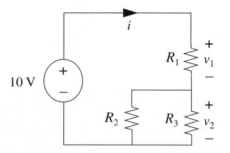

1.22 Determine the currents through each branch in the circuit of Example 1.6. Verify that the solution obtained satisfies both KVL and KCL.

1.23 Figure 1.42(c) shows the equivalent circuit of a 1.5-V and a 9-V battery wired in parallel, including the internal series resistance of each. Assume that the internal resistances are 0.3 Ω for the 1.5-V battery and 2.0 Ω for the 9-V battery. Determine the voltage V which appears across the terminals of the wired pair. Why is this experiment not a good idea?

1.24 In the circuit below, $R = 1\ \mathrm{k\Omega}$ and $V_S = 3\ \mathrm{V}$.

(a) Assume the switch is open. Determine the equivalent resistance between nodes **A** and **B**.

(b) Solve for the current I when the switch is closed.

c) With the switch closed, what is the node voltage at **C**? What is it when the switch is open?

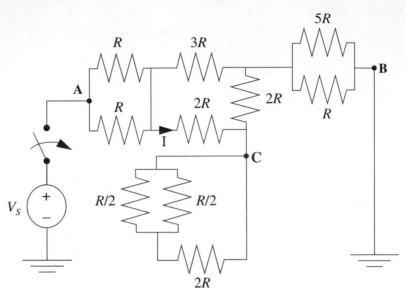

1.25 In the following circuit, solve for all node voltages. (The nodes are labeled alphabetically on the circuit diagram.) Is the 5-V voltage source absorbing or delivering power? Is the current source absorbing or delivering power? Calculate the power absorbed in all the resistors, and the net power delivered to the circuit by the sources. Comment on the relationship between the two quantities.

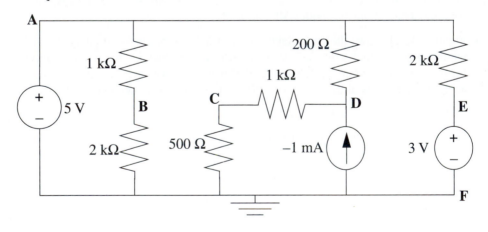

1.26 In the circuit below, $V_{S1} = 5$ V and $V_{S2} = 2$ V; determine I_1, I_2, and I_3. Is the V_{S2} source delivering or absorbing power?

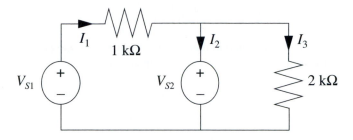

1.27 In the circuit below, we wish to determine the current i after the switch closes. Reduce the circuit into an equivalent form containing one equivalent resistance and capacitance, and determine their values.

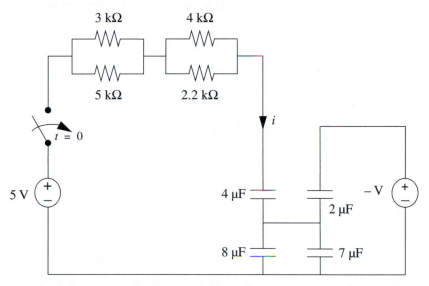

1.28 Reduce the circuit below to one equivalent resistor R_{eq} and determine the value of this equivalent resistance.

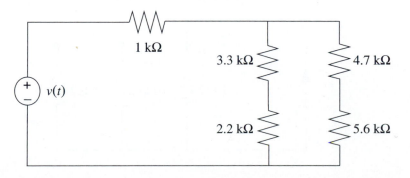

1.29 Show that equivalent capacitance of two capacitors in series can be expressed as

$$C_{eq} = \frac{C_1 C_2}{C_1 + C_2}$$

1.30 A load resistor, R_L, is connected to the two free terminals in the circuit below. Determine the maximum amount of power which can be dissipated in R_L.

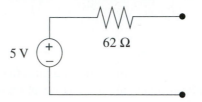

1.31 Repeat Exercise 1.30, replacing the 62-Ω resistor with an arbitrary resistance R_S and the 5-V source with an arbitrary source V_S.

1.32 Determine all branch voltages and currents in the circuit of Figure 1.45 (Example 1.5).

1.33 Determine v_S and the currents i_1 and i_2 in the circuit below, given that

$$v(t) = 2.3\cos(2\pi 10^3 t + 0.4)$$

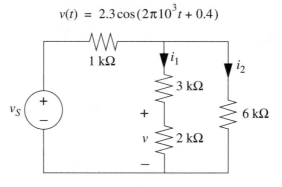

1.34 Determine v and the currents i, i_1, and i_3 in the circuit below, given that

$$i_2(t) = 0.055\cos 2\pi 10^3 t$$

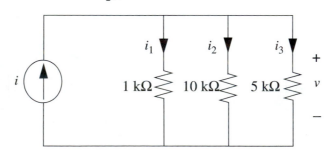

1.35 Determine the currents i_1 and i_2 in the circuit below.

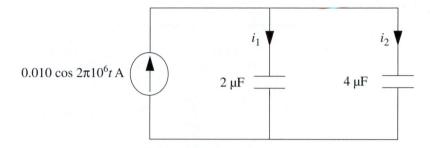

1.36 Determine the values of V_1, V_2, and V_3 in the circuit below.

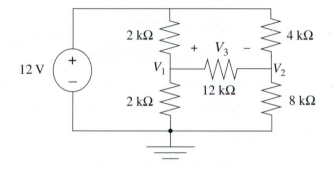

1.37 Write a set of equations in matrix form which determine the values of V_1 and V_2 in the circuit below.

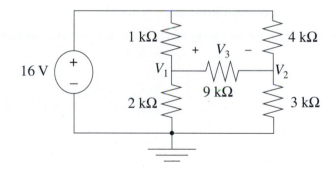

1.38 Determine the value of R in the circuit below which will result in a current of 4.5 mA through the 1-kΩ resistor.

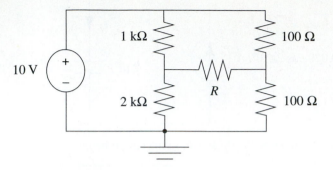

1.39 The current through a 200-μF capacitor is given by

$$i_C(t) = 4.0 \times 10^{-6} \cos 2\pi 10^6 t$$

Determine the voltage across the capacitor, if $v_C(0) = 1.0$ V.

1.40 In Exercise 1.39, write an expression for the energy stored in the capacitor as a function of time.

1.41 The current through an inductor (1 μH) in an underdamped RLC circuit is given by

$$i_L(t) = e^{-\alpha t}(I_1 \cos \omega_d t + I_2 \sin \omega_d t)$$

where $\alpha = 500 \times 10^6$ s^{-1}, $\omega_d = 800 \times 10^6$ s^{-1}, $I_1 = I_2 = 10$ mA.

(a) Determine the energy stored in the inductor at $t = 0$.

(b) Write an expression for the energy stored in the inductor as a function of time.

Gate Delay and *RC* Circuits

2

2.1 Introduction: Delays in Logic Circuits

Digital systems, including computers, make extensive use of **logic gates**, which are circuits that implement logical functions like AND, OR, NAND, NOR, and NOT. Logical variables can take on one of two values, 0 and 1, and logical functions act on these values in a specified way. For example, the logic function B = NOT (A) means that B = 0 if A = 1, and B = 1 if A = 0. Other logical functions depend on more than one input variable.

In designing a computer, the first consideration is functionality: does the design implement the desired function? At this level, delays between logic transitions are generally not of concern. Unfortunately, there is an unavoidable delay between the time an input signal enters a gate and the time something happens at the output. It is the cumulative effect of these delayed logic transitions which limits the speed of computers and other digital systems.

Consider the combinational logic circuit of Figure 2.1. Suppose we were to change the value of input A from zero to one. This change, which we refer to as a *logic transition*, affects the values of other logic signals throughout the circuit:

$$A: 0 \rightarrow 1$$
$$\Rightarrow \quad E: 1 \rightarrow 0$$
$$\Rightarrow \quad G: 0 \rightarrow 1$$

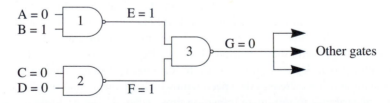

Figure 2.1 Two-stage NAND-gate logic circuit with logic signals A through G.

As an aid to interpreting the specific details of the logic transitions we plot the logic signals A through G as functions of time in a *timing diagram*. The timing diagram for the logic circuit of Figure 2.1 is illustrated in Figure 2.2.

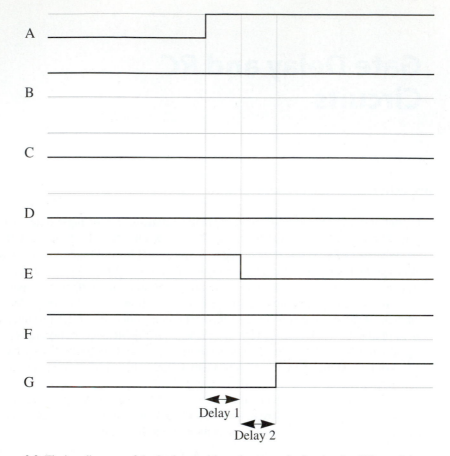

Figure 2.2 Timing diagram of the logic transitions that occur in the circuit of Figure 2.1.

As shown in Figure 2.2, the logic transitions E: $1 \to 0$ and G: $0 \to 1$ do not happen instantaneously with A: $0 \to 1$; delays are incurred from one logic transition to the next. The time delay between logic transitions is called the transition time or **gate delay**. Delay 1 is the time required for the logic transition in gate 1 of Figure 2.1 to occur; Delay 2 is the time required for the logic transition in gate 3 to occur. When signal E reaches logic zero, the transition G: $0 \to 1$ begins. Thus, the total transition time for this combinational logic circuit is given by the sum of Delay 1 and Delay 2. Decreasing the transition times increases the speed with which circuits containing logic gates can operate. Engineers seek to minimize transition times when designing logic gates, since gate delay limits overall system speed.

Timing diagrams, like that shown in Figure 2.2, are a high-level abstraction of the logic circuit. In designing a computer it is useful to treat logic signals as digital, with two discrete values we call logic low (0) and logic high (1). However, to understand the fundamental limitations on speed which are imposed by the physical realization of the circuit, this abstraction is insufficient. We need to treat the discrete logic transitions as continuous voltage transitions, rather than instantaneous and abrupt changes. For example, in a real computer, the signal A might transition from 0 V to 5 V. If we were to view the signals in the circuit of Figure 2.1 with an oscilloscope, we would observe waveforms like those shown in Figure 2.3.

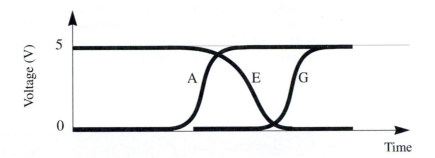

Figure 2.3 Actual voltage transitions of the circuit depicted in Figure 2.1. The signals inside a "digital" computer are actually continuous-time, analog signals.

Figure 2.3 illustrates the fact that the logic transitions are actually continuous functions that vary between 0 V and 5 V. (In fact, the signals can go outside the design range because of noise, interference from other sources, and circuit inductance. We'll talk more about these phenomena later.) We will see that instantaneous, abrupt changes in the signals are not possible, as these would violate physical principles. Therefore, we can conceptualize a logic transition as a gradual process that occurs over a period of time. The length of the transition time varies, depending on the family of logic gates used in the design. Furthermore, it is necessary to precisely define what we mean by "transition time," since the signals are continuous.

2.2 Transition Times in CMOS

The most widely used families of logic are built using a semiconductor technology called CMOS, which is an acronym for *C*omplementary *M*etal *O*xide *S*emiconductor. In this chapter, and throughout the text, we will develop circuit-analysis techniques to predict the speed of computers and other digital systems. As a vehicle, we will use the example of delays in interconnected CMOS inverters and CMOS logic gates, and relate these delays to overall computer performance. We will see that even simple models of gate switching will provide reasonable estimates of system limitations.

As a start, we now introduce the operation of a CMOS inverter. The description here, which treats the transistors inside the inverter as switches, is very useful in understanding gate delays. Although this model does not treat the details of how transistors operate at the level of holes and electrons, it is more than adequate for our purposes. Much more sophisticated models are, of course, necessary for other purposes, such as MOS transistor design.

First, we consider what we mean by "logic low" and "logic high" in the context of a CMOS inverter which is designed to operate with a 5-V power supply. Typically, in

CMOS logic gates any input voltage between roughly 80% and 100% of the maximum voltage is recognized as logic high, while any input voltage between 0% and about 20% of the maximum voltage is recognized as logic low. That is, for 5 V of "logic swing" we could say that a logic transition will occur when the input voltage has increased above 4 V or fallen below 1 V. In this example an input voltage of 4 V is referred to as the **high voltage threshold** (V_{th}) and one of 1 V is the **low voltage threshold** (V_{tl}). The region between the low and high voltage thresholds is logically undefined. An illustration of the threshold voltages and voltage ranges which define the two logic levels is shown in Figure 2.4.

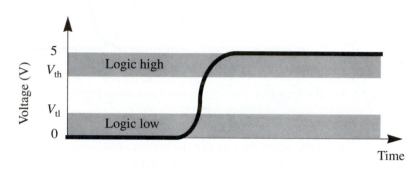

Figure 2.4 Illustration of the high and low threshold voltages, and the concept of voltage ranges for logic high and low.

In Figure 2.3, the transition times for signals E and G are not the same. We distinguish such cases with the concepts of rise and fall times. The transition time from a logic low to a logic high is called the **rise time**; that from logic high to logic low is called the **fall time**. In the example illustrated in Figure 2.3 we observe that the fall time is slightly greater than the rise time.

Our goal is to accurately approximate the rise and fall times of a CMOS inverter through the construction of a simple circuit model. To develop a valid approximation of the characteristics of a CMOS inverter we next look at the components inside the inverter and develop a functional understanding of their operation.

2.3 Inside the CMOS Inverter

The construction of a CMOS inverter is illustrated in Figure 2.5. Recall from Chapter 1 that the arrow pointing upward is a node held at a constant potential, in this case 5 V, with respect to the ground node. This upward-pointing arrow is equivalent to drawing a 5-V dc voltage source with its negative terminal tied to ground.

The circuit symbols inside the inverter in Figure 2.5 represent p-channel (PMOS) and *n*-channel (NMOS) MOS transistors. (These symbols may be unfamiliar to you, but there is no need to worry, as we don't have to understand the operation of the transistors in any detail.) A highly simplified model of the operation of PMOS and NMOS transistors is shown in Figure 2.6. The idea is that the position of the switch, which is connected

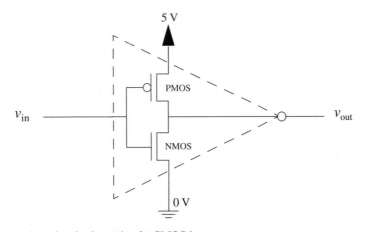

Figure 2.5 Transistor-level schematic of a CMOS inverter.

between the two terminals on the right, is a function of the value of the voltage v_{in} (measured with respect to ground). For example, if v_{in} rises above V_{th}, the two terminals on the right of the PMOS transistor look like an open circuit, while the two terminals on the right of the NMOS transistor are connected together through a closed switch. If v_{in} drops below V_{th}, nothing at all happens to the switches, until v_{in} drops below V_{tl}. At this point the switch associated with the PMOS transistor closes, and the switch associated with the NMOS transistor opens.

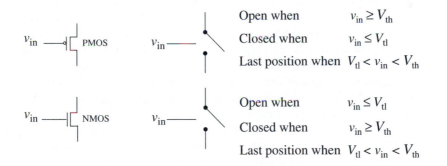

Figure 2.6 Left: symbols for PMOS and NMOS transistors. Right: zeroth-order models for the transistors. If the input voltage is between the two threshold values, the switch remains in its last known position.

    The simple switch models for the PMOS and NMOS transistors can be combined into a dual-switch model for the CMOS inverter. This model is illustrated in Figure 2.7. It provides a rudimentary yet functionally accurate view of the operation of a CMOS inverter, insofar as it outputs a digital "0" for an input of digital "1," and vice versa.

    The two switches in Figure 2.7 move simultaneously and are said to be *ganged*. These ganged switches are also presumed to move instantaneously from one position to

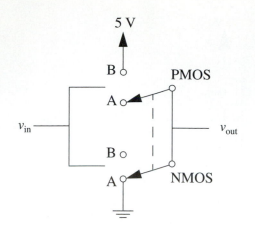

Figure 2.7 Zeroth-order dual-switch model of a CMOS inverter. This model is obtained by inserting the models for the NMOS and PMOS transistors (Figure 2.6) into the CMOS inverter circuit (Figure 2.5). The switches are ganged (meaning that they switch together); this is symbolized by the dashed line. When $v_{in} \geq V_{th}$ (i.e., logic high), both switches are in positions A, and $v_{out} = 0$ V (logic low). Similarly, when $v_{in} \leq V_{tl}$, both switches are in positions B, and $v_{out} = 5$ V.

the other whenever the input voltage crosses a threshold value. Figure 2.7 illustrates another important point: there is no connection between the input and output of the inverter. The value of the voltage at the input simply determines the position of the ganged switch.

It would be wonderful if CMOS inverters could be accurately modeled using the circuit of Figure 2.7. Unfortunately, several aspects of MOS transistors, not included in Figure 2.6, significantly limit the performance of integrated circuits. So, instead of using this zeroth-order model of a CMOS inverter, we will include the effects of two important transistor properties to establish a first-order model. This model will contain sufficient detail to model gate and interconnect delay using straightforward circuit-analysis techniques.

Although MOS transistors do behave as switches, they are not ideal switches. The path between the two "output" terminals (i.e., the two terminals on the right in Figure 2.6) is not a perfect conductor. For our purposes this path can be modeled in terms of an average resistance, typically a few hundred ohms. The dual-switch model for the CMOS inverter incorporating output resistances R_p and R_n for the PMOS and NMOS transistors is shown in Figure 2.8. These resistances, at the outputs of gates, are one factor limiting the transition speed of digital logic circuits.

The other important aspect of MOS transistors not present in the zeroth-order model is the input capacitance. This capacitance, while small, is extremely important in determining the gate delay, because it takes a finite time to change the amount of charge on a capacitor. We will account for this effect by including in the CMOS inverter two capacitors, C_p and C_n, which represent the input capacitances of the PMOS and NMOS transistors, respectively. Inclusion of these capacitors results in a model for the CMOS inverter as shown in Figure 2.9.

We recognize that the ganged switch in Figure 2.9 can be replaced with a single switch. This results in our first-order model for the CMOS inverter, Figure 2.10.

The CMOS inverter model shown in Figure 2.10 operates as follows. When the input voltage v_{in} rises above the high voltage threshold (V_{th}), the switch instantaneously moves down to position D; when v_{in} falls below the low voltage threshold (V_{tl}), the switch instantaneously moves up to position U. R_p represents the average resistance of the PMOS transistor channel, which, when turned on by an input voltage falling below V_{tl}, ultimately

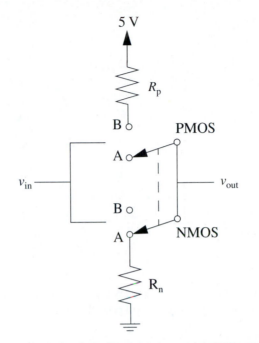

Figure 2.8 Model of a CMOS inverter including average output resistances of the MOS transistors.

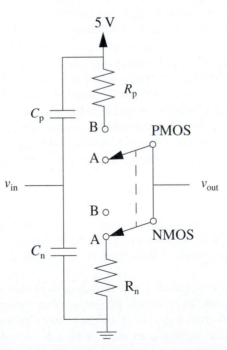

Figure 2.9 The CMOS inverter model incorporating input capacitance. C_p represents the capacitance of the PMOS transistor; C_n is the capacitance of the NMOS transistor.

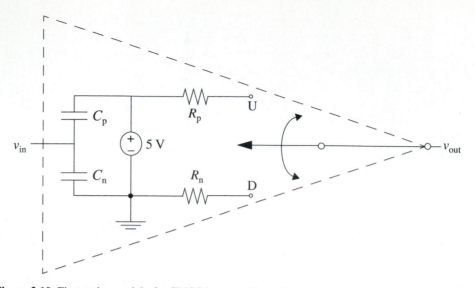

Figure 2.10 First-order model of a CMOS inverter. The voltage source is the supply voltage which sets the value of the logic swing; we will generally use a value of 5 V.

pulls up the output voltage v_{out} to 5 V. Similarly, R_n represents the average resistance of the NMOS transistor channel, which, when turned on by an input voltage rising above V_{th}, ultimately pulls the output voltage down to 0 V.

Logic families other than CMOS may operate differently but almost all can be reasonably represented, for timing purposes, with this same basic circuit model. A more complete inverter model would include many other components; for example, it would have capacitors representing the charge stored in the channels of the MOS transistors. While such a model might bring a bit more realism to bear, it would unnecessarily complicate the subsequent analysis without providing much additional insight. The simple circuit model of Figure 2.10 is sufficient to explain many of the limitations of logic-circuit switching. (In Chapter 5 we will consider one additional improvement in the inverter model: the parasitic inductance of the wires connecting the gate to its power supply.)

2.3.1 Interconnected CMOS Inverters

In our CMOS inverter model, the input voltage v_{in} determines the position of the switch. We have made the idealization that the switch moves *instantaneously* whenever v_{in} crosses a logic threshold boundary. Why, then, is there a gate delay if the switch moves instantaneously?

To answer this question, it is important to realize that inverters (and gates) do not operate in isolation. To be useful, the output of the inverter must be connected to something: another inverter, a gate, or an output circuit which performs some function like displaying a pixel or triggering an alarm. (We call such devices *loads*.) Even though the open-circuited output appears to switch instantaneously, once the inverter is connected to something else the situation changes dramatically.

Most of the gates inside a computer are connected to other gates. To approximate gate delay, we will look at the connection of two or more cascaded inverters, as shown in Figure 2.11. It is the combination of the output resistance (R_p or R_n) of one inverter, and the input capacitance (C_p and C_n) of the following inverter that results in a noninstantaneous logic transition of the node voltage v_C at the connection of the two inverters. The voltage v_C is the output voltage of the driver as well as the input voltage of the driven (load) inverter.

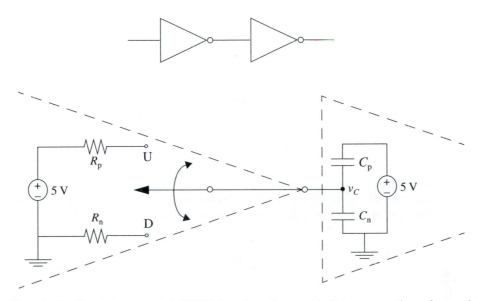

Figure 2.11   Two interconnected CMOS inverters. *Top:* symbolic representation of cascaded inverter pair. The inverter on the left is the driving inverter, or driver; the inverter on the right is the load. *Bottom:* Those parts of the inverter models necessary to solve for $v_C(t)$ are shown in this diagram.

We have not shown the input capacitance of the driving inverter in Figure 2.11. We are interested in calculating the time difference between the "switching event" and the change in logic level of the node v_C. Since we are assuming that the switching event follows the gate input, the driver input capacitance has no place in our calculation. (It *will* affect the delay of preceding gates, though.) Similarly, the output circuitry of the load inverter, containing the output resistors and switch, has no effect whatsoever on v_C. For these reasons, only those portions of the cascaded inverter circuit necessary to solve for the gate delay are shown in Figure 2.11.

The circuit shown in Figure 2.11 can be analyzed by considering the two possible switching events: the *pull-down* case which occurs when the switch moves down to position D, and the *pull-up* case which occurs when the switch moves up to position U. The two simplified circuits corresponding to each case are shown in Figure 2.12 and Figure 2.13.

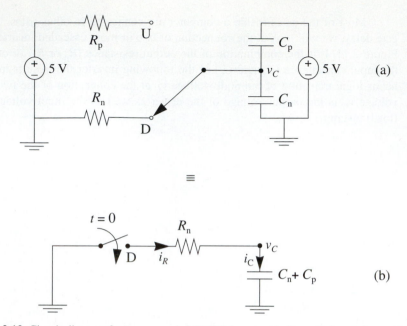

Figure 2.12 Circuit diagram for two cascaded CMOS inverters in the pull-down phase. (a) The circuit extracted directly from Figure 2.11. (b) Simplified version of (a), with the "virtually parallel" capacitances C_n and C_p combined into a single equivalent capacitance (as described in Chapter 1).

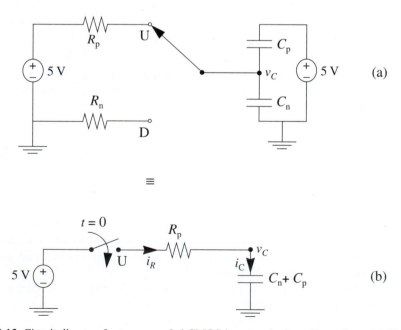

Figure 2.13 Circuit diagram for two cascaded CMOS inverters in the pull-up phase. (a) The circuit extracted directly from Figure 2.11. (b) Simplified version of (a).

The circuits illustrated in Figure 2.12(b) and Figure 2.13(b) both contain one resistor and one capacitor. Circuits that contain only resistors and capacitors, along with sources, are called **RC circuits**; these particular circuits, containing only one capacitor, are known as **first-order** *RC* circuits. In general, when a circuit contains an energy-storage element (capacitance or inductance), or a switch, the circuit currents and voltages are functions of time. (This is in contrast to circuits with only resistors and dc sources; the currents and voltages in such circuits are also dc.) To determine the circuit current and voltages ("waveforms") we will apply Kirchhoff's laws and our knowledge of circuit elements to write differential equations that characterize the behavior of the circuit.

For the pull-down circuit in Figure 2.12(b), application of KCL at the node labeled v_C yields

$$-i_R + i_C = 0 \tag{2.1}$$

After substituting the appropriate branch relations for the currents into Eq. (2.1) we obtain

$$\frac{v_C}{R_n} + (C_n + C_p)\frac{dv_C}{dt} = 0 \tag{2.2}$$

Rearranging Eq. (2.2),

$$(C_n + C_p)\frac{dv_C}{dt} = -\frac{v_C}{R_n} \tag{2.3}$$

Similarly, for the circuit of Figure 2.13(b) the application of KCL at the node labeled v_C yields the same result as Eq. (2.1). Substituting the appropriate branch relations for the currents we obtain

$$-\frac{5 - v_C}{R_p} + (C_n + C_p)\frac{dv_C}{dt} = 0 \tag{2.4}$$

Rearranging Eq. (2.4), we get

$$(C_n + C_p)\frac{dv_C}{dt} = -\frac{v_C}{R_p} + \frac{5}{R_p} \tag{2.5}$$

Letting $C \equiv C_n + C_p$ and $R = R_n$ or R_p (as appropriate), both Eq. (2.3) and Eq. (2.5) can be expressed with a single expression:

$$C\frac{dv_C}{dt} = -\frac{1}{R}(v_C - V_S) \tag{2.6}$$

In this equation, if we set V_S equal to zero we get Eq. (2.3); setting V_S equal to 5 V yields Eq. (2.5). Finally, we will rearrange Eq. (2.6) slightly:

$$RC\frac{dv_C}{dt} + v_C = V_S \tag{2.7}$$

This equation describes the series RC circuit shown in Figure 2.14.

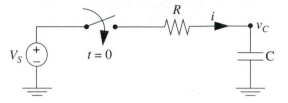

Figure 2.14 Generic series RC circuit characterized by Eq. (2.6) can represent either the pull-down case of Figure 2.12 (with $V_S = 0$, $R = R_n$, and $C = C_p + C_n$) or the pull-up case of Figure 2.13 (with $V_s = 5$, $R = R_p$ and $C = C_p + C_n$).

It is important to note that in deriving Eq. (2.7), we have taken the case where the switch is closed. Thus this equation is valid only for $t > 0$.

The voltage and current waveforms of the circuit in Figure 2.14 can be obtained using a variety of techniques. In the next section we present one technique which will yield the voltage at node v_C as a function of time. Once we know $v_C(t)$, we can calculate the length of the logic-gate delay time.

2.4 Solving First-Order *RC* Circuits

2.4.1 The General Case

In the previous section we derived the differential equation describing this circuit, Eq. (2.7), by using KCL. There is nothing sacred about using one technique or another; we must get the same result no matter how we go about it. To illustrate this, we will derive Eq. (2.7) by formally applying KVL instead of KCL.

In Figure 2.15 we have redrawn the series RC circuit along with some additional notation. Again we will assume the switch is closed; this restricts the solution we obtain to times $t > 0$. By Ohm's law the voltage across the resistor is given by

$$v_R = iR \tag{2.8}$$

Applying KVL around the loop in the direction shown, starting from the leftmost ground node, yields

$$-V_S + v_R + v_C = 0$$

which can be written

$$v_R + v_C = V_S$$

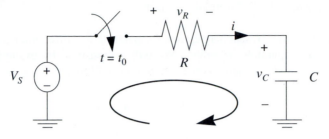

Figure 2.15 The series RC circuit model of a coupled CMOS inverter pair. The loop indicates the direction of travel we use in applying KVL.

After substitution of Eq. (2.8),

$$iR + v_C = V_S$$

Since the capacitor and the resistor are in series, the same current flows through them. We can therefore use the capacitor relation for the current,

$$i = C \frac{dv_C}{dt}$$

which results in

$$RC \frac{dv_C}{dt} + v_C = V_S \qquad (2.9)$$

which is the same as Eq. (2.7).

Equation (2.9) describes how the capacitor voltage varies as a function of time, but not in an obvious way. This equation is an example of a **first-order, linear ordinary differential equation**, so called because:

- only the first derivative of the variable v_C is present (hence, first order);

- the variable v_C occurs as a linear term, not as an argument of another function (hence, linear);

- there are no partial derivatives (therefore, an "ordinary" rather than a "partial" differential equation);

- instead of an explicit expression for the variable, v_C is expressed in terms of its derivatives (so it is a differential equation).

We must figure out a way to find $v_C(t)$ from Eq. (2.9). There are many ways to accomplish this; we describe a simple method in the next section.

2.4.2 Solving the Differential Equation

The method we present here involves separating the differential equation into expressions which are separately integrable. We will demonstrate by example. First, we perform a series of straightforward algebraic steps on Eq. (2.9), as shown in the steps below:

$$RC\frac{dv_C}{dt} = V_S - v_C$$

$$RC\,dv_C = (V_S - v_C)dt$$

$$\frac{dv_C}{V_S - v_C} = \frac{dt}{RC}$$

$$\frac{dv_C}{v_C - V_S} = -\frac{dt}{RC}$$

The objective here is to put all of the terms containing the variable v_C on the left and all of the terms containing the variable t on the right.

The next step is to integrate both sides of this equation. Since v_C is a function of t, we need to have the limits of integration on both sides correspond. This is accomplished by integrating the right side from t_0 to t; the corresponding limits on the left side are $v_C(t_0)$ and $v_C(t)$:

$$\int_{v_C(t_0)}^{v_C(t)} \frac{dv_C}{v_C - V_S} = -\frac{1}{RC}\int_{t_0}^{t} dt$$

After integration and evaluation at the limits of integration,

$$\ln\left(v_C(t) - V_S\right) - \ln\left(v_{C0} - V_S\right) = -\frac{1}{RC}(t - t_0) \tag{2.10}$$

where we have introduced the notation for the *initial condition*:

$$v_{C0} \equiv v_C(t_0) \tag{2.11}$$

Evaluating Eq. (2.10) gives us

$$\ln\left(\frac{v_C(t) - V_S}{v_{C0} - V_S}\right) = -\frac{1}{RC}(t - t_0) \tag{2.12}$$

We get an explicit expression for $v_C(t)$ by exponentiating both sides of Eq. (2.12) and performing some algebra. The result is the solution to the first-order differential equation in Eq. (2.9):

$$v_C(t) = (v_{C0} - V_S)e^{-\frac{1}{RC}(t - t_0)} + V_S \tag{2.13}$$

The denominator of the argument of the exponential function, as we will see, has a great deal of significance; we call it the **time constant** of the circuit. Time constant is denoted by the Greek letter τ (tau) and has units of seconds. For a series RC circuit $\tau = RC$. Therefore, the general solution given in Eq. (2.13) can be equivalently written

$$v_C(t) = (v_{C0} - V_S)e^{-\frac{t-t_0}{\tau}} + V_S \qquad (2.14)$$

$$\tau = RC$$

The value of the time constant dictates how rapidly the voltage $v_C(t)$ reacts to the change in the switch position. The larger τ is, the slower the response; i.e., v_C changes from its initial to final value over a longer period of time.

Example 2.1

A certain set of cascaded CMOS inverter gates have $C_{in} = 1$ nF, and $R_n = R_p = 1$ kΩ. Determine the time constant.

Since the pull-up resistance (R_n) and pull-down resistance (R_p) are identical, both the pull-up and pull-down time constants are the same and are given by $\tau = RC$:

$$\tau = RC$$

$$= 10^3 10^{-9}$$

$$= 10^{-6} \text{ s}$$

The time constant is 1 μs.

Equation (2.14) is known as the **total response** of the circuit. It describes the response of the circuit from time t_0 onward. The solution is composed of two terms, of which one depends on t, and the other is independent of t:

$$(v_{C0} - V_S)e^{-\frac{t-t_0}{\tau}} \qquad \text{and} \qquad V_S$$

When written in this form, the solution illustrates some general features about all first-order circuits. The first term of the general solution is given by

$$(v_{C0} - V_S)e^{-\frac{t-t_0}{\tau}}$$

This term is called the **transient response** of the circuit. The transient response of all first order circuits is a decaying exponential function, with a characteristic decay time of τ. We call this part of the solution a "transient" because it decays away to insignificance after the passage of a certain amount of time. This can be seen by examining the limiting value of the transient term as t approaches infinity:

$$\lim_{t \to \infty}\left[(v_{C0} - V_S)e^{-\frac{t-t_0}{\tau}}\right] = 0$$

The transient response bridges the gap in voltage between the initial capacitor voltage and the second part of the general solution. In the early stages, just after the switching event, the transient response dominates the total response of the circuit. It then decays to zero as t becomes large. (A quantitative description of "large" is given in the next section.)

The second part of the solution, which does not depend on t, dominates the total response at large values of t. The time-independent term is called the **dc steady-state response**. In this example the dc steady-state response is equal to the value of the voltage source, V_S. This term is always given by

$$\lim_{t \to \infty} v_C(t);$$

hence you will sometimes see the notation $v_C(\infty)$ for the dc steady-state capacitor voltage.

The general solution of the first-order circuit is encountered many times in electronic analysis. We will apply the solution to the problem of switching speed of cascaded inverter pairs.

2.5 *RC* Delays in Integrated Circuits

Using a first-order *RC* model of interconnected inverters, we see that the voltage at the input of the second inverter does not change instantaneously when the first inverter switches states. By quantifying the delay associated with the two inverters, we can make a first estimate of the speed of a computer.

We will generally take the origin of the time axis to be at the instant of switching. This results in $t_0 = 0$ and a considerable simplification in notation.

2.5.1 Inverter Pull-Down

In Figure 2.16 we have drawn the circuit for two cascaded inverters where the second inverter is being pulled down from logic high (5 V) to logic low (0 V). (In engineering jargon this situation is referred to as an *undriven* circuit, because there is no active voltage or current source in the circuit.) The input capacitance of the second gate has been given the explicit notation C_{in} ($= C_p + C_n$) to distinguish it from other capacitances we will consider later.

Example 2.2

Determine the capacitor voltage as a function of time, given the circuit parameters:

$$C_{in} = 20 \text{ pF}$$
$$R_n = 10 \text{ k}\Omega$$
$$v_{C0} = 5 \text{ V}$$

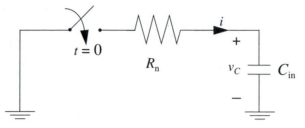

Figure 2.16 *RC* circuit model for the pull-down (undriven) case of two cascaded inverters.

The solution is found by inserting the appropriate numbers into the general solution

$$v_C(t) = (v_{C0} - V_S)e^{-\frac{t}{\tau}} + V_S \tag{2.15}$$

$$\tau = R_n C_{in}$$

Recall that the short-circuit connection between the left-hand terminal of the switch and the ground node in Figure 2.16 can be considered a voltage source of zero volts; therefore $V_S = 0$. The numerical values are thus

$$\tau = (10 \times 10^3)(20 \times 10^{-12}) = 200 \times 10^{-9} = 200 \text{ ns}$$
$$V_S = 0$$
$$v_{C0} = 5$$

Substituting these parameters into Eq. (2.15) yields:

$$v_C(t) = (5 - 0)e^{-\frac{t}{200 \times 10^{-9}}} + 0$$

which simplifies to

$$v_C(t) = 5e^{-\frac{t}{200 \times 10^{-9}}} \tag{2.16}$$

Eq. (2.16) is plotted in Figure 2.17. Note that the capacitor voltage has nearly reached the dc steady-state value of 0 V after 1 μs; this amount of time is equal to 5τ.

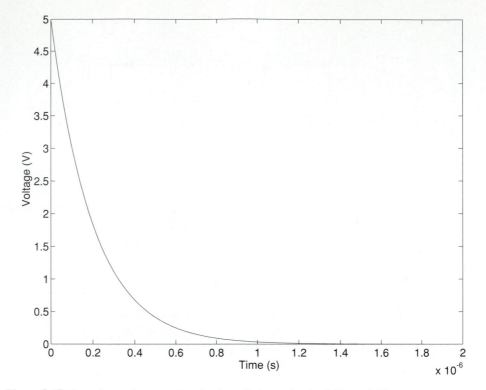

Figure 2.17 Capacitor voltage vs. time for the pull-down circuit of Figure 2.16.

2.5.2 Inverter Pull-Up

Consider the circuit of Figure 2.18, which models the pull-up case of the coupled inverter pair. The response of the circuit in Figure 2.18 is an example of a *driven* or *forced* response, so-called because the source voltage V_S is nonzero.

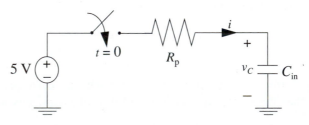

Figure 2.18 RC circuit model for the pull-up (driven) case of two cascaded inverters.

Example 2.3

Determine the capacitor voltage in Figure 2.18 as a function of time, given the circuit parameters:

$$C_{in} = 200 \text{ pF}$$
$$R_n = 75 \text{ } \Omega$$
$$v_{C0} = 1 \text{ V}$$

The solution is found by inserting the appropriate numbers into the general solution

$$v_C(t) = (v_{C0} - V_S)e^{-\frac{t}{\tau}} + V_S$$
$$\tau = R_p C_{in}$$

This yields

$$v_C(t) = (1-5)e^{-\frac{t}{15\times10^{-9}}} + 5$$

which simplifies to

$$v_C(t) = -4e^{-\frac{t}{15\times10^{-9}}} + 5 \tag{2.17}$$

Equation (2.17) is plotted in Figure 2.19. Again, note that the capacitor voltage has nearly reached the dc steady-state value (5 V in this case) after 5τ.

In both the pull-down and pull-up cases it takes an infinite amount of time for the capacitor voltage to reach the steady-state value. We observe, though, that there is little change after a period of approximately 5τ. This is investigated further in the next section. However, it is not necessary for the voltage to completely reach the steady-state value before the next inverter switches. Recall that the important parameters involved in gate switching are the threshold voltages: V_{th} (for the pull-up) and V_{tl} (for the pull-down). The gate delay is defined by the time it takes for the voltage to reach the appropriate threshold voltage, which is considerably shorter than 5τ in a well-designed system. Notice that the closer the thresholds are to the extreme values of the logic levels (V_H and V_L) the longer the gate delay.

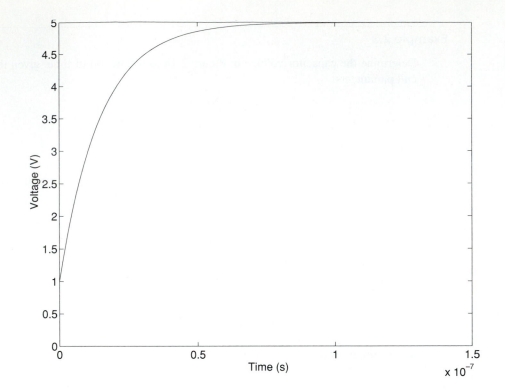

Figure 2.19 Capacitor voltage as a function of time for the pull-up case illustrated in Figure 2.18.

2.6 Significance of the Time Constant τ

In both the pull-down and pull-up cases we observed that the capacitor voltage reaches its dc steady-state value after an interval of approximately 5τ. This can be demonstrated quantitatively by evaluating the transient term of the general solution at $t = 5\tau$:

$$(v_{C0} - V_S)e^{-\frac{t}{\tau}}\bigg|_{t\,=\,5\tau} \approx 0.0067(v_{C0} - V_S)$$

In other words, the magnitude of the transient term has dropped to less than 0.7% of the value it had at the switching event. The time constant, then, represents the time scale over which the capacitor voltage (or other first-order circuit variable) decays toward its final value.

Plotted in Figures 2.20 and 2.21 are universal exponential decay curves corresponding to the pull-down ($V_S = 0$) and pull-up ($V_S > 0$) cases just considered. They are "universal" in the sense that the time axis is marked off in units of τ rather than seconds. With proper interpretation of the y-axis, these curves represent the solution for any first-order transient problem. [In fact the only substantive difference between the two figures is that $v_{C0} > v_C(\infty)$ in Figure 2.20 and $v_{C0} < v_C(\infty)$ in Figure 2.21.]

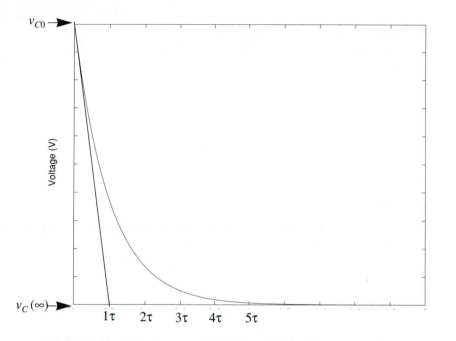

Figure 2.20 Graph of the time response of a first-order circuit where the initial value is greater than the final value. The straight line represents the initial slope of the curve.

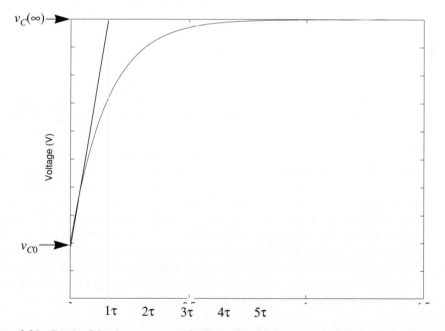

Figure 2.21 Graph of the time response of a first-order circuit where the initial value is less than the final value. The straight line represents the initial slope of the curve.

The straight lines in these two figures represent the initial slope of the response. If we sketch a line tangent to the response at $t = 0$, it intersects a horizontal line representing the final value $v_C(\infty)$ at $t = \tau$. We can show this formally by differentiating the general solution and evaluating at $t = 0$:

$$\left.\frac{dv_C}{dt}\right|_{t=0} = -\left(\frac{v_{C0} - v_C(\infty)}{\tau}\right) = \frac{v_C(\infty) - v_{C0}}{\tau} \tag{2.18}$$

This derivative is equal to the slope of a straight line tangent to the response; it is given in geometrical terms as the line connecting the points $(0, v_{C0})$ and $(\tau, v_C(\infty))$.

Example 2.4

Determine the initial rate of change of the voltage in the pull-up case of Example 2.3.

The inverter voltage is given by Eq. (2.17):

$$v_C(t) = -4e^{-\dfrac{t}{15\times10^{-9}}} + 5$$

We could find the initial slope of this curve by differentiating, but it is easier to form the ratio given by Eq. (2.18):

$$\left.\frac{dv_C}{dt}\right|_{t=0} = \frac{v_C(\infty) - v_{C0}}{\tau}$$

$$= \frac{4}{15 \times 10^{-9}}$$

$$= 2.67 \times 10^8$$

The voltage rises toward the final value with an initial slope of 267 MV/s.

We can also examine the magnitude of the transient term as a percentage of its initial value. At the switching time $t = 0$ the transient term is at its maximum value (since $e^{-(0/\tau)} = 1$). As t increases, the transient response decays exponentially. At $t = \tau$ the exponential part of the transient term has decayed to $e^{-1} = 0.37$; therefore, the transient term as a whole has decayed to approximately 37% of its initial value. At $t = 2\tau$ and $t = 3\tau$ the transient term has decayed to 14% and 5% of its initial value. The approximation of a "completely" decayed transient after five time constants (where the transient term is 99.3% decayed) is a common engineering approximation, but one that must be applied intelligently, as some systems (such as high-precision analog-to-digital converters) require an even more complete decay, and thus a longer time.

2.7 Maximum Inverter-Pair Switching Speed

We are now ready to calculate the maximum frequency at which a pair of cascaded CMOS inverter gates can be switched. We will assume that the switch inside the driving inverter moves up and down abruptly at frequency f; when this frequency is too high, some failure occurs, which prevents the next inverter from switching. Our task is to determine the nature of this failure, and the frequency f_{max} at which it occurs.

The state of the switch inside the driving inverter (up or down) is a consequence of the logic state at the gate input. Since we are assuming the ideal case of completely abrupt switch transitions, the input signal takes the form of a **square wave**. This signal is typical of that generated by the clock circuitry of a computer which is used to synchronize all of the fundamental operations. For maximum speed it is necessary that the clock signal be **symmetric**. A symmetric square wave, Figure 2.22, spends an equal amount of time at logic high and logic low. (Engineering jargon for the ratio of time at the high level to the period is a percentage called the **duty cycle**. The symmetric clock signals we will deal with have a **50% duty cycle**.)

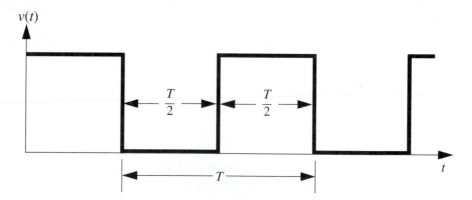

Figure 2.22 Symmetric square-wave clock signal. The frequency f of a periodic signal is given by the reciprocal of the period T.

We will calculate the maximum switching frequency in two ways. First, we will assume that the inverter pair has not made a transition for many clock cycles preceding $t = 0$. As a result, the effect of transients which may have occurred for $t < 0$ have died away, which sets an initial condition for the capacitor voltage (i.e., the input of the load inverter) equal to either 0 V (logic low) or 5 V (logic high). This assumption, which we call Case I, represents the situation of random logic inside a computer, since one cannot predict when a transition will take place. This estimate will provide a worst-case switching frequency.

The second method of calculating f_{max} assumes that the driving inverter is being switched continuously by the clock signal. Although this situation is not often encountered in a computer, it is a close approximation to the method by which integrated circuits are tested for maximum speed. We will call this scenario Case II.

In both examples we will assume the following:

$V_{th} = 3.7$ V (therefore v_C is logic high if $3.7 \leq v_C \leq 5.0$)

$V_{tl} = 1.0$ V (therefore v_C is logic low if $0.0 \leq v_C \leq 1.0$)

$R_n = 1.0$ kΩ

$R_p = 1.5$ kΩ

$C_{in} = 10$ pF

2.7.1 Case I: Inverter Pair Initially at Rest

To find the maximum switching frequency, we need to compute the transition time from 0 V to V_{th} (pull-up case) and the transition time from 5 V to V_{tl} (pull-down case), since the output of the driving inverter may be at either 0 V or 5 V at $t = 0$. Once the output of the driving inverter has reached a threshold value, the load inverter will switch, resulting in a forward propagation of the logic signal. If the clock signal were to switch before this occurred, the logic signal would not trigger the next inverter.

The pull-up and pull-down transition times for coupled inverters (or logic gates) are not equal because of differences in output resistance (i.e., $R_n \neq R_p$) and the difference in relative threshold voltage. The switching speed will be limited by the *larger* of these two transition times where t' is the longer of the two transition times. Since the clock signal is symmetric, we can approximate f_{max} from

$$f_{max} = \frac{1}{2t'} \tag{2.19}$$

Pull-Up Transition Time

The circuit model for the pull-up transition is shown in Figure 2.23; the resulting voltage across the capacitor is sketched in Figure 2.24. We wish to determine the transition time, t_1, defined as the time required for v_C to rise from 0 V to V_{th}.

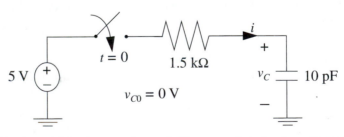

Figure 2.23 Circuit model for the example coupled inverter pair during the pull-up transition.

The voltage v_C is given by

$$v_C(t) = (v_{C0} - V_S)e^{-\frac{t}{\tau}} + V_S$$

Substituting the given circuit values yields:

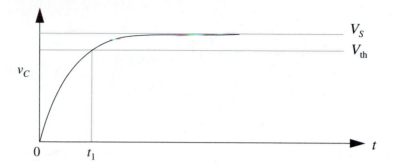

Figure 2.24 Capacitor voltage during pull-up transition.

$$v_C(t) = 5 + (0-5)e^{-\frac{t}{(1.5\times10^3)(10\times10^{-12})}} = 5-5e^{-\frac{t}{15\times10^{-9}}} \qquad (2.20)$$

At $t = t_1$ the capacitor voltage is at the logic high threshold. We express this symbolically as $v_C(t_1) = V_{th} = 3.7$. Inserting this condition into Eq. (2.20):

$$3.7 = 5-5e^{-\frac{t_1}{15\times10^{-9}}}$$

Solving for t_1,

$$-1.3 = -5e^{-\frac{t_1}{15\times10^{-9}}}$$

$$\frac{1.3}{5} = e^{-\frac{t_1}{15\times10^{-9}}}$$

$$\ln\left(\frac{1.3}{5}\right) = \frac{-t_1}{15\times10^{-9}}$$

$$t_1 = (-15\times10^{-9})\ln\left(\frac{1.3}{5}\right) = 20.2 \text{ ns} \qquad (2.21)$$

Thus, it requires 20.2 ns for the inverter to switch from 0 V to the logic high threshold of 3.7 V.

Pull-Down Transition Time

The circuit model for the pull-down transition is shown in Figure 2.25; the resulting voltage across the capacitor is sketched in Figure 2.26. We will now determine the transition time t_2, defined as the time required for v_C to fall from 5 V to V_{tl}.

The general solution, as always, is

$$v_C(t) = (v_{C0}-V_S)e^{-\frac{t}{\tau}} + V_S$$

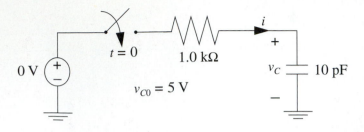

Figure 2.25 Circuit model for example coupled inverter pair during pull-down transition.

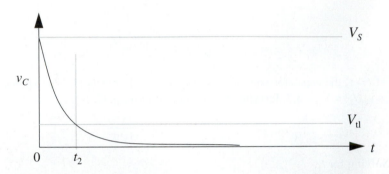

Figure 2.26 Capacitor voltage during the pull-down transition.

where $V_S = 0$. Substituting the appropriate values:

$$v_C(t) = 0 + (5 - 0)e^{-\frac{t_2}{(10 \times 10^3)(10 \times 10^{-12})}} = 5e^{-\frac{t_2}{10 \times 10^{-9}}}$$

The transition time t_2 is defined by $v_C(t_2) = V_{tl} = 1.0$. Solving for t_2:

$$1 = 5e^{-\frac{t_2}{10 \times 10^{-9}}}$$

$$\frac{1}{5} = e^{-\frac{t_2}{10 \times 10^{-9}}}$$

$$\ln\left(\frac{1}{5}\right) = -\frac{t_2}{10 \times 10^{-9}}$$

$$-10 \times 10^{-9}\ln\left(\frac{1}{5}\right) = t_2$$

$$t_2 = 10 \times 10^{-9} \ln(5) = 16.1 \text{ ns} \tag{2.22}$$

The pull-up transition time, 20.2 ns, is greater than the pull-down transition time, which is only 16.1 ns. In designing a digital system we must allow for the worst case; hence, the clock period must be at least twice the longer delay:

$$T = 2t_1 = 2 \times 20.2 \text{ ns} = 40.4 \text{ ns}$$

The clock frequency is equal to the reciprocal of the period; this results in a maximum clock frequency of

$$f_{max} = \frac{1}{40.4 \times 10^{-9}} = 24.8 \times 10^6 = 24.8 \text{ MHz} \tag{2.23}$$

This calculated maximum clock frequency is only an *approximation* of the coupled inverter pair's true operating limit. We have used many assumptions in deriving Eq. (2.23); e.g., we used a linear circuit model to characterize the responses of the nonlinear inverters. However, the solution obtained is in fact quite reasonable and provides a great deal of insight into the important parameters which limit the speed of digital systems. Throughout this text we will improve upon the model so that it more closely approximates the actual system; this will lead to continuous improvements in our ability to estimate the maximum switching speed.

2.7.2 Case II: Inverter Pair Switching Continuously

A second way to calculate the maximum switching speed is to visualize the following experiment. Suppose the cascaded inverter pair are driven by a variable-frequency clock generator. The inverters will be driven continuously and will switch every half cycle of the clock. If the frequency is sufficiently low, then the voltage v_C will come within a few milli-volts of 0 V and 5 V on every cycle. Lowering the frequency further will cause v_C to approach these two asymptotic values even closer, within microvolts of the extrema.

Now, suppose the clock frequency is increased by a small amount. After a few cycles, we will find that v_C still comes up to a maximum value somewhat less than 5 V at the end of the pull-up; it then enters the pull-down phase and reaches a minimum value slightly greater than 0 V. We will call the maximum and minimum values that v_C reaches at the switching times v_h and v_l, as illustrated in Figure 2.27. (Don't confuse these with V_{th} and V_{tl}, which are particular values of threshold voltage that depend on the inverter's internal construction.)

Continuing to increase the clock frequency, we find that v_h drops more and more below 5 V, and v_l rises more and more above 0 V. At some frequency we find that the output of the second inverter no longer switches states. What is the value of this frequency?

This problem illustrates an easy way to test the performance of integrated circuits. It is quite straightforward to set up a cascade of inverters and experimentally determine the frequency at which the logic signals no longer pass through. The maximum frequency found from this test is a good relative measure of the performance of different IC technol-

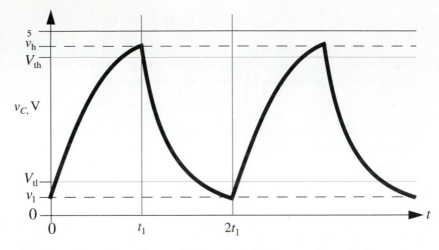

Figure 2.27 Capacitor voltage of a coupled inverter pair driven continuously. The system operates reliably as long as $v_h \geq V_{th}$ and $v_l \leq V_{tl}$.

ogies and is easier to perform than the one-shot timing measurement described in the last section.[1]

If we examine v_C just before the clock frequency has been raised to the point where the logic signals no longer propagate through the cascaded inverters, we will find it looks like Figure 2.28. The maximum value of capacitor voltage, v_h, is just equal to the threshold high voltage V_{th}. If the frequency were increased beyond this point, v_h would drop below V_{th}, which is insufficiently high to cause the output of the second inverter to switch states.

In Figure 2.28 we see that v_l is still below the threshold low value, V_{tl}, The fact that the upper limiting voltage v_h decreases to V_{th} before v_l increases to V_{tl}, as frequency is increased, is a consequence of the particular values of time constants and threshold voltages chosen in this example. These inverters, as we saw in the last section, are limited by the pull-up transition time. The limiting frequency is thus set by the point where the voltage at the end of the pull-up half-cycle *just reaches* V_{th}. In the following half-cycle, the voltage "overshoots" the low threshold V_{tl}.

To find the maximum frequency, we need to determine the half-cycle switching time t_1 as shown in Figure 2.28. However, we do not know the value of v_l, so we cannot simply solve for t_1 using the general solution of the series RC circuit. We need to find both t_1 and v_l by simultaneously solving the constraint at the end of the pull-up transient given by

$$v_C(t_1) = V_{th} = 5 + (v_1 - 5)e^{-\dfrac{t_1}{15 \times 10^{-9}}} \tag{2.24}$$

and the constraint at the end of the pull-down transient,

---

[1] Many IC manufacturers include on their chips a circuit known as a *ring oscillator*, which is simply an odd number of inverters cascaded together into a closed loop. The inverters in this circuit will switch continuously, without any external clock signal, at a frequency which depends strongly on how well the chip was made. The ring frequency is thus used as a rough measure of process quality and circuit speed.

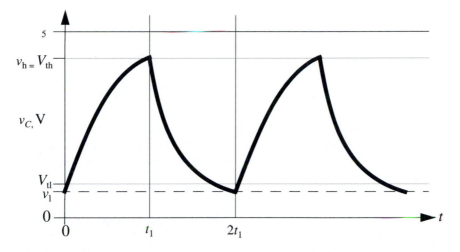

Figure 2.28 Capacitor voltage of example coupled inverter pair operating continuously at the maximum clock frequency. Because the pull-up transition is more sluggish than the pull-down, the maximum voltage v_h just equals the threshold high level.

$$v_C(2t_1) = v_1 = 0 + (V_{th} - 0)e^{-\frac{2t_1 - t_1}{10 \times 10^{-9}}} \tag{2.25}$$

In this system of equations, we can eliminate one of the variables by cross-substitution. This will lead to a transcendental equation which expresses t_1 implicitly. Another way is find t_1 by solving Eqs. (2.24) and (2.25) iteratively; we will use this technique.

We will use an arbitrarily chosen value[2] of 1.0 V as an initial estimate of v_1. Substituting this value in Eq. (2.24):

$$3.7 = 5 + (1 - 5)e^{-\frac{t_1}{15 \times 10^{-9}}}$$

Solving for t_1:

$$t_1 = (15 \times 10^{-9})\ln\left(\frac{5 - 1}{5 - 3.7}\right) = 16.86 \text{ ns}$$

Now we take this first estimate of $t1$ and substitute it into Eq. (2.25); this will yield an approximation for vl:

$$v_1 = 3.7e^{-\frac{16.86 \times 10^{-9}}{10 \times 10^{-9}}} = 0.6855 \text{ V}$$

We now use this estimate of v_1 to find a more accurate value for t_1, by substituting $v_1 = 0.6855$ in Eq. (2.24):

[2.] "Arbitrarily chosen value" is engineering lingo for "guess"; if someone is paying you to do an analysis of this sort, it is inadvisable to use words like "guess."

$$3.7 = 5 + (0.6855 - 5)e^{-\dfrac{t_1}{15 \times 10^{-9}}}$$

Solving for t_1:

$$t_1 = (15 \times 10^{-9})\ln\left(\frac{5 - 0.6855}{5 - 3.7}\right) = 17.99 \text{ ns}$$

This estimate of t, 17.99 ns, differs from the previous estimate, 16.86 ns. We must continue the iteration until successive estimates are equal, to within the desired degree of accuracy. (When this happens, we say that the variable has **converged** to a stable value.) The next approximation for v_1 is:

$$v_1 = 3.7e^{-\dfrac{17.99 \times 10^{-9}}{10 \times 10^{-9}}} = 0.6122 \text{ V}$$

which is used to find the next estimate of t_1:

$$t_1 = (15 \times 10^{-9})\ln\left(\frac{5 - 0.6122}{5 - 3.7}\right) = 18.25 \text{ ns}$$

Continuing this process:

$$v_1 = 3.7e^{-\dfrac{18.25 \times 10^{-9}}{10 \times 10^{-9}}} = 0.5965 \text{ V}$$

$$t_1 = (15 \times 10^{-9})\ln\left(\frac{5 - 0.5965}{5 - 3.7}\right) = 18.30 \text{ ns}$$

$$v_1 = 3.7e^{-\dfrac{18.30 \times 10^{-9}}{10 \times 10^{-9}}} = 0.5935 \text{ V}$$

$$t_1 = (15 \times 10^{-9})\ln\left(\frac{5 - 0.5935}{5 - 3.7}\right) = 18.31 \text{ ns}$$

After five iterations, t_1 has converged to a limiting value of 18.3 ns (to three significant figures).

The period of a clock driving the cascaded inverters at their maximum continuous switching speed is

$$T = 2t_1 = 2 \times 18.3 \text{ ns} = 36.6 \text{ ns}$$

resulting in a maximum he clock frequency of

$$f_{max} = \frac{1}{36.6 \times 10^{-9}} = 27.3 \times 10^6 = 27.3 \text{ MHz} \qquad (2.26)$$

The estimate of maximum switching speed found using this procedure is somewhat greater than before. This is to be expected, since the logic swing, $v_h - v_l$, is larger in the first case than in the second. In other words, with all circuit and system parameters equal, it takes longer to drive a capacitor voltage through a larger range.

In this example the switching speed is limited by the pull-up transition. It is just as possible that the pull-down transition could be the limiting factor; this could be the case if the pull-down resistance R_n were larger than R_p, or if the threshold low voltage V_{tl} were closer to 0 V. Under these conditions, the inverter pair would switch between V_{tl} and v_h (where $v_h > V_{th}$), as shown in Figure 2.29.

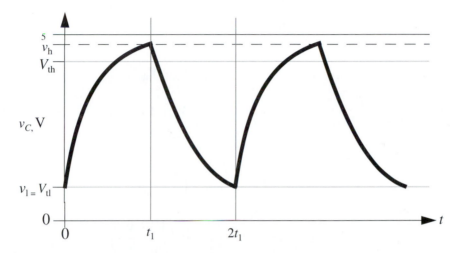

Figure 2.29 Capacitor voltage of a pull-down-limited coupled inverter pair operating continuously at the maximum clock frequency. In this case the pull-down transition is more sluggish than the pull-up, therefore the maximum voltage v_h is above the threshold high level.

2.8 Algebraic Analysis of Inverter-Pair Switching Speed

Not all inverters operate between 0 V and 5 V, nor do they all have the values of threshold voltages, output resistance, etc., given in the last section. We can generalize the analysis of inverter-pair switching speed by performing the previous analysis entirely with symbolic quantities. In addition to the quantities already defined, we will use the following symbols:

$$V_H = \text{maximum voltage at logic high}$$

$$V_L = \text{minimum voltage at logic low}$$

$$\tau_u = R_p C_{in} \text{ (pull-up time constant)}$$

$$\tau_d = R_n C_{in} \text{ (pull-down time constant)}$$

$$t_{low \rightarrow high} = \text{low-to-high transition time (} v_C \text{ rises from } V_L \text{ to } V_{th}\text{)}$$

$$t_{high \rightarrow low} = \text{high-to-low transition time (} v_C \text{ drops from } V_H \text{ to } V_{tl}\text{)}$$

For the pull-up transition from low to high we have

$$v_C(0) = V_L$$
$$v_C(t_{\text{low}\to\text{high}}) = V_{\text{th}}$$
$$v_C(\infty) = V_H$$

Inserting these into the general solution of the first-order series-*RC* circuit gives

$$V_{\text{th}} = (V_L - V_H)e^{-\dfrac{t_{\text{low}\to\text{high}}}{\tau_u}} + V_H$$

Solving for the low-to-high transition time:

$$(V_H - V_L)e^{-\dfrac{t_{\text{low}\to\text{high}}}{\tau_u}} = V_H - V_{\text{th}}$$

$$e^{-\dfrac{t_{\text{low}\to\text{high}}}{\tau_u}} = \dfrac{V_H - V_{\text{th}}}{V_H - V_L}$$

$$e^{\dfrac{t_{\text{low}\to\text{high}}}{\tau_u}} = \dfrac{V_H - V_L}{V_H - V_{\text{th}}}$$

$$\ln\left(e^{\dfrac{t_{\text{low}\to\text{high}}}{\tau_u}}\right) = \ln\left(\dfrac{V_H - V_L}{V_H - V_{\text{th}}}\right)$$

$$\dfrac{t_{\text{low}\to\text{high}}}{\tau_u} = \ln\left(\dfrac{V_H - V_L}{V_H - V_{\text{th}}}\right)$$

$$t_{\text{low}\to\text{high}} = \tau_u\ln\left(\dfrac{V_H - V_L}{V_H - V_{\text{th}}}\right) \qquad (2.27)$$

The transition time from high to low is calculated in a similar manner. For the pull-down we have

$$v_C(0) = V_H$$
$$v_C(t_{\text{high}\to\text{low}}) = V_{\text{tl}}$$
$$v_C(\infty) = V_L$$

Substituting these into the general solution:

$$V_{\text{tl}} = (V_H - V_L)e^{-\dfrac{t_{\text{high}\to\text{low}}}{\tau_d}} + V_L$$

Solving for the transition time from high to low,

$$(V_L - V_H)e^{-\frac{t_{\text{high} \to \text{low}}}{\tau_d}} = V_L - V_{\text{tl}}$$

$$e^{-\frac{t_{\text{high} \to \text{low}}}{\tau_d}} = \frac{V_L - V_{\text{tl}}}{V_L - V_H}$$

$$e^{\frac{t_{\text{high} \to \text{low}}}{\tau_d}} = \frac{V_L - V_H}{V_L - V_{\text{tl}}}$$

$$\ln\left(e^{\frac{t_{\text{high} \to \text{low}}}{\tau_d}}\right) = \ln\left(\frac{V_L - V_H}{V_L - V_{\text{tl}}}\right)$$

$$\frac{t_{\text{high} \to \text{low}}}{\tau_d} = \ln\left(\frac{V_L - V_H}{V_L - V_{\text{tl}}}\right)$$

$$t_{\text{high} \to \text{low}} = \tau_d \ln\left(\frac{V_L - V_H}{V_L - V_{\text{tl}}}\right) \tag{2.28}$$

Equations (2.27) and (2.28) give explicit expressions for the logic transition times for the case where the capacitor voltage has reached a steady-state value of V_L or V_H. Analogous expressions can be derived for the continuous-switching case.

2.8.1 Increasing Switching Speed

How can we increase switching speed in digital systems? Several possibilities are suggested by Eqs. (2.27) and (2.28).

One way is to decrease the time constants τ_u and τ_d. This means that the transistor parameters C_{in}, R_p, and R_n, must be made smaller. Decreasing C_{in} is possible, but the charge stored in the capacitance C_{in} is the physical realization of the digital data; it cannot be reduced beyond a certain point without compromising the noise immunity[3] of the circuit. Reduction of R_p and R_n has the effect of making the transistors wider, so fewer devices can be packed into a given space, which reduces functionality. (It also increases the input capacitance.)

In fact, we will see in future chapters that the limiting factor in state-of-the-art digital systems is not the devices themselves, but the wires connecting them together. Up to now we have treated the interconnects as perfect conductors; using a more realistic model will show that they are important factors in limiting system speed. Although device designers continue to shrink transistors to ever-decreasing dimensions, the effect of this effort on overall system performance is not as large as one would expect from Eqs. (2.27) and (2.28).

---

[3] "Noise immunity" is engineering lingo for the property of computers that resists interpreting external events, like switching on a toaster or electric shaver, as data.

Another approach to increasing speed is to reduce the voltage swing, given by $V_H -$ V_L. Instead of driving our system with a 5-V power supply, we could use 3 V or even less. Pushing the threshold voltages closer together, by lowering the high logic threshold voltage V_{th} and raising the low logic threshold voltage V_{tl}, will also help to speed things up. Again, the price for this improvement is a reduction is noise immunity. An unwanted voltage spike, which can come from many different sources, can more easily be interpreted as data when the logic levels are spaced close together. Another problem is that as threshold voltages are changed in this direction, the transistor does not look as much like an open circuit when it is supposed to be off. This results in leakage of charge when we are expecting it to remain constant, which affects the integrity of our data.

It is possible to overcome these difficulties to some degree. The direction of the microelectronics industry is to lower power-supply voltages and logic swings so that overall system speed is improved. Also, other logic families, which use different kinds of transistors and have logic swings of less than 1 V, are used in demanding applications (such as supercomputers) where speed is essential. (Such systems are limited by how fast the heat can be removed before the transistors enter a state known as *thermal runaway*.)

We will return to the question of how to best increase system speed after we have investigated the problem of interconnect delay.

2.9 Energy and Power Dissipation in Digital Systems

If you have ever used a high-performance laptop computer on your lap, you know that it can get hot enough to cause serious damage. However, a pocket calculator, which presumably operates using the same types of resistors, capacitors, and electrons as a laptop, does not get hot, or even warm. Why?

2.9.1 Energy Dissipation

Data in computers and other digital systems is charge. Capacitors store charge; the process of computation involves moving charge around. Every time a logic zero is processed into a logic one, or vice versa, a capacitor is charged or discharged, resulting in a current through a series *RC* circuit. The resistance in this circuit converts the motion of electrical charge (i.e., the process of computation) into heat.

We can calculate the energy dissipated during a logic transition by integrating the power dissipated in the resistor. We saw in Chapter 1 that the power is given by

$$p_R(t) = Ri^2(t) \tag{2.29}$$

The current in the series *RC* circuit can be found from the capacitor voltage using

$$i(t) = C\frac{d}{dt}v_C(t) \tag{2.30}$$

Consider a logic transition from low to high. The capacitor voltage has an initial value of $v_C(0) = V_L$ and a dc steady-state value of $v_C(\infty) = V_H$; therefore the capacitor voltage as a function of time can be expressed as

$$v_C(t) = (V_L - V_H)e^{-\frac{t}{\tau}} + V_H \tag{2.31}$$

Inserting this expression in Eq. (2.30) yields the current:

$$i(t) = \frac{C(V_H - V_L)}{\tau} e^{-\frac{t}{\tau}} \tag{2.32}$$

The energy dissipated in the resistor during the logic transition is found by integrating Eq. (2.29) with the above expression for current inserted:

$$E_R = R \int_0^{t_s} \left[\frac{C(V_H - V_L)}{\tau} e^{-\frac{t}{\tau}} \right]^2 dt \tag{2.33}$$

where t_s is the switching time of the transition. Since the logic levels are ranges, t_s can be anywhere from $t_{\text{low} \to \text{high}}$ to ∞.

Carrying out the integration, with the substitution of supply voltage V_S for $V_H - V_L$:

$$E_R = \frac{RC^2}{\tau^2}(V_H - V_L)^2 \int_0^{t_s} e^{-\frac{2t}{\tau}} dt$$

$$E_R = -\frac{\tau RC^2}{2 \tau^2} V_S^2 e^{-\frac{2t}{\tau}} \Big|_0^{t_s}$$

$$E_R = -\frac{C}{2} V_S^2 \left[e^{-\frac{2t_s}{\tau}} - 1 \right]$$

$$E_R = \frac{1}{2} C V_S^2 \left[1 - e^{-\frac{2t_s}{\tau}} \right] \tag{2.34}$$

We know that the transition time t_s is the range $t_{\text{low} \to \text{high}} \leq t_s < \infty$. For the lower limit, we can use the expression for $t_{\text{low} \to \text{high}}$ given in Eq. (2.27):

$$\min(E_R) = \frac{1}{2} C V_S^2 \left[1 - e^{-\frac{2t_s}{\tau}} \right] \Big|_{t_s = t_{\text{low} \to \text{high}}}$$

$$\min (E_R) = \frac{1}{2} CV_S{}^2 \left[1 - e^{-\frac{2\tau \ln\left(\frac{V_H - V_L}{V_H - V_{th}}\right)}{\tau}} \right]$$

$$\min (E_R) = \frac{1}{2} CV_S{}^2 \left[1 - e^{-\ln\left(\frac{V_S}{V_H - V_{th}}\right)^2} \right]$$

$$\min (E_R) = \frac{1}{2} CV_S{}^2 \left[1 - \left(\frac{V_H - V_{th}}{V_S}\right)^2 \right] \qquad (2.35)$$

The maximum energy dissipated corresponds to a transition all the way from V_L to V_H; letting $t_s \to \infty$ in Eq. (2.34) yields:

$$\max (E_R) = \frac{1}{2} CV_S{}^2 \qquad (2.36)$$

These two expressions are the same, except for the second (bracketed) term in Eq. (2.35), which accounts for the energy dissipated by the resistor as v_C rises from V_{th} to V_H. We can compare the magnitude of this term to that of the first term by inserting the values we used previously: $V_H = 5$ V, $V_S = 5$ V, $V_{th} = 3.7$ V:

$$\frac{1}{2} CV_S{}^2 \left[1 - \left(\frac{V_H - V_{th}}{V_S}\right)^2 \right] = \frac{1}{2} CV_S{}^2 \left[1 - \left(\frac{5 - 3.7}{5}\right)^2 \right]$$

$$= \frac{1}{2} CV_S{}^2 \left[1 - \left(\frac{1.3}{5}\right)^2 \right]$$

$$= \frac{1}{2} CV_S{}^2 [0.932]$$

We see that ignoring the second term in Eq. (2.35) results in an error of less than 7%. Thus it is convenient to take the maximum energy dissipated in an ideal (complete) transition, given by Eq. (2.36), as a measure of the energy dissipated in a real transition:

$$E_R = \frac{1}{2} CV_S{}^2 \qquad (2.37)$$

Example 2.5

Cascaded CMOS inverter gates operating from a 5-V supply have $C_{in} = 5$ pF, $R_n = 1$ kΩ, and $R_p = 1.2$ kΩ. Determine the energy dissipated by the driving gate during a pull-up and pull-down transition.

From Eq. (2.37), the energy dissipated by the driving-gate resistance in a complete logic transition is given by

$$E_R = \frac{1}{2}CV_S^2$$

$$= \frac{1}{2} \times 5 \times 10^{-12} \times 5^2$$

$$= 62.5 \times 10^{-12}$$

The transition dissipates 62.5 pJ of energy. Because Eq. (2.37) does not include a term representing the gate output resistance, the same energy is dissipated in pull-up and pull-down transitions, even though the two resistances which dissipate the energy have different values.

In Chapter 1 we learned that a capacitor charged from 0 to V_S stores energy:

$$E_C = \frac{1}{2}CV_S^2 \qquad (2.38)$$

During the logic transition from V_L to V_H, the source delivers a total amount of energy to the circuit given by

$$E = \int_0^\infty V_S i(t)dt$$

Evaluating:

$$E = \int_0^\infty V_S \frac{C(V_H - V_L)}{\tau} e^{-\frac{t}{\tau}} dt$$

$$= \frac{CV_S^2}{\tau} \int_0^\infty e^{-\frac{t}{\tau}} dt$$

$$= \frac{CV_S^2}{\tau} \left(-\tau e^{-\frac{t}{\tau}}\right)\Big|_0^\infty$$

$$= CV_S^2$$

Thus the energy supplied by the source is

$$E = E_R + E_C = \frac{1}{2}CV_S^2 + \frac{1}{2}CV_S^2 = CV_S^2 \qquad (2.39)$$

Equation (2.39) tells us that, in storing energy in the capacitor, we also dissipate an equal amount in the resistor. The process of computation, which in a digital system is implemented by the transfer of charge among capacitors, results in an unavoidable energy loss in the form of heat. Interestingly, the amount of heat E_R lost to the environment is independent of the value of the resistance.

2.9.2 Power Dissipation by the Computer

We can now estimate the amount of power turned into heat for various digital systems. We know that each logic transition results in a conversion of $\frac{1}{2}CV_S^2$ joules of energy into heat. There can be, at most, two logic transitions per clock cycle; typically there are far fewer. Let's call the average number of logic transitions per clock cycle the *activity* 2α. There are f clock cycles per second. Finally, suppose there are N logic gates in the system. Combining, the estimated power dissipation for the system is given by

$$P = \frac{1}{2}CV_S^2 \times 2\alpha \times N \times f = CV_S^2\alpha Nf \tag{2.40}$$

A typical circa-1994 microprocessor containing 10^7 gates has gate capacitances of about 10 fF, a logic swing of approximately 5 V, and runs at approximately 40 MHz. If we estimate the activity at one logic transition every other clock cycle ($2\alpha = 0.5$):

$$P = CV_S^2\alpha Nf$$

$$= 10 \times 10^{-15}5^2 10^7 (0.25)40 \times 10^6$$

$$= 25 \text{ W}$$

which is healthy amount of power. (Think about pressing a lit 25-W light bulb into your lap.) A pocket calculator has far fewer gates and runs at a considerably slower speed, so it does not warm perceptibly.

Getting rid of this heat is a serious engineering challenge. Equation (2.40) tells us that, as we increase the number of gates and the clock frequency in future systems, even more heat will be produced. Reduction of this power requires reducing the capacitance and voltage swing; the latter is much more important, since the power dissipation varies as the square of V_S. Note that reductions of both of these parameters are also important in increasing the speed of the computer.

2.10 Other First-Order *RC* Circuits

Up to now we have considered only one circuit: a series connection of a voltage source, resistor, and capacitor. We showed that the general solution of the series *RC* circuit can be written as

$$v_C(t) = [v_C(0) - v_C(\infty)]e^{-\frac{t}{\tau}} + v_C(\infty) \tag{2.41}$$

which is the solution to the differential equation

$$\tau \frac{d}{dt} v_C(t) + v_C(t) = v_C(\infty) \qquad (2.42)$$

describing the circuit.

It turns out that Eq. (2.41) describes the response of **any** first-order RC circuit with dc sources. That is, any RC circuit, containing only dc voltage and current sources and a single capacitor, can be solved by inspection. One has only to determine the values of $v_C(0)$, $v_C(\infty)$, and τ and insert them into Eq. (2.41).

Consider the slightly more complicated RC circuit of Figure 2.30. Applying KCL at the upper capacitor node:

$$\frac{V_S - v_C}{R_1} = \frac{v_C}{R_2} + C \frac{dv_C}{dt}$$

Rearranging:

$$C \frac{dv_C}{dt} + \left(\frac{1}{R_2} + \frac{1}{R_1} \right) v_C = \frac{V_S}{R_1}$$

$$R_1 R_2 C \frac{dv_C}{dt} + (R_1 + R_2) v_C = R_2 V_S$$

$$\frac{R_1 R_2}{R_1 + R_2} C \frac{dv_C}{dt} + v_C = \frac{R_2}{R_1 + R_2} V_S \qquad (2.43)$$

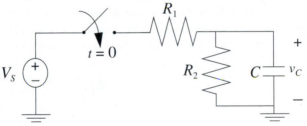

Figure 2.30 First-order RC circuit with two resistors.

Note that Eq. (2.43) is identical in form to the differential equation Eq. (2.42), provided we identify the variables as follows:

$$v_C(\infty) = \frac{R_2}{R_1 + R_2} V_S$$

$$\tau = \frac{R_1 R_2}{R_1 + R_2} C$$

Therefore, the capacitor voltage in this circuit is given by

$$v_C(t) = \left[v_C(0) - \frac{R_2}{R_1 + R_2}V_s\right]\exp\left(-\frac{t}{\dfrac{R_1 R_2}{R_1 + R_2}C}\right) + \frac{R_2}{R_1 + R_2}V_s$$

The last term on the right, the dc steady-state response, can be found directly from the circuit. Recall that the dc steady-state response is given by

$$\lim_{t \to \infty} v_C(t) \, ;$$

it is what's left after the transient exponential term has decayed away. If we examine the limit of Eq. (2.42) as $t \to \infty$, we will discover an important fact:

$$\lim_{t \to \infty}\left[\tau\frac{d}{dt}v_C(t) + v_C(t) = v_C(\infty)\right] \Rightarrow \lim_{t \to \infty}\left[\frac{d}{dt}v_C(t) = 0\right] \qquad (2.44)$$

Equation (2.44) states that the derivative of capacitor voltage is zero in the steady state limit, long after the switch is thrown. But, since capacitor current is directly proportional to

$$\frac{d}{dt}v_C(t) \, ,$$

this means that there is **zero current through the capacitor in steady state**.

We can therefore determine the dc steady-state conditions in a circuit containing sources, resistors, and capacitors by replacing all capacitors with their steady-state equivalent model: an open circuit. This relationship can be summarized with the rule: **The dc steady-state equivalent model of a capacitor is an open circuit.**

Applying this rule to the circuit of Figure 2.30 results in the dc equivalent circuit shown in Figure 2.31. Using this equivalent circuit, it is trivial to solve for the steady state voltage across the capacitor using the voltage divider rule.

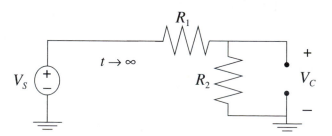

Figure 2.31 The dc steady-state equivalent circuit of the RC circuit of Figure 2.30.

Example 2.6

The second-order two-RC ladder circuit shown in Figure 2.32 will be introduced in the next chapter. Determine the dc steady-state voltages V_{C1} and V_{C2}.

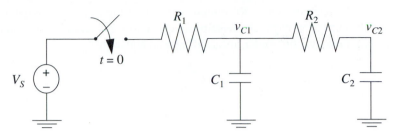

Figure 2.32 Second-order *RC* ladder circuit with two resistors and two capacitors.

Replacing the two capacitors with their dc equivalents results in the circuit of Figure 2.33. By inspection, there is no current flowing in this circuit, therefore the voltage drops across the two resistors are zero. This yields dc steady-state capacitor voltages

$$V_{C1} = V_S$$
$$V_{C2} = V_S$$

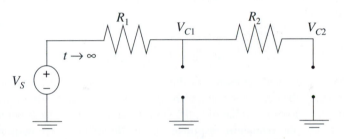

Figure 2.33 The dc steady-state equivalent of the second-order circuit shown in Figure 2.32.

2.11 Summary

This chapter has introduced several concepts useful in understanding the delays inherent in coupled logic gates. A logic transition, from low to high or high to low, is not an instantaneous digital event, but a continuous swing in voltage over time. Logic levels likewise are not discrete values but are a range in voltage (V_L to V_{tl} for logic low and V_{th} to V_H for logic high). Approximate values of the transition times of coupled inverter pairs can be determined using a first-order *RC* circuit model of an inverter, as shown in Figure 2.10.

This model was solved for two specific cases: the pull-down case, which characterizes a high-to-low logic transition, and the pull-up case, which characterizes a low-to-high logic transition. Both cases are modeled in terms of an output resistance (i.e., a resistor at the output of the driving inverter) in series with an input capacitance (a capacitor at the input of the load inverter) and a voltage source. In an *RC* circuit the currents and voltages are functions of time.

The resulting series RC circuit is shown in Figure 2.34. Applying Kirchhoff's laws and the current-voltage relationships for capacitors and resistors, we found that this circuit is described by a first-order linear differential equation:

$$RC\frac{dv_C}{dt} + v_C = V_S$$

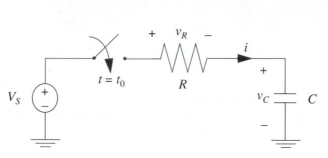

Figure 2.34 Series RC circuit approximating the behavior of a coupled pair of CMOS inverters.

The general solution of this differential equation (also known as the **total response** of the circuit) is:

$$v_C(t) = (v_{C0} - V_S)\, e^{-\frac{t - t_o}{\tau}} + V_S$$

where τ is the **time constant** of the circuit, given by $\tau = RC$. τ determines the time scale over which the capacitor voltage changes from the **initial value** v_{C0} to the **dc steady-state value**; the larger the value of τ, the slower the response. The general solution has two distinct terms: the **transient response** and the **dc steady-state response**. The transient part bridges the gap between the initial capacitor voltage and the steady-state capacitor voltage; the steady-state response is the response of the circuit as time t approaches infinity.

Problems

2.1   A **truth table** is a table listing the output of a logic gate for all combinations of inputs. For example, the truth table for an OR gate where C = A OR B is:

A	B	C
0	0	0
0	1	1
1	0	1
1	1	1

Determine the truth table for D = A NOR B and E = A NAND B.

2.2 Write the truth table for D = NOT((A NAND B) OR C).

2.3 A set of cascaded CMOS inverter gates have $C_{in} = 0.8$ nF and $R_n = R_p = 1.2$ kΩ. Determine the pull-up and pull-down time constants.

2.4 A set of cascaded CMOS inverter gates have $C_n = 0.8$ nF, $C_p = 1.2$ nF, $R_n = 1.2$ kΩ, and $R_p = 2.4$ kΩ. Determine the pull-up and pull-down time constants.

2.5 Each of the CMOS inverter gates in the circuit below has $C_{in} = 1$ nF and $R_n = R_p = 800$ Ω. Determine the pull-up and pull-down time constants.

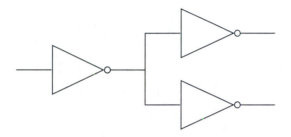

2.6 The circuit below has three CMOS inverter gates with different characteristics. Inverter "A" has $C_{in} = 1$ nF, $R_n = 800$ Ω, and $R_p = 1.2$ kΩ; inverter "B" has $C_n = 0.5$ nF, $C_p = 0.7$ nF, $R_n = 600$ Ω, and $R_p = 1.3$ kΩ; inverter "C" has $C_n = 0.6$ nF, $C_p = 0.6$ nF, $R_n = 900$ Ω, and $R_p = 950$ Ω; Determine the pull-up and pull-down time constants.

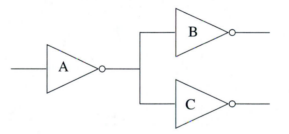

2.7 Two cascaded inverter gates powered with a 5-V source have the following parameters: $C_{in} = 1$ nF, $R_n = 750$ Ω, and $R_p = 1.1$ kΩ. Solve for the voltage on the input of the second inverter as a function of time for the pull-up and pull-down cases, assuming the initial voltages are at logic low and logic high, respectively. Sketch the time behavior of the voltages over a time period of 5 time constants.

2.8 In Exercise 2.7 determine the gate delay times (both pull-up and pull-down), assuming that the logic threshold voltages $V_{tl} = 0.7$ V and $V_{th} = 4.1$ V.

2.9 Two cascaded inverter gates powered with a 5-V source have $C_{in} = 10$ pF, $R_n = 80$ Ω, $R_p = 120$ Ω, and $V_{tl} = 0.7$ V. What is the logic high threshold voltage V_{th} if the pull-up and pull-down logic transition times are equal?

2.10 In Exercise 2.9 what is the initial slope of the second inverter input voltage in pull-up and pull-down? Do these values depend on the threshold voltages?

2.11 Two cascaded inverter gates powered with a 5-V source have $C_{in} = 10$ pF, $R_n = 80$ Ω, $R_p = 120$ Ω, and $V_{th} = 4.0$ V. What is the range of logic low threshold voltage V_{tl} if the pull-up and pull-down logic transition times are equal to within 10%?

2.12 Assume the switch has been in position A for a long time before it switches to B at $t = 0$. Determine $v_C(t)$ for t > 0.

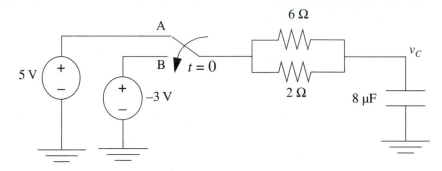

2.13 For each of the two circuits below, determine (a) the capacitor voltage as a function of time and (b) the initial rate of change of the capacitor voltage. Assume the switches have been in the up position for a long time before moving downward.

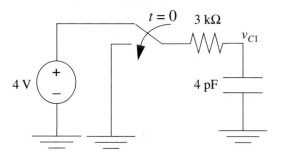

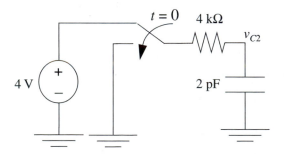

2.14 Assume that the following circuit switches continuously, driven by a symmetric square wave that transitions from 0 to 5 V. Determine the maximum switching speed of the circuit to three significant figures. The circuit parameters are:

$C_n = 4.5 \text{ pF}$ $C_p = 4.5 \text{ pF}$

$R_n = 1.6 \text{ k}\Omega$ $R_p = 1.6 \text{ k}\Omega$

$V_L = 0.0 \text{ V}$ $V_H = 5.0 \text{ V}$

$V_{t1} = 1.5 \text{ V}$ $V_{th} = 3.7 \text{ V}$

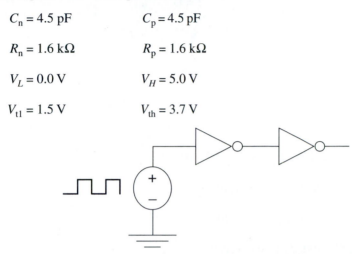

2.15 The circuit below is driven by a symmetric square wave that is at logic high for $t < 0$. At $t = 0$ the input switches instantaneously to 0 V; thereafter it transitions between 0 and 5 V at a frequency of 36.0 MHz. Determine the elapsed time before the second inverter switches state. The circuit parameters are:

$C_n = 4.5 \text{ pF}$ $C_p = 5.5 \text{ pF}$

$R_n = 1.0 \text{ k}\Omega$ $R_p = 1.0 \text{ k}\Omega$

$V_L = 0.0 \text{ V}$ $V_H = 5.0 \text{ V}$

$V_{t1} = 1.0 \text{ V}$ $V_{th} = 4.0 \text{V}$

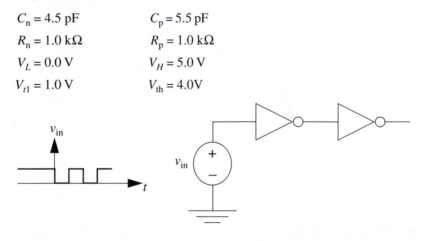

2.16 Determine $v_C(t)$ and $i_C(t)$ in the RC circuit below for each of the four sets of parameters. Also determine the time interval required for $i_C(t)$ to reach 0.01% of its value at $t = 0$.

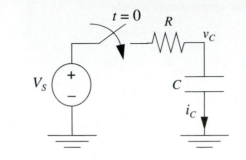

(a) $R = 3.0$ kΩ, $C = 10$ pF, $V_S = 5$ V, $v_C(0) = 0$ V

(b) $R = 1.5$ kΩ, $C = 8$ pF, $V_S = 4$ V, $v_C(0) = 2$ V

(c) $R = 1.0$ kΩ, $C = 5$ nF, $V_S = 0$ V, $v_C(0) = 4$ V

(d) $R = 300$ Ω, $C = 100$ pF, $V_S = 0$ V, $v_C(0) = 6$ V

2.17 Redraw the circuit below with a single equivalent R and a single equivalent C. Solve for $v_C(t)$, assuming $v_C(0) = 3.2$ V.

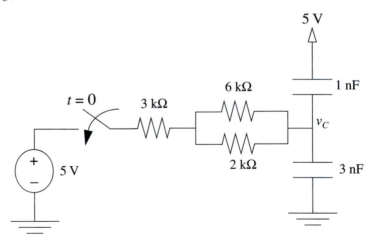

2.18 Using your solution to Exercise 2.17, write an expression for the voltage across the 1-nF capacitor.

2.19 In the RC circuit below, $R = 3$ kΩ. The capacitor voltage is given by

$$v_C = -3.4\, e^{-\frac{t}{2\times10^{-6}}} + 5$$

where t is in seconds and v_C is in volts.

(a) Determine $v_C(0)$.

(b) Determine C.

(c) Determine V_S.

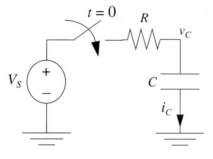

2.20 In the circuit below, a load capacitance $C_L = 1$ pF is driven by a CMOS inverter with the following parameters:

$C_n = 10$ fF	$C_p = 20$ fF
$R_n = 2.0$ kΩ	$R_p = 2.5$ kΩ
$V_L = 0.0$ V	$V_H = 5.0$ V
$V_{t1} = 1.0$ V	$V_{th} = 4.0$V

(a) Draw the RC model of the circuit in pull-up. Determine the pull-up transition time. Plot v_{out} over a time range of five time constants.

(b) Repeat for the pull-down transition.

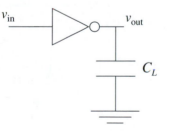

2.21 Coupled inverter pairs are built in a technology with the following parameters:

$C_n = 1.0$ pF	$C_p = 5.0$ pF
$R_n = 1.6$ kΩ	$R_p = 8.0$ kΩ
$V_L = 0.0$ V	$V_H = 5.0$ V
$V_{t1} = 0.8$ V	

During the manufacturing process it is possible to adjust the value of V_{th}. Determine V_{th} such that the pull-up and pull-down times (starting from V_L and V_H) are equal.

2.22 Repeat Exercise 2.21 so that the second inverter input reaches V_{t1} and V_{th} when switched continuously at the maximum operating frequency.

2.23 The circuit below has CMOS inverters built with these parameters:

$$C_n = 10 \text{ fF} \qquad\qquad C_p = 20 \text{ fF}$$
$$R_n = 2.0 \text{ k}\Omega \qquad\qquad R_p = 2.5 \text{ k}\Omega$$
$$V_L = 0.0 \text{ V} \qquad\qquad V_H = 5.0 \text{ V}$$
$$V_{tl} = 1.0 \text{ V} \qquad\qquad V_{th} = 4.0 \text{ V}$$

Determine the maximum switching speed of this configuration when the inverters have been quiet for a long time.

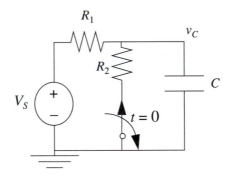

2.24 Repeat Exercise 2.23, assuming the inverters are switched continuously with a symmetric square-wave clock signal.

2.25 In the circuit below, the switch has been closed a long time. At $t = 0$, the switch opens. The component values are: $V_S = 5 \text{ V}$, $R_1 = 1 \text{ k}\Omega$, and $R_2 = 470 \text{ }\Omega$.

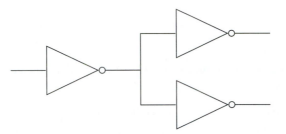

(a) Determine $v_C(0^+)$.

(b) Determine the steady-state capacitor voltage.

(c) Given that $v_C = 2.3 \text{ V}$ at $t = 5.3 \text{ ms}$, determine the capacitance value C.

2.26 In the circuit below, $v_C(0) = 0$ V, $V_S = 12$ V, $R_1 = 4$ kΩ, $R_2 = 2$ kΩ, $R_3 = 6$ kΩ, $R_4 = 6$ kΩ, and $C = 1$ nF. Determine $v_C(t)$.

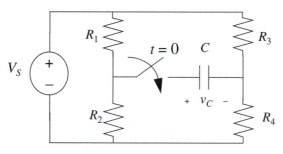

2.27 The circuit below represents a "forbidden combination" of elements. Explain the unphysical nature of this circuit.

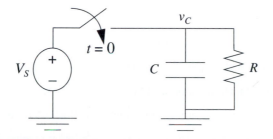

2.28 Many practical systems are described as being accurate to N bits. For example, a digital telephone system is accurate to about $N = 8$ bits, while a high-quality digital audio system is accurate to $N = 18$ bits. If we model the transient response of such an N-bit system using a first order RC circuit, "accurate to N bits" means that transition between the initial voltage $v(0)$ and a final voltage $v(\infty)$ must be $(1 - 2^{-N}) \times 100\%$ complete. If the system can be approximated by a first-order RC circuit, determine the number of time constants required to obtain:

(a) 6-bit accuracy

(b) 12-bit accuracy

(c) 18-bit accuracy.

2.29 In Section 2.9.2 we calculated the approximate power dissipation of a microprocessor chip. Using the parameters of this chip, determine the number of fundamental unit charges which represent a digital "1" in the computer.

2.30 Future computers will have ever-smaller transistors. Two factors which limit this minimum size are: (1) a digital bit must be represented by at least one fundamental charge, and (2) the minimum voltage swing between a "0" and "1" should be at least the thermal energy (about 26 mV at room temperature). Given these constraints, what is the approximate minimum size of a data-storage capacitor?

2.31 Determine the current $i(t)$ in the circuit below for each set of parameters. $R = 1.0$ kΩ.

(a) $C_1 = 10$ pF, $C_2 = 10$ pF, $V_1 = 5$ V, $V_2 = -5$ V, $v_1(0^-) = 0$ V, $v_2(0^-) = 0$ V

(b) $C_1 = 10$ pF, $C_2 = 20$ pF, $V_1 = 10$ V, $V_2 = -5$ V, $v_1(0^-) = -8$ V, $v_2(0^-) = -6$ V

(c) $C_1 = 10$ pF, $C_2 = 15$ pF, $V_1 = -5$ V, $V_2 = -50$ V, $v_1(0^-) = 3$ V, $v_2(0^-) = 3$ V

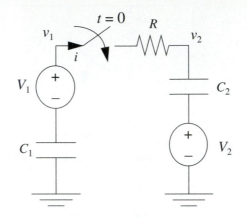

2.32 In the circuit below, $C_1 = 1$ F, $C_2 = 1$ F, $v_1(0) = 10$ V, $v_2(0) = 2$ V.

(a) Determine the total charge stored on the two capacitors before the switch closes.

(b) Determine the total energy stored in the circuit before the switch closes.

(c) Determine the total charge stored on the two capacitors after the circuit reaches steady state.

(d) Determine the total energy stored in the circuit when it reaches steady state.

(e) Determine the total energy dissipated in the resistor.

(f) Would the previous result be valid if the resistance were exactly zero? Explain.

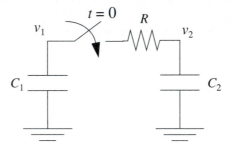

2.33 Coupled inverter pairs are made from two competing integrated-circuit technologies A and B, whose characteristics are listed below. Which technology has the greater switching speed? Justify your answer.

	A	B
R_n	1.0 kΩ	1.2 kΩ
R_p	1.2 kΩ	800 Ω
C_n	2.5 pF	2.4 pF
C_p	2.5 pF	2.8 pF
V_L	−5.0 V	0.0 V
V_{tl}	−4.0 V	0.7 V
V_{th}	−0.5 V	3.5 V
V_H	0.5 V	3.9 V

2.34 Calculate the energy dissipation per logic transition for the two technologies in Exercise 2.33. Which technology is more power hungry?

2.35 The circuit below is driven by a symmetric square wave that is at logic high for $t < 0$. At $t = 0$ the input switches instantaneously to 0 V; thereafter it transitions between 0 V and 5 V at a frequency of 36.0 MHz. Determine the energy dissipated in the first inverter per logic transition. The circuit parameters are:

$C_n = 4.5 \text{ pF}$ $\qquad$ $C_p = 5.5 \text{ pF}$

$R_n = 1.0 \text{ k}\Omega$ $\qquad$ $R_p = 1.0 \text{ k}\Omega$

$V_L = 0.0 \text{ V}$ $\qquad$ $V_H = 5.0 \text{ V}$

$V_{tl} = 1.0 \text{ V}$ $\qquad$ $V_{th} = 4.0\text{V}$

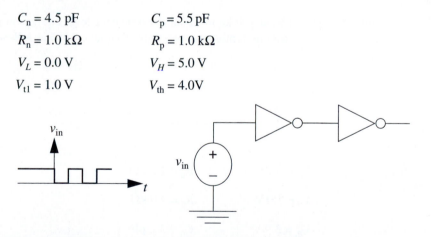

2.36 Repeat Exercise 2.35 for a square-wave frequency of 30 MHz.

2.37 Coupled inverter pairs are switched continuously at frequency f. The second inverter is loaded by a parallel combination of resistance and capacitance ($R_L = 10 \text{ k}\Omega$, $C_L = 10 \text{ pF}$). Determine the power consumption of the system for $f = 10$ MHz and 25 MHz. The circuit parameters are:

$C_n = 2$ pF $\qquad\qquad$ $C_p = 2$ pF

$R_n = 2.0$ kΩ $\qquad\quad$ $R_p = 2.5$ kΩ

$V_L = 0.0$ V $\qquad\quad$ $V_H = 5.0$ V

$V_{tl} = 1.0$ V $\qquad\quad$ $V_{th} = 4.0$ V

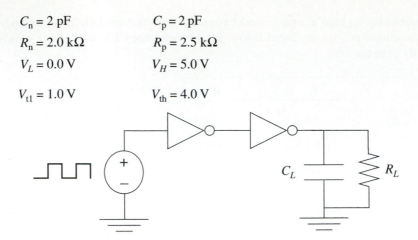

2.38 The internal construction of a CMOS NAND gate is shown below:

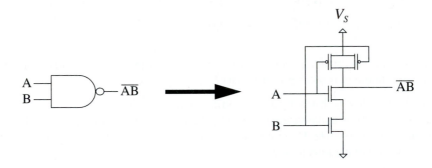

Given this information and what you know about the transistor-level equivalent of a CMOS inverter, determine the maximum continuous switching frequency of the following circuit.

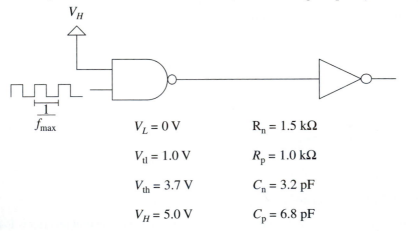

$$V_L = 0 \text{ V} \qquad\qquad R_n = 1.5 \text{ k}\Omega$$

$$V_{tl} = 1.0 \text{ V} \qquad\qquad R_p = 1.0 \text{ k}\Omega$$

$$V_{th} = 3.7 \text{ V} \qquad\qquad C_n = 3.2 \text{ pF}$$

$$V_H = 5.0 \text{ V} \qquad\qquad C_p = 6.8 \text{ pF}$$

2.39 A CMOS NOR gate, like the NAND gate, also uses two NMOS and two PMOS transistors. Determine the internal construction of a NOR gate.

2.40 An exclusive-OR (XOR) gate is constructed using five CMOS NAND gates:

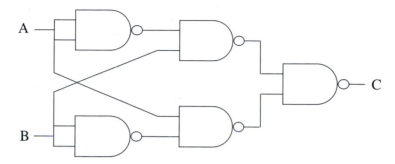

(a) Draw the transistor-level schematic diagram for this circuit.

(b) Suppose the B input is held low, and A is instantaneously switched from high to low. Calculate the delay time before the output C switches from V_H to V_L. Assume the output C is loaded by a 10-pF capacitance and that A and B were in their initial states for a long time before the switching event. Use the following parameters (R and C values are for *each* transistor).

$$V_L = 0 \text{ V}$$
$$V_{tl} = 1.0 \text{ V}$$
$$V_{th} = 3.7 \text{ V}$$
$$V_H = 5.0 \text{ V}$$
$$R_p = 1.5 \text{ k}\Omega$$
$$R_n = 1.0 \text{ k}\Omega$$
$$C_{in} = 10 \text{ pF}$$

Interconnects and *RC* Ladder Circuits

3

3.1 Introduction

The behavior of cascaded CMOS inverter gates in integrated circuits was approximated in Chapter 2 with single-lump *RC* circuits. In this approximation, we ignored the electrical characteristics of the connections between the output of the driving gate and the input of the load gate. However, the conducting paths between the MOS transistors in integrated circuits are not ideal; they have resistance and capacitance.

These conducting paths between devices are called **interconnects**. Interconnects are usually made of aluminum alloys but can also be constructed of polysilicon or other less conductive materials. Viewed from above the chip, interconnects often appear as long paths which can be decomposed into rectangles, as illustrated by the shaded regions of Figure 3.1.

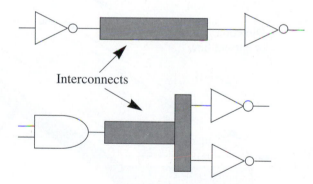

Figure 3.1 Examples of interconnect paths. The "wires" connecting devices in an integrated circuit have finite resistance and capacitance; these must be taken into consideration when calculating circuit speed.

In the past, interconnect resistance and capacitance were often ignored because the internal transistor resistance and capacitance dominated switching speed. In recent years, IC technology has progressed to the point where the electrical characteristics of the interconnect must be taken into account. The simultaneous decrease in MOS transistor size and increase in chip area have combined to make interconnects both thinner and longer; they are therefore responsible for much of the switching delay in modern integrated circuits.

In this chapter we will modify our single-lump RC model of interconnected inverter gates to include interconnect resistance and capacitance. Using increasingly accurate models of the interconnect, we will find that the coupled inverter pairs can be represented by a circuit known as the RC ladder circuit. In solving this circuit you will learn about techniques for approaching multiple-capacitor circuits and about the concept of natural frequencies.

3.2 Resistance and Capacitance of Interconnects

Consider the run of interconnect illustrated in Figure 3.2. The lightly shaded region is part of a metal trace connecting two components of an integrated circuit. The metal path, l units long, w units wide, and h units high, is situated over an insulating layer of SiO_2 with thickness d. The interconnect carries current along the path indicated by the arrow in Figure 3.2. The insulator isolates the metal interconnect from the silicon substrate; we will assume that the potential everywhere in the silicon under the interconnects is equal and identified with the ground node of the circuit. Although metals are good conductors, they have a nonzero resistivity, which results in a finite interconnect resistance. Also, the interconnect/insulator/substrate sandwich comprises a capacitor. In effect, the interconnect structure is a four-terminal device exhibiting both resistance and capacitance.

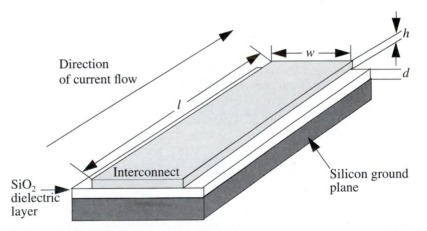

Figure 3.2 Perspective view of an integrated-circuit interconnect (not drawn to scale). The current flows in the direction of the arrow inside the interconnect layer, along a path l units long and wh units in area. The interconnect also has a parallel-plate capacitance with the ground plane, with area wl and dielectric thickness d. The silicon substrate is approximated as a perfectly conducting ground plane.

We can approximate the values of the interconnect resistance and capacitance using the formulas introduced in Chapter 1:

$$R_{\text{int}} = \rho \frac{l}{A_R}$$

and

$$C_{\text{int}} = \varepsilon \frac{A_C}{d}$$

Note that the "areas" in these expressions for resistance and capacitance are different. Resistance R_{int} is inversely proportional to A_R, the cross-sectional area of the resistor, and capacitance C_{int} is directly proportional to A_C, the area overlapping both plates of the capacitor. (Since capacitor plates, in general, can have different shapes and sizes, we use the overlap area as a measure of the capacitor area A_C. In the present case A_C is simply the interconnect area, since the metal layer overlaps the substrate everywhere.)

In Figure 3.3, we illustrate the interconnect structure as viewed from the side, orthogonal to the direction of current flow. Note that the interconnect resistance, which originates from the finite resistivity of the interconnect metal, is along the path of the current. We can denote this by superposing the symbol of a resistor along the length of the interconnect, as shown in the lower left of Figure 3.3. To visualize the capacitance, we first ignore the fact that the interconnect has any resistance; by pretending that it is a perfectly conducting plane, we see a parallel plate capacitance formed between the interconnect and the substrate. This is denoted by the capacitor superposed on the structure as shown in the lower right of Figure 3.3.

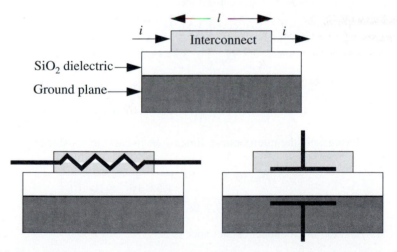

Figure 3.3 *Top:* Side view of an interconnect showing the current path along the length l. *Bottom left:* The interconnect resistance R_{int} appears along the length of the interconnect. *Bottom right:* The interconnect capacitance C_{int} appears between the interconnect and the ground plane; in this case we approximate the interconnect as an equipotential.

Treating the interconnect resistance and capacitance as independent properties of the structure, we see that the appropriate areas for A_R and A_C are given by

$$A_R = wh$$

and

$$A_R = lw$$

Hence, the interconnect resistance is given by:

$$R_{int} = \rho \frac{l}{wh} \qquad (3.1)$$

and the interconnect capacitance is approximately

$$C_{int} = \varepsilon \frac{lw}{d} \qquad (3.2)$$

While these formulas are only approximations,[1] they are quite useful in calculating the effect of interconnect material and geometry on circuit performance.

Example 3.1

An aluminum integrated-circuit interconnect, 2 mm long, 1.0 μm thick, and 1.5 μm wide, connects two gates over a 100-nm SiO_2 dielectric layer. Determine the interconnect resistance and capacitance.

The permittivity of silicon dioxide is

$$\varepsilon_{SiO_2} = 3.37 \times 10^{-13} \text{ F/cm}$$

and the resistivity of aluminum is

$$\rho_{Al} = 2.8 \text{ μΩ-cm}$$

Expressing the interconnect dimensions in cm, the resistance is given by

$$R_{int} = 2.8 \times 10^{-6} \frac{0.2}{1.5 \times 10^{-4} \times 1 \times 10^{-4}}$$

$$= 37.3 \text{ Ω}$$

and the interconnect capacitance is

[1.] In Section 3.4.2 we will see that incorrect conclusions can be drawn regarding the effects of interconnect if these approximate formulas are taken too seriously. In particular, Eq. (3.2) is valid only when $w > d$; it should therefore not be applied if this condition does not hold.

$$C_{int} = 3.37 \times 10^{-13} \left(\frac{0.2 \times 1.5 \times 10^{-4}}{100 \times 10^{-7}} \right)$$

$$= 1.1 \times 10^{-12}$$

$$= 1.1 \text{ pF}$$

3.3 Interconnect Models

Because it has both resistance and capacitance, the presence of an IC interconnect in a circuit affects the electrical response. There are several ways to approximate the effects of interconnect; we will describe two here.

In the previous section, we derived approximate expressions for the interconnect resistance and capacitance using convenient but conflicting assumptions. When calculating the resistance (from end to end), we assumed a current flow which results in a voltage drop along the length of the line; when finding the capacitance (from line to ground), we assumed no current flow, so that the line could be treated as an equipotential. In Figure 3.4 we see how these two properties of an interconnect would be independently modeled in a simple circuit. Figure 3.4 (a) illustrates the physical layout of a series connection of a voltage source, resistor, and load capacitor joined by an interconnect. Figures 3.4 (b) and (c) show how the circuit is affected by each of the interconnect properties, independent of the other.

The obvious question is: what model is appropriate which *simultaneously* takes into account the **distributed** nature of the interconnect resistance and capacitance? If we were to map the physical construction of the interconnect into a schematic symbol, it would look like the device in Figure 3.5. Unfortunately, there is no simple and useful expression which describes the current-voltage relations for this "device."

The simplest useful model, which exhibits both resistance and capacitance, is shown in Figure 3.6. In this model we lump together all of the distributed resistance into one element, and all of the distributed capacitance into another. For simplicity and ease of analysis, the lumped capacitance is placed at the right end of the line. This model is known as a **single-*RC*-lump model** of an interconnect. (The circuits we studied in Chapter 2 are also known as single-lump-*RC* circuits, since they have the same topology as Figure 3.6.)

While the single-lump *RC* model is simple, it is not sufficiently accurate for many problems. It is clearly an approximation, since in an actual interconnect the capacitance is not really "lumped" at the end of the line but is distributed along the length. A more accurate model will reflect this. For example, we could split the lumped capacitance into two equal parts and place them at either end of the interconnect (Figure 3.7). Although this is a better approximation than the single-lump model, analyses incorporating this model are more difficult.

Another approach to increasingly accurate *RC* interconnect models is illustrated in Figure 3.8. By subdividing the interconnect into smaller portions, each of which is modeled by a single *RC* lump, we obtain a better approximation of the real device. If we use the *n*-lump model at the bottom of Figure 3.8, and let $n \to \infty$, then the model becomes

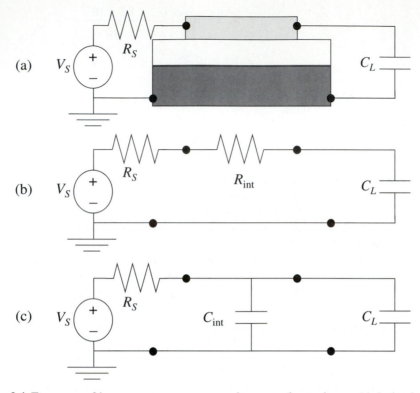

Figure 3.4 Treatment of interconnect system as a resistance and capacitance. (a) A simple circuit consisting of a voltage source, series resistance, and load capacitance connected with the interconnect and silicon ground plane. (b) Circuit model, ignoring interconnect capacitance. Note that the resistance originates from the resistivity of the upper metal interconnect layer. (c) Circuit model, ignoring interconnect resistance. In this case, the parallel-plate capacitance of the interconnect/dielectric/substrate gives rise to a capacitance in parallel with the load. In both (b) and (c) we treat the silicon substrate as a perfect conductor at ground potential.

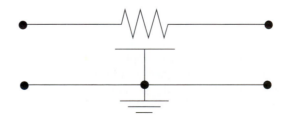

Figure 3.5 Schematic symbol for a distributed RC line. Although this symbol simply captures the physical reality of an interconnect, it does not have a correspondingly simple current-voltage relationship.

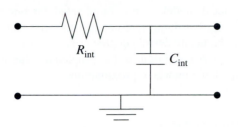

Figure 3.6 Single-RC-lump model of an interconnect.

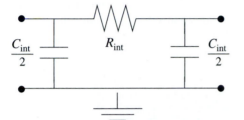

Figure 3.7 Another possible model for an IC interconnect.

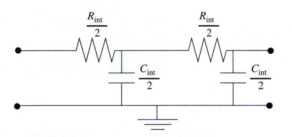

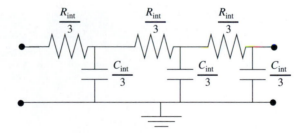

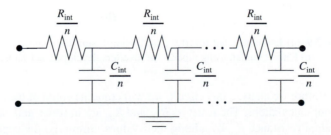

Figure 3.8 Two-lump, three-lump, and n-lump models of an RC interconnect.

exact. However, the analysis of this ideal model is too complicated for our purposes. Therefore, we will use the one-lump and two-lump interconnect models, which can be generalized to higher-order models to obtain any desired degree of accuracy.

In the following sections, we will derive solutions to the coupled inverter gate delay problem, using one-lump and two-lump interconnect approximations.

3.4 Single-*RC*-Lump Approximation of an Interconnect

We begin our analysis by approximating the physical interconnect with a single-lump *RC* model. Inserting this model into the cascade connection of two inverter gates results in the circuit illustrated in Figure 3.9.

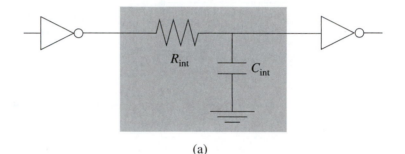

(a)

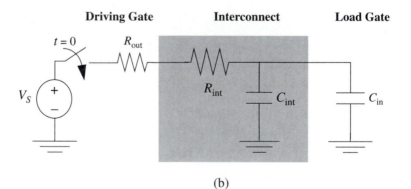

(b)

Figure 3.9 (a) Interconnected inverter gates, using a single-*RC*-lump interconnect model (shaded portion). (b) Circuit model, using models for inverters developed in Chapter 2.

Before we analyze this circuit, we will combine some of the circuit elements to simplify our calculations. The resistors R_{out} and R_{int} are in series and can be combined into an equivalent resistance R_{eq} by adding their values. Similarly, the capacitors C_{in} and C_{int} are in parallel, and can be combined into an equivalent capacitance C_{eq}:

$$R_{eq} = R_{out} + R_{int}$$
$$C_{eq} = C_{in} + C_{int}$$

Substituting the equivalent component values into the circuit results in the equivalent circuit illustrated in Figure 3.10.

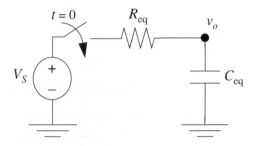

Figure 3.10 Simplified equivalent circuit for a pair of cascaded CMOS inverters using a single-lump interconnect model.

Example 3.2

Determine the output voltage $v_o(t)$ using a single lump model for the interconnect.

Assume the following:

$$V_S = 5 \text{ V (pull-up case)}$$
$$R_{out} = 500 \ \Omega$$
$$C_{in} = 1.00 \text{ pF}$$
$$l = 1.00 \text{ mm}$$
$$w = 1.25 \ \mu\text{m}$$
$$h = 0.5 \ \mu\text{m}$$
$$d = 0.2 \ \mu\text{m}$$
$$\rho = \rho_{Al} = 2.8 \times 10^{-6} \ \Omega\text{-cm}$$

We apply Eqs. (3.1) and (3.2) to calculate the interconnect resistance and capacitance. Expressing all dimensions in cm, we find:

$$R_{int} = \rho \frac{l}{wh}$$

$$= 2.8 \times 10^{-6} \frac{0.1}{1.25 \times 10^{-4} \times 0.5 \times 10^{-4}}$$

$$= 44.8 \ \Omega$$

and

$$C_{int} = \varepsilon \frac{lw}{d}$$

$$= 3.36 \times 10^{-13} \times \frac{0.1 \times 1.25 \times 10^{-4}}{0.2 \times 10^{-4}}$$

$$= 0.210 \times 10^{-12} = 0.210 \text{ pF}$$

These values are used to determine the equivalent resistance and capacitance values in Figure 3.10:

$$R_{eq} = 500 + 44.8 = 544.8 \ \Omega$$

$$C_{eq} = 1.00 + 0.210 = 1.21 \ \text{pF}$$

Note that the circuit of Figure 3.10 is exactly the same as the first-order *RC* circuits examined in Chapter 2. By inserting the appropriate values into the general solution, we will obtain an expression for the output voltage as a function of time.

Since we are considering the pull-up case, the initial voltage across the capacitor is assumed to be zero volts at $t = 0$. The time constant is found by multiplying the equivalent resistance and capacitance; we will denote the time constant including the effect of interconnect as $\tau_{w \ int}$:

$$\tau_{w \ int} = R_{eq}C_{eq}$$

$$= (544.8)(1.21 \times 10^{-12}) = 659 \times 10^{-12} \ s$$

where the subscript "w int" means "with interconnect included." The general solution is

$$v_o(t) = (v_o(0) - V_S)e^{-\left(\frac{t}{\tau_{w \ int}}\right)} + V_S$$

After substitution and simplification:

$$v_o(t) = -5e^{-\left(\frac{t}{659 \times 10^{-12}}\right)} + 5 \tag{3.3}$$

Now that we have an expression for the inverter voltage as a function of time, we can find the logic transition times using the methods outlined in Chapter 2. Instead of repeating these calculations, we will show numerically why it is important to include the effect of interconnects in our estimates of system speed.

In Section 2.8, we showed that the pull-up transition time is given symbolically by

$$t_{low \to high} = \tau_u \ln\left(\frac{V_H - V_L}{V_H - V_{th}}\right)$$

and the pull-down transition time by

$$t_{high \to low} = \tau_d \ln\left(\frac{V_L - V_H}{V_L - V_{tl}}\right)$$

Note that in both cases any effects of interconnect resistance and capacitance are contained in the time constant. R_{int} and C_{int} have no effect whatsoever on the logic levels V_L and V_H; nor can they affect the logic threshold values V_{tl} and V_{th}, since these parameters are internal to the inverters. Since the logic transition time is linearly proportional to the time constant, we can compare the effect of interconnect on transition time by examining the effect on the time constant alone. In other words, we can determine the ratio of switching times, with and without interconnect included, from

$$\frac{t_{w\ int}}{t_{w/o\ int}} = \frac{\tau_{w\ int}}{\tau_{w/o\ int}} \tag{3.4}$$

where the subscript "w/o int" refers to the case where the effects of interconnect are ignored.

As a general engineering approximation, we will consider a change to be "significant" if it results in a numerical change in an important output parameter (e.g., transition time) of at least 5%. Therefore, we will consider interconnect properties important if they change the switching time of our coupled inverter pair by 5% or more. Conversely, the effect of interconnect will be considered unimportant if

$$\frac{t_{w\ int}}{t_{w/o\ int}} < 1.05 \tag{3.5}$$

Example 3.3

Determine whether the effect of interconnect in Example 3.2 is significant.

In determining whether the interconnect has a significant effect, we must first ask, effect on *what*? Our figure of merit is system speed, so we will compare logic transition times with and without interconnect. From Eq. (3.4) we know that the transition times are linearly proportional to time constant; therefore, we will compare the circuit time constants with and without interconnect.

We first calculate $\tau_{w/o\ int.}$. From the original CMOS parameters we have

$$R_{out} = 500\ \Omega$$

and

$$C_{in} = 1.0\ pF$$

The time constant is given by their product:

$$\tau_{w/o\ int} = R_{out}C_{in} = (500)(1 \times 10^{-12})$$
$$= 500 \times 10^{-12}\ s$$

or 500 ps. We previously calculated the time constant including the effect of interconnect as $\tau_{w\ int} = 659$ ps. Inserting these values into Eq. (3.4), we have

$$\frac{\tau_{w\ int}}{\tau_{w/o\ int}} = \frac{6.59 \times 10^{-10}}{5.0 \times 10^{-10}} = 1.318 > 1.05$$

We see that the time constant, and hence the switching time, is degraded by 31.8% when the interconnect properties are taken into consideration. This is considerably more than the 5% level at which we deem an effect "significant." Hence, the interconnect properties in this example are significant in comparison to the CMOS inverter parameters and should be included in calculations of gate delay.

In these examples we assumed that the interconnect material is aluminum, a common choice due to its relatively low value of resistivity ($< 3 \times 10^{-6}$ Ω-cm). Suppose instead that two inverters were connected with a polysilicon interconnect of the same dimensions. The resistivity of polysilicon is approximately 5×10^{-4} Ω-cm, considerably higher than that of aluminum. This results in an interconnect resistance

$$R_{int} = 8000\ \Omega$$

We expect this large increase in resistance to have a significant effect on gate delay, compared to an aluminum interconnect with a resistance of 44.8 Ω.

The interconnect capacitance is the same as before, since it does not depend on the resistivity:

$$C_{int} = 0.210\ \text{pF}$$

Example 3.4

Determine the effect on switching time if the interconnect in Example 3.2 is fabricated from polysilicon.

The time constant, excluding the interconnect, is the same as before:

$$\tau_{w/o\ int} = R_{out}C_{in} = (500)(1.0 \times 10^{-12}) = 500 \times 10^{-12}\ \text{s}$$

Including the interconnect, the time constant is

$$\tau_{w\ int} = R_{eq}C_{eq}$$
$$= (500 + 8000) \times ((1.0 + 0.210) \times 10^{-12})$$
$$= 10.3\ \text{ns}$$

Taking the ratio of the two time constants:

$$\frac{\tau_{w\ int}}{\tau_{w/o\ int}} = \frac{10.3 \times 10^{-9}}{500 \times 10^{-12}} = 20.6$$

Using a polysilicon interconnect results in a delay time more than twenty times greater (i.e., 2000%) than the ideal case (ignoring the interconnect). For this reason, long runs of polysilicon interconnect are avoided in IC design. It is convenient for technological reasons to use polysilicon as a local interconnect, but the length of these runs is limited to a few microns. Major interconnections, which may be many millimeters long, are made of metal (aluminum alloys, usually) to avoid horrendous gate delays.

Figure 3.11 illustrates the inverter voltage v_o as a function of time after the switching event. The two curves represent the response ignoring and including the effect of a polysilicon interconnect. Inclusion of the polysilicon resistance and capacitance dramatically slows down the circuit response.

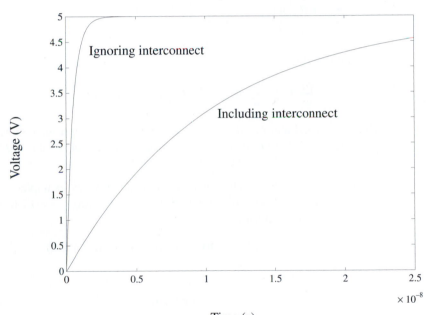

Figure 3.11 Inverter output voltage waveforms, ignoring the effects of polysilicon interconnect (top line) and including them (bottom line). Failure to take the interconnect properties into account results in a severe underestimation of the inverter switching time.

3.4.1 Maximum Interconnect Length

Recall that both interconnect resistance and capacitance are proportional to the length l. In designing an IC, one wants to retain maximum flexibility over the layout, but the problem of interconnect delay sets a constraint on where cascaded stages may be placed. Using the single-lump approximation for the interconnect, we can determine the maximum length of interconnect which will not result in "significant" delays (i.e., the transition time is degraded by less than 5% compared to the ideal case without interconnect).

Example 3.5

Determine the maximum interconnect length l_{max} such that the interconnect may be neglected.

We start with the inequality given in Eq. (3.5), which can be written as:

$$\frac{\tau_{w\ int}}{\tau_{w/o\ int}} = \frac{(R_{out} + R_{int})(C_{in} + C_{int})}{R_{out}C_{in}} < 1.05$$

Replacing R_{int} and C_{int} by Eq. (3.1) and Eq. (3.2), respectively, we get

$$\frac{(R_{out} + R_{int})(C_{in} + C_{int})}{R_{out}C_{in}} = \frac{\left(R_{out} + \rho\dfrac{l}{wh}\right)\left(C_{in} + \varepsilon\dfrac{lw}{d}\right)}{R_{out}C_{in}} < 1.05$$

Expanding terms:

$$\frac{\rho\varepsilon}{hd}\ l^2 + \left(\frac{\varepsilon w}{d}\ R_{out} + \frac{\rho}{wh}\ C_{in}\right)l - 0.05R_{out}C_{in} < 0$$

which is a quadratic inequality. We can find the maximum interconnect length by setting the left side to zero and solving for the positive root; a value of l less than this root will result in a less than significant delay.

Inserting the parameter values that we used in previous examples ($R_{out} = 500\ \Omega$, $C_{in} = 1.00$ pF, $l = 1.00$ mm, $w = 1.25\ \mu m$, $h = 0.5\ \mu m$, $d = 0.2\ \mu m$, $\rho = \rho_{Al} = 2.8 \times 10^{-6}\ \Omega$-cm) results in:

$$(942 \times 10^{-12})l^2 + (1.50 \times 10^{-9})l - (25 \times 10^{-12}) = 0$$

which has two roots:

$$l = \begin{cases} -1.61 \times 10^0 \\ 16.5 \times 10^{-3} \end{cases}$$

of which only the positive root is physically significant. The critical interconnect length is therefore $l_{max} = 16.5 \times 10^{-3}$ cm, or 165 μm. Aluminum interconnect runs which are shorter than 165 μm will result in less than a 5% increase in interconnect delay for the example parameters.

We can reconsider the same problem applied to a polysilicon interconnect of the same dimensions and connecting the same gates (R_{out} and C_{in} remain unchanged). The only change is the resistivity ρ, which increases to $\rho_{poly} = 5 \times 10^{-4}\ \Omega$-cm. Solving for the critical interconnect length, we find that $l_{max} = 3.08$ μm.

Notice the striking difference in maximum lengths of interconnect. The maximum length of a polysilicon interconnect is much shorter, owing to its significantly higher resistivity; this leads to a longer delay than that for a similar aluminum interconnect. Thus, polysilicon is ill suited to all but the shortest connections between devices.

3.4.2 Interconnect Scaling

We have seen that there is a practical upper bound on interconnect length. Interconnects longer than this limiting value result in significant increases in gate delay. It is also true that, as IC technology continues to evolve, the problem of gate delay is increasingly dominated by the interconnects, rather than the gates. This can be seen from the fact that both interconnect resistance R_{int} and capacitance C_{int} are proportional to length l. Thus, as new generations of ICs increase in area, which results in longer maximum interconnect lengths, the resistance and capacitance of interconnects both increase.

The first-order interconnect approximation yields a solution time constant given by

$$\tau_{w\,int} = R_{eq}C_{eq} = (R_{out} + R_{int})(C_{in} + C_{int})$$

When expanded, the last term is given by

$$R_{int}C_{int} = \rho\varepsilon\,\frac{l^2}{dh} \tag{3.6}$$

There are two important observations to be made from Eq. (3.6). First, we see that the interconnect delay scales as the *square* of the interconnect length. (The other terms are either independent of l or linear in l.) This means that, once the contribution of interconnect to the total delay exceeds a certain level, further increases in interconnect length have an extremely bad effect on the delay. Thus, the issue of interconnect length minimization is crucial in designing ICs.

Another point is that the interconnect width w does not appear in Eq. (3.6). This would appear to imply that technological improvements which enable narrower features on the IC, and thus smaller interconnect widths, will have no effect on interconnect delay. The apparent independence of delay on w is a consequence of the fact that R_{int} scales as the reciprocal of w, while C_{int} scales linearly with w; therefore the product is independent of w. In fact, this is not true; once w decreases below a certain point, the interconnect delay begins to increase with further shrinkages in the width. The reason for this inconsistency is that Eq. (3.2), the expression for interconnect capacitance, is not a very good approximation once $w < d$ (i.e., if interconnect width is less than the dielectric thickness). The approximate formula in Eq. (3.2) ignores the effect of fringing fields (electric field not directly between the capacitor plates), which dominate the capacitance when w is narrow (Figure 3.12). In practice, we can make a reasonable approximation of the interconnect capacitance from

$$C_{int} = k_A A_C + k_P P_C$$

where the first term is proportional to the interconnect area A_C, and the second term is proportional to the interconnect perimeter P_C.

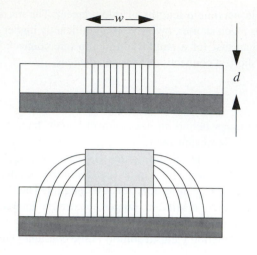

Figure 3.12 Electric-field distribution around an interconnect. *Top:* The parallel-plate capacitor approximation takes into account only those field lines directly under the interconnect. This approximation is reasonable when the interconnect width w is considerably larger than the dielectric thickness d. *Bottom:* The actual field distribution includes fringing field around the perimeter of the interconnect. The contribution to the capacitance from the fringing field becomes more significant as w is scaled downward.

The failure of Eq. (3.6) to correctly predict the effect of narrow linewidths on gate delay is a consequence of the limits of validity of the models used to derive it. An extremely important aspect of engineering practice that the range of validity of all approximations be understood, so that invalid conclusions may be avoided when these ranges are exceeded.

3.5 Two-*RC*-Lump Interconnect Approximation

The single-*RC*-lump approximation of the interconnect which we used in Section 3.4 to analyze the switching delay in cascaded inverter gates lets us make quantitative estimates of maximum computer speed. We have also seen the relative importance of the effects of interconnect properties on ICs. This approach provides useful first-order estimates of the speed-limiting factors and approximate values of maximum clock speed.

Several questions naturally arise. How accurate is this model? Is the distributed nature of the interconnect accurately modeled by the single-*RC*-lump approximation? How do the shortcomings of this approximation, if any, affect the estimate of interconnect delay?

As we will see in this section, the answers to these questions are: not very; no; and considerably. That is, the single-lump interconnect approximation, while useful for understanding gross effects, is insufficiently accurate when economic decisions are at stake.[2]

A better approximation for the interconnect is the two-*RC*-lump model, shown inserted between two cascaded inverters in Figure 3.13. An even better approximation is the *n*-lump model, where *n* is a sufficiently large number. As a first step, though, we will consider the two-lump interconnect model and learn how to analyze the resulting ladder circuit.

---

[2] For example, how much should we charge for this IC? Should we build 1,000,000 or 10,000,000 per year? Should we reassign the engineer who gave us a lousy estimate of the maximum clock frequency? To answer these questions, we must not only fill out the back of the envelope, but the sheet inside as well.

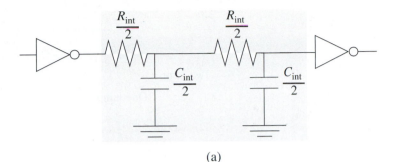

(a)

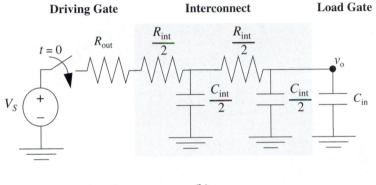

(b)

Figure 3.13 (a) Interconnected inverter gates, using a two-lump-*RC* interconnect model (shaded portion). (b) Corresponding circuit model, using models for inverters developed in Chapter 2.

The mathematics needed to analyze the two-section *RC* ladder circuit is considerably more involved than for the single-lump *RC* circuit. Before we embark on the detailed analysis, let us examine the results. Assuming inverter and interconnect parameters of $R_{out} = R_{int} = 1000\ \Omega$ and $C_{int} = C_{in} = 1.0$ pF, the single-lump interconnect approximation yields an inverter output voltage v_o given by

$$v_o(t) = -5e^{-250\times10^6 t} + 5 \tag{3.7}$$

whereas the two lump interconnect approximation predicts

$$v_o(t) = -5.23e^{-278\times10^6 t} + 0.23e^{-6.38\times10^9 t} + 5 \tag{3.8}$$

[A detailed explanation of the steps required to derive Eq. (3.8), the solution for the two-section *RC* ladder circuit, is presented later in this chapter.]

These two solutions are graphed in Figure 3.14 over different time scales. We see that there are significant differences: the single-lump approximation has the output voltage rising more steeply at the beginning of the transient, but more slowly than the two-lump model after approximately 1.6 ns. The single-lump model predicts a significantly longer

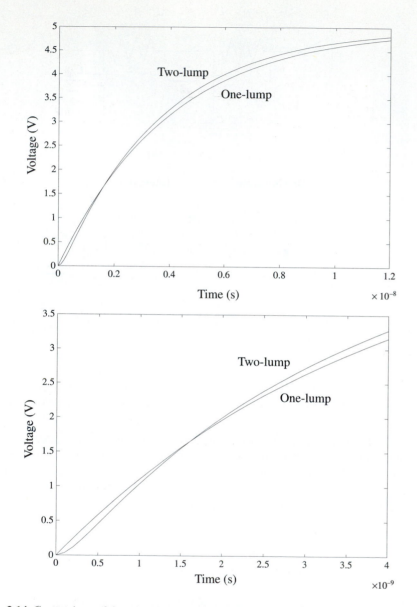

Figure 3.14 Comparison of the cascaded inverter pull-up transient using single-lump and two-lump models for the interconnect. (a) Time scale from zero to 12 ns. (b) Early part of the transient, from zero to 4 ns.

gate delay than the more accurate two-lump model. To see this, the two responses have been replotted in Figure 3.15, along with a horizontal line representing a high threshold voltage value of $V_{th} = 4.0$ V. The low-to-high transition time is predicted to be 6.4 ns by the single-lump interconnect model, but the two-lump model predicts that the logic transition takes only 5.9 ns. In other words, by using a more accurate model, we have realized an increase of system speed of more than 7%, which certainly meets the 5% significance test.

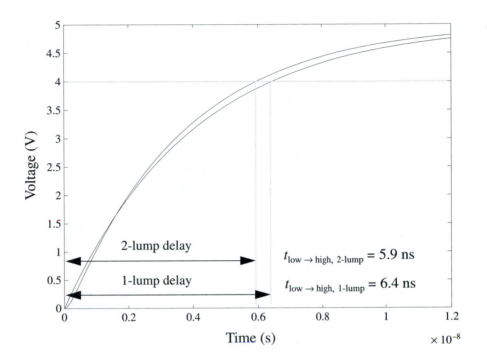

Figure 3.15 Transient response of the coupled inverter pair using one-lump and two-lump interconnect models. The more accurate two-lump model predicts a significantly shorter switching delay.

Clearly, the single-lump interconnect approximation has resulted in a too-conservative estimate of the maximum system speed. Thus, we need to develop techniques to solve circuits containing multiple-lump interconnect models.

3.6 Analysis of the Two-Section *RC* Ladder Circuit

The comparison of the single-lump and two-lump approximations, Figure 3.15, illustrates the importance of accurate interconnect models. We will now develop a general method to solve the two-lump (and *n*-lump) *RC* ladder circuits.

We begin with the two-lump circuit as shown in Figure 3.13. After combining resistors in series and capacitors in parallel, we have the equivalent circuit of Figure 3.16. In this circuit, the equivalent component values are

$$R_1 = R_{out} + \frac{R_{int}}{2} \tag{3.9}$$

$$R_2 = \frac{R_{\text{int}}}{2} \tag{3.10}$$

$$C_1 = \frac{C_{\text{int}}}{2} \tag{3.11}$$

$$C_2 = C_{\text{in}} + \frac{C_{\text{int}}}{2} \tag{3.12}$$

We have labeled the two capacitor voltages v_{C1} and v_{C2}; the latter is the same as the voltage v_o appearing at the input of the load inverter.

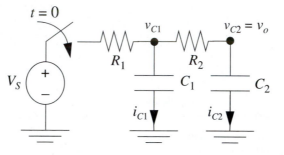

Figure 3.16 Equivalent circuit for the cascaded inverter pair, using a two lump interconnect model.

3.6.1 Derivation of the Differential Equations

At time $t = 0$, the switch closes instantaneously, whereupon the capacitors begin to charge. Applying KCL at the v_{C1} node:

$$\frac{v_{C1} - V_S}{R_1} + i_{C1} + \frac{v_{C1} - v_{C2}}{R_2} = 0 \tag{3.13}$$

and since

$$i_{C1} = C_1 \frac{dv_{C1}}{dt} \tag{3.14}$$

Equation (3.13) becomes

$$\frac{v_{C1} - V_S}{R_1} + C_1 \frac{dv_{C1}}{dt} + \frac{v_{C1} - v_{C2}}{R_2} = 0 \tag{3.15}$$

Solving Eq. (3.15) for the derivative of v_{C1} yields

$$\frac{dv_{C1}}{dt} = \frac{V_S - v_{C1}}{R_1 C_1} - \frac{v_{C1} - v_{C2}}{R_2 C_1}$$

$$\frac{dv_{C1}}{dt} = \frac{1}{R_1 C_1} V_S - \frac{1}{R_1 C_1} v_{C1} - \frac{1}{R_2 C_1} v_{C1} + \frac{1}{R_2 C_1} v_{C2}$$

Combining terms:

$$\frac{dv_{C1}}{dt} = -\frac{1}{C_1}\left(\frac{1}{R_1} + \frac{1}{R_2}\right) v_{C1} + \left(\frac{1}{R_2 C_1}\right) v_{C2} + \frac{1}{R_1 C_1} V_S \qquad (3.16)$$

Applying KCL at the v_{C2} node:

$$\frac{v_{C2} - v_{C1}}{R_2} + i_{C2} = 0 \qquad (3.17)$$

and since

$$i_{C2} = C_2 \frac{dv_{C2}}{dt} \qquad (3.18)$$

Equation (3.17) becomes

$$\frac{v_{C2} - v_{C1}}{R_2} + C_2 \frac{dv_{C2}}{dt} = 0 \qquad (3.19)$$

Solving Eq. (3.19) for the derivative of v_{C2} yields

$$\frac{dv_{C2}}{dt} = \left(\frac{1}{R_2 C_2}\right) v_{C1} + \left(-\frac{1}{R_2 C_2}\right) v_{C2} \qquad (3.20)$$

Equations (3.16) and (3.20) are a pair of coupled first order differential equations. Expressing Eqs. (3.16) and (3.20) in matrix form:

$$\begin{bmatrix} \dfrac{dv_{C1}}{dt} \\[2mm] \dfrac{dv_{C2}}{dt} \end{bmatrix} = \begin{bmatrix} -\dfrac{1}{C_1}\left(\dfrac{1}{R_1} + \dfrac{1}{R_2}\right) & \dfrac{1}{R_2 C_1} \\[3mm] \dfrac{1}{R_2 C_2} & -\dfrac{1}{R_2 C_2} \end{bmatrix} \begin{bmatrix} v_{C1} \\[2mm] v_{C2} \end{bmatrix} + \begin{bmatrix} \dfrac{V_S}{R_1 C_1} \\[2mm] 0 \end{bmatrix} \qquad (3.21)$$

Our goal is to find functions $v_{C1}(t)$ and $v_{C2}(t)$ which satisfy Eq. (3.21). As was the case with the first-order circuit, the capacitor voltages can be expressed as the sum of two terms—a transient part which depends on t, and a dc steady-state value which is reached as $t \to \infty$:

$$v_{C1}(t) = v_{Ct1}(t) + V_{C1} \qquad (3.22)$$

$$v_C(t) = v_{Ct2}(t) + V_{C2} \tag{3.23}$$

In this notation, the total response is given by $v_{C1}(t)$; the transient response is $v_{Ct1}(t)$; the dc steady-state response is $V_{C1} = v_{C1}(\infty)$; and similarly for $v_{C2}(t)$.

We can determine the steady-state response in two ways; the first is strictly mathematical, while the second takes advantage of our physical intuition about the circuit.

3.6.2 Determining the dc Steady-State Response

The straightforward mathematical way to determine the steady-state values is to use the fact that "dc steady state" means that the time derivatives are all zero. After a long time, the transient terms have died away, leaving constant voltages V_{C1} and V_{C2} across the two capacitors. Since the capacitor voltages are constant, we can write

$$\frac{d}{dt}V_{C1} = 0 \qquad \frac{d}{dt}V_{C2} = 0$$

Substituting these relationships into Eq. (3.21), with the expressed understanding that we are dealing with the limit as $t \to \infty$, results in

$$\left.\begin{bmatrix} \dfrac{dV_{C1}}{dt} \\ \dfrac{dV_{C2}}{dt} \end{bmatrix}\right|_{t \to \infty} = \begin{bmatrix} 0 \\ 0 \end{bmatrix} = \begin{bmatrix} -\dfrac{1}{C_1}\left(\dfrac{1}{R_1} + \dfrac{1}{R_2}\right) & \dfrac{1}{R_2 C_1} \\ \dfrac{1}{R_2 C_2} & -\dfrac{1}{R_2 C_2} \end{bmatrix}\begin{bmatrix} V_{C1} \\ V_{C2} \end{bmatrix} + \begin{bmatrix} \dfrac{V_S}{R_1 C_1} \\ 0 \end{bmatrix} \tag{3.24}$$

Expanding the second row of Eq. (3.24) yields

$$\frac{1}{R_2 C_2}V_{C1} - \frac{1}{R_2 C_2}V_{C2} = 0$$

$$V_{C1} = V_{C2}$$

Substitution of this result into an expansion of the first row of Eq. (3.24) gives

$$-\frac{1}{R_1 C_1}V_{C2} - \frac{1}{R_2 C_1}V_{C2} + \frac{1}{R_2 C_1}V_{C2} + \frac{1}{R_1 C_1}V_S = 0$$

$$V_{C2} = V_S$$

which produces the final result:

$$V_{C1} = V_{C2} = V_S \tag{3.25}$$

While this method works, it does not have much to recommend it. It is much easier to use a physical approach to find the steady-state capacitor voltages. In Chapter 2, we learned that the steady-state equivalent model for a capacitor is an open circuit. (This follows from the fact that capacitor current is proportional to the time derivative of capacitor voltage; in steady state, all time derivatives are zero, therefore the dc steady-state current through a capacitor is zero.) To solve the circuit for $t \to \infty$, we simply replace all of our components with their equivalent steady-state models. Replacing the capacitors in Figure 3.16 with open circuits results in the equivalent circuit of Figure 3.17. From this diagram it is immediately clear that the two steady-state voltages V_{C1} and V_{C2} are equal to V_S.

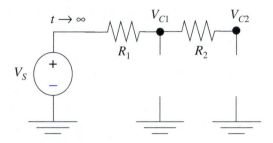

Figure 3.17 Equivalent dc steady-state circuit for the cascaded inverter pair, valid for $t \to \infty$. The capacitors have been replaced by their dc equivalent models, which are open circuits.

3.6.3 The Initial Conditions

Recall that the transient response of the first-order RC circuit contained a term which reflected our knowledge of the capacitor voltage at the instant the switch closed. We used this *initial condition* as a lower limit of a definite integral which was encountered in our solution method. In general, we must have an initial condition for each capacitor in our circuit in order to have a complete, unambiguous solution. This is because a single differential equation can have many transient solutions for different initial conditions; this corresponds to the physical reality of different voltage responses for different initial capacitor voltages, even though the circuit (and hence the governing differential equation) is the same.

This point can be illustrated using the circuit of Figure 3.16. If we have only the circuit, *and no other information*, it is impossible to determine unique initial conditions. Any arbitrary values of $v_{C1}(0)$ and $v_{C2}(0)$ will result in valid solutions. Only if we know something about the history of the circuit prior to or at the switch closure can we find initial conditions. For example, if we know that the switch has been open for a long time prior to $t = 0$, then we can reason out that $v_{C1}(0) = v_{C2}(0)$. This is a start, but it does not tell us whether the common initial capacitor voltage is 10 mV or 10 kV; either of these values is possible.

If a circuit is presented in the abstract, then we will generally assume that the initial values are given. For applied problems, such as the estimation of interconnect delay, the initial conditions are part of the problem context.

We know that capacitor voltages cannot change instantaneously. Thus, the voltages across the capacitors immediately after the switch is closed, $t = 0^+$, are equal to the voltages across the capacitors immediately before the switch is closed, $t = 0^-$. We can therefore write the initial conditions as:

$$v_{C1}(0^+) = v_{C1}(0^-) = v_{C1}(0) \tag{3.26}$$

$$v_{C2}(0^+) = v_{C2}(0^-) = v_{C2}(0) \tag{3.27}$$

Equations (3.26) and (3.27) are often useful in determining the initial conditions. For example, if we know that the circuit has reached a steady-state condition before the switch closure, we can use these steady-state values (from the appropriate circuit model) as the initial conditions at $t = 0^-$. The capacitor voltages will then be the same the instant after the switch is thrown.

Often, the initial conditions can be determined from the complete problem statement. For example, when we solved for the continuous switching case of two coupled inverters in Chapter 2, we reasoned that the initial conditions are given by the extreme values v_h and v_l at the switching times; further reasoning led us to conclude that, at the maximum switching speed, one of these must be a threshold logic level, such as V_{th} or V_{tl}.

3.6.4 Determining the Transient Solution

To determine the transient solution, we insert the dc steady-state values into Eqs. (3.22) and (3.23):

$$v_{C1}(t) = v_{Ct1}(t) + V_S \tag{3.28}$$

$$v_{C2}(t) = v_{Ct2}(t) + V_S \tag{3.29}$$

Substituting these expressions into Eq. (3.21) yields

$$
\begin{bmatrix} \dfrac{d}{dt}(v_{Ct1} + V_S) \\[2ex] \dfrac{d}{dt}(v_{Ct2} + V_S) \end{bmatrix}
=
\begin{bmatrix} -\dfrac{1}{C_1}\left(\dfrac{1}{R_1} + \dfrac{1}{R_2}\right) & \dfrac{1}{R_2 C_1} \\[2ex] \dfrac{1}{R_2 C_2} & -\dfrac{1}{R_2 C_2} \end{bmatrix}
\begin{bmatrix} (v_{Ct1} + V_S) \\[2ex] (v_{Ct2} + V_S) \end{bmatrix}
+
\begin{bmatrix} \dfrac{V_S}{R_1 C_1} \\[2ex] 0 \end{bmatrix}
$$

Expanding this expression:

$$
\begin{bmatrix} \dfrac{dv_{Ct1}}{dt} + \dfrac{dV_S}{dt} \\[2ex] \dfrac{dv_{Ct2}}{dt} + \dfrac{dV_S}{dt} \end{bmatrix}
=
\begin{bmatrix} -\dfrac{1}{C_1}\left(\dfrac{1}{R_1} + \dfrac{1}{R_2}\right) & \dfrac{1}{R_2 C_1} \\[2ex] \dfrac{1}{R_2 C_2} & -\dfrac{1}{R_2 C_2} \end{bmatrix}
\begin{bmatrix} v_{Ct1} \\[2ex] v_{Ct2} \end{bmatrix}
$$

$$
+
\begin{bmatrix} -\dfrac{1}{C_1}\left(\dfrac{1}{R_1} + \dfrac{1}{R_2}\right) & \dfrac{1}{R_2 C_1} \\[2ex] \dfrac{1}{R_2 C_2} & -\dfrac{1}{R_2 C_2} \end{bmatrix}
\overset{\textstyle 1}{\begin{bmatrix} V_S \\[2ex] V_S \end{bmatrix}}
+
\begin{bmatrix} \dfrac{V_S}{R_1 C_1} \\[2ex] 0 \end{bmatrix}
$$

Evaluating the second and third terms on the right side, we find

$$
\begin{bmatrix} -\dfrac{1}{C_1}\left(\dfrac{1}{R_1}+\dfrac{1}{R_2}\right) & \dfrac{1}{R_2 C_1} \\[3mm] \dfrac{1}{R_2 C_2} & -\dfrac{1}{R_2 C_2} \end{bmatrix} \begin{bmatrix} V_S \\[3mm] V_S \end{bmatrix} + \begin{bmatrix} \dfrac{V_S}{R_1 C_1} \\[3mm] 0 \end{bmatrix} = \begin{bmatrix} 0 \\[3mm] 0 \end{bmatrix}
$$

which yields the following:

$$
\begin{bmatrix} \dfrac{dv_{Ct1}}{dt} \\[4mm] \dfrac{dv_{Ct2}}{dt} \end{bmatrix} = \begin{bmatrix} -\dfrac{1}{C_1}\left(\dfrac{1}{R_1}+\dfrac{1}{R_2}\right) & \dfrac{1}{R_2 C_1} \\[3mm] \dfrac{1}{R_2 C_2} & -\dfrac{1}{R_2 C_2} \end{bmatrix} \begin{bmatrix} v_{Ct1} \\[4mm] v_{Ct2} \end{bmatrix}
\tag{3.30}
$$

Equation (3.30) is similar to Eq. (3.21), except that the constant term present in the right side of Eq. (3.21) is no longer here, and the equation variables are the transient voltages rather than the total voltage. The reason for rewriting the differential equations in this form is that the solution is easier to find; the two forms are, of course, completely equivalent.

With the passage of time, the transient response approaches zero, so it is a reasonable conjecture that it might be composed of decaying exponential terms. Based on our knowledge of first-order differential equations which we developed from the series RC circuit, we assume transient response solutions of the form

$$
v_{Ct1} = V_1 e^{st}
\tag{3.31}
$$

$$
v_{Ct2} = V_2 e^{st}
\tag{3.32}
$$

We will see shortly that this approach is correct, but the solution expressions Eqs. (3.31) and (3.32) are incomplete. Also, we need to determine appropriate values of s which satisfy the differential equations of Eq. (3.30). This can be accomplished by substituting the assumed solutions and their derivatives into Eq. (3.30). Differentiating the assumed solutions:

$$
\frac{dv_{Ct1}}{dt} = sV_1 e^{st}
$$

$$
\frac{dv_{Ct2}}{dt} = sV_2 e^{st}
$$

Substituting these and the expressions for $v_{Ct1}(t)$ and $v_{Ct2}(t)$ into Eq. (3.30):

$$
\begin{bmatrix} sV_1 e^{st} \\[4mm] sV_2 e^{st} \end{bmatrix} = \begin{bmatrix} -\dfrac{1}{C_1}\left(\dfrac{1}{R_1}+\dfrac{1}{R_2}\right) & \dfrac{1}{R_2 C_1} \\[3mm] \dfrac{1}{R_2 C_2} & -\dfrac{1}{R_2 C_2} \end{bmatrix} \begin{bmatrix} V_1 e^{st} \\[4mm] V_2 e^{st} \end{bmatrix}
\tag{3.33}
$$

Canceling out the common exponential terms and subtracting the right side from both sides yields

$$\begin{bmatrix} sV_1 \\ sV_2 \end{bmatrix} - \begin{bmatrix} -\dfrac{1}{C_1}\left(\dfrac{1}{R_1} + \dfrac{1}{R_2}\right) & \dfrac{1}{R_2C_1} \\[2ex] \dfrac{1}{R_2C_2} & -\dfrac{1}{R_2C_2} \end{bmatrix} \begin{bmatrix} V_1 \\ V_2 \end{bmatrix} = \begin{bmatrix} 0 \\ 0 \end{bmatrix} \tag{3.34}$$

The leftmost term can be expanded as

$$\begin{bmatrix} sV_1 \\ sV_2 \end{bmatrix} = \begin{bmatrix} s & 0 \\ 0 & s \end{bmatrix} \begin{bmatrix} V_1 \\ V_2 \end{bmatrix} = s \begin{bmatrix} 1 & 0 \\ 0 & 1 \end{bmatrix} \begin{bmatrix} V_1 \\ V_2 \end{bmatrix} = s\mathbf{I} \begin{bmatrix} V_1 \\ V_2 \end{bmatrix}$$

where $\mathbf{I}$ is the identity matrix (main diagonal elements unity, all other elements zero). Eq. (3.34) can be written as

$$s\mathbf{I} \begin{bmatrix} V_1 \\ V_2 \end{bmatrix} - \begin{bmatrix} -\dfrac{1}{C_1}\left(\dfrac{1}{R_1} + \dfrac{1}{R_2}\right) & \dfrac{1}{R_2C_1} \\[2ex] \dfrac{1}{R_2C_2} & -\dfrac{1}{R_2C_2} \end{bmatrix} \begin{bmatrix} V_1 \\ V_2 \end{bmatrix} = \begin{bmatrix} 0 \\ 0 \end{bmatrix}$$

$$\left(s\mathbf{I} - \begin{bmatrix} -\dfrac{1}{C_1}\left(\dfrac{1}{R_1} + \dfrac{1}{R_2}\right) & \dfrac{1}{R_2C_1} \\[2ex] \dfrac{1}{R_2C_2} & -\dfrac{1}{R_2C_2} \end{bmatrix} \right) \begin{bmatrix} V_1 \\ V_2 \end{bmatrix} = \begin{bmatrix} 0 \\ 0 \end{bmatrix} \tag{3.35}$$

$$\begin{bmatrix} s + \dfrac{1}{R_1C_1} + \dfrac{1}{R_2C_1} & -\dfrac{1}{R_2C_1} \\[2ex] -\dfrac{1}{R_2C_2} & s + \dfrac{1}{R_2C_2} \end{bmatrix} \begin{bmatrix} V_1 \\ V_2 \end{bmatrix} = \begin{bmatrix} 0 \\ 0 \end{bmatrix} \tag{3.36}$$

Although it might not be immediately obvious, Eq. (3.36) contains enough information to find the correct values of s which are needed to solve for the voltages in the ladder circuit. While the procedure is actually quite simple, the explanation requires several steps, involving careful interpretation of Eq. (3.36).

We begin by noting that the right side of Eq. (3.36) is a column vector equal to zero. By inspection, one possible solution to Eq. (3.36) is the trivial solution:

$$\begin{bmatrix} V_1 \\ V_2 \end{bmatrix} = \begin{bmatrix} 0 \\ 0 \end{bmatrix}$$

However, if this were the only solution, it would mean that the transient voltages $v_{Ct1}(t)$ and $v_{Ct2}(t)$ are also identically zero. Since we know that the transient solution must bridge the gap between the initial and dc steady-state capacitor voltages, this result is unsatisfactory; there must be nontrivial (i.e., nonzero) solutions to Eq. (3.36).

The 2×2 matrix on the left side of Eq. (3.36) defines a system of linear algebraic equations. For there to be a nontrivial solution to such a system of linear equations, **the determinant of the matrix must equal zero**. The determinant of a 2×2 matrix is given by

$$\det \begin{bmatrix} a & b \\ c & d \end{bmatrix} = ad - bc$$

It follows that, for the nontrivial solution to exist,

$$ad - bc = 0$$

If the determinant of a 2×2 matrix equals zero, then the rows are linearly dependent; i.e., the second row is a multiple of the first.

To see why this is true, consider the following example:

$$\begin{bmatrix} a & b \\ a & c \end{bmatrix} \begin{bmatrix} x \\ y \end{bmatrix} = 0, \qquad b \neq c$$

where $a \neq b \neq c$. Clearly, the determinant $(ac - ab)$ is not equal to zero, which means that the first row and the second row are independent and the matrix is **nonsingular**. If we subtract the first row from the second and substitute the result for the second row, we obtain

$$\begin{bmatrix} a & b \\ 0 & c-b \end{bmatrix} \begin{bmatrix} x \\ y \end{bmatrix} = 0$$

Examining the second row, we find that

$$(c-b)y = 0$$
$$y = 0$$
$$x = 0$$

which is the only solution for x and y in this case. There are therefore no nontrivial solutions if the matrix is nonsingular.

However, if c and b are equal, nonzero solutions for x and y can be found. Assuming $b = c$, the system can be written

$$\begin{bmatrix} a & b \\ a & b \end{bmatrix} \begin{bmatrix} x \\ y \end{bmatrix} = 0$$

The second row is linearly dependent on the first; in fact, the two rows are the same. The determinant of this matrix, $(ab - ab)$, equals zero. Subtracting the first row from the second and substituting the result for the second row yields

$$\begin{bmatrix} a & b \\ 0 & 0 \end{bmatrix} \begin{bmatrix} x \\ y \end{bmatrix} = 0$$

$$ax + by = 0$$

It is now possible to have nontrivial solutions for x and y. For example, one solution is $x = b$, $y = -a$.

Returning to the circuit problem, we wish to find the values of s which result in non-trivial solutions to Eq. (3.36). To do so, we set the determinant of the matrix equal to zero:[3]

$$\det \begin{bmatrix} s + \dfrac{1}{R_1 C_1} + \dfrac{1}{R_2 C_1} & -\dfrac{1}{R_2 C_1} \\[2ex] -\dfrac{1}{R_2 C_2} & s + \dfrac{1}{R_2 C_2} \end{bmatrix} = 0 \qquad\qquad (3.37)$$

Expanding terms:

$$\left(s + \frac{1}{R_1 C_1} + \frac{1}{R_2 C_1} \right)\left(s + \frac{1}{R_2 C_2} \right) - \left(-\frac{1}{R_2 C_1} \right)\left(-\frac{1}{R_2 C_2} \right) = 0$$

$$s^2 + \left(\frac{1}{R_1 C_1} + \frac{1}{R_2 C_1} + \frac{1}{R_2 C_2} \right) s + \frac{1}{R_1 R_2 C_1 C_2} + \frac{1}{R_2^2 C_1 C_2} - \frac{1}{R_2^2 C_1 C_2} = 0$$

$$s^2 + \left(\frac{1}{R_1 C_1} + \frac{1}{R_2 C_1} + \frac{1}{R_2 C_2} \right) s + \frac{1}{R_1 R_2 C_1 C_2} = 0$$

Applying the quadratic formula:

$$s_1 = -\frac{1}{2}\left(\frac{1}{R_1 C_1} + \frac{1}{R_2 C_1} + \frac{1}{R_2 C_2} \right) - \frac{1}{2}\sqrt{\left(\frac{1}{R_1 C_1} + \frac{1}{R_2 C_1} + \frac{1}{R_2 C_2} \right)^2 - \frac{4}{R_1 R_2 C_1 C_2}}$$

$$s_2 = -\frac{1}{2}\left(\frac{1}{R_1 C_1} + \frac{1}{R_2 C_1} + \frac{1}{R_2 C_2} \right) + \frac{1}{2}\sqrt{\left(\frac{1}{R_1 C_1} + \frac{1}{R_2 C_1} + \frac{1}{R_2 C_2} \right)^2 - \frac{4}{R_1 R_2 C_1 C_2}}$$

We call s_1 and s_2 the **natural frequencies** of the circuit.[4] It is possible to show that, for the RC ladder circuit we are considering, they are both real, negative numbers. Although they are called "frequencies" in order to be consistent with terminology and

[3.] Note that this can also be written as $\det\left[s\mathbf{I} - \begin{bmatrix} -\dfrac{1}{C_1}\left(\dfrac{1}{R_1} + \dfrac{1}{R_2} \right) & \dfrac{1}{R_2 C_1} \\[2ex] \dfrac{1}{R_2 C_2} & -\dfrac{1}{R_2 C_2} \end{bmatrix} \right] = 0.$

[4.] In linear algebra they are known as **eigenvalues**. If you have had formal training in linear algebra, you have probably learned about eigenvalues (and eigenvectors too) but might not have been told what they are good for. This is one of the many things they are good for.

notation we develop later on, they do not have the usual interpretation of "cycles per second" which we associate with the frequency of a periodic signal. As such, it is appropriate to express numerical values of s_1 and s_2 using the unit of inverse seconds: s^{-1}.

The concept of natural frequencies of a system is one of the most significant and important ideas in engineering. The natural frequencies relate a fundamental property of the circuit: the rate at which the circuit changes from one state to another. The greater the value of $|s|$, the faster the circuit reaches dc steady state. The symbolic expressions we derived show that both natural frequencies depend on the values of both capacitors and resistors; however, they do **not** depend on the source voltage V_S, nor on the initial conditions. This result is quite general; in all of the circuits we consider, the natural frequencies depend only on their component values.

Example 3.6

The two-lump RC ladder circuit approximation of cascaded inverters having the response shown in Figure 3.15 has component values $R_1 = 1500\ \Omega$, $R_2 = 500\ \Omega$, $C_1 = 0.5\ \text{pF}$, and $C_2 = 1.5\ \text{pF}$. Determine the natural frequencies of the response.

The natural frequencies are found from Eq. (3.37):

$$\det \begin{bmatrix} s + \dfrac{1}{R_1 C_1} + \dfrac{1}{R_2 C_1} & -\dfrac{1}{R_2 C_1} \\[2ex] -\dfrac{1}{R_2 C_2} & s + \dfrac{1}{R_2 C_2} \end{bmatrix} = 0$$

Substituting values,

$$\det \begin{bmatrix} s + \dfrac{1}{1500 \times 0.5 \times 10^{-12}} + \dfrac{1}{500 \times 0.5 \times 10^{-12}} & -\dfrac{1}{500 \times 0.5 \times 10^{-12}} \\[2ex] -\dfrac{1}{500 \times 1.5 \times 10^{-12}} & s + \dfrac{1}{500 \times 1.5 \times 10^{-12}} \end{bmatrix} = 0$$

$$\det \begin{bmatrix} s + 5.33 \times 10^9 & -4.00 \times 10^9 \\ -1.33 \times 10^9 & s + 1.33 \times 10^9 \end{bmatrix} = 0$$

$$(s + 5.33 \times 10^9)(s + 1.33 \times 10^9) - (-4.00 \times 10^9)(-1.33 \times 10^9) = 0$$

$$s^2 + 6.66 \times 10^9 s + 1.78 \times 10^{18} = 0$$

Applying the quadratic formula:

$$s = -\frac{6.66 \times 10^9}{2 \times 1} \pm \sqrt{\left(\frac{6.66 \times 10^9}{2 \times 1}\right)^2 - \frac{1.78 \times 10^{18}}{1}}$$

$$s_1 = -278 \times 10^6 \text{ s}^{-1}$$

$$s_2 = -6.38 \times 10^9 \text{ s}^{-1}$$

These are the values of natural frequencies found in Eq. (3.8) describing the two-*RC*-lump circuit response.

Another significant feature of this procedure is that it has resulted in *two* values of *s*. Recall that we originally assumed transient solutions of the form:

$$v_{Ct1} = V_1 e^{st}$$

$$v_{Ct2} = V_2 e^{st}$$

These solutions will satisfy the differential equation [Eq. (3.30)] with either of the two values of *s*. More generally, since Eq. (3.30) is a linear equation, we can construct linear combinations of $e^{s_1 t}$ and $e^{s_2 t}$ which will also solve the system of differential equations. Therefore, the transient solutions are given by

$$v_{Ct1} = V_{11} e^{s_1 t} + V_{12} e^{s_2 t}$$
$$v_{Ct2} = V_{21} e^{s_1 t} + V_{22} e^{s_2 t} \tag{3.38}$$

For mathematical completeness, we could take these assumed solutions and repeat the procedure which led to the two natural frequencies. This is a straightforward algebraic exercise (Exercise 3.24) with the result that these assumed solutions (and the derived values of s_1 and s_2) are correct.

To recap, the procedure for determining the natural frequencies for a system described by

$$\begin{bmatrix} \dfrac{dv_{Ct1}}{dt} \\ \dfrac{dv_{Ct2}}{dt} \end{bmatrix} = \mathbf{A} \begin{bmatrix} v_{Ct1} \\ v_{Ct2} \end{bmatrix}$$

is to determine values of *s* which satisfy

$$\det[s\mathbf{I} - \mathbf{A}] = 0 \tag{3.39}$$

Now that we have found a procedure for determining the natural frequencies, we can dispense with the formulation of the problem in terms of the transient solution. From now on we will work directly with the total solution; with the dc steady-state values of V_S inserted, they are:

$$v_{C1}(t) = V_{11} e^{s_1 t} + V_{12} e^{s_2 t} + V_S$$

$$v_{C2}(t) = V_{21}e^{s_1 t} + V_{22}e^{s_2 t} + V_S$$

To complete our solution of the two-section RC ladder circuit, we need to determine six constants: s_1 and s_2 (already done); V_{11}, V_{12}, V_{21}, and V_{22}. Although finding the remaining four constants is straightforward, the symbolic expressions are quite unattractive, even worse than the expressions for s_1 and s_2. Therefore, instead of deriving general symbolic expressions for the other four constants, we will illustrate the general procedure by using a a specific numerical example.

Example 3.7

Determine the voltages $v_{C1}(t)$ and $v_{C2}(t)$ in the circuit of Figure 3.18, given the following parameters:

$$R_1 = \tfrac{1}{3}\ \Omega$$
$$R_2 = \tfrac{1}{2}\ \Omega$$
$$C_1 = C_2 = 1\ F$$
$$v_{C1}(0) = 0\ V$$
$$v_{C2}(0) = 0\ V$$

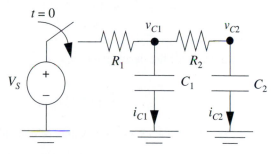

Figure 3.18 Two-section RC ladder circuit.

The component values in this example are highly unrealistic; they have been chosen so that the arithmetic does not obscure the solution method. Also, we are assuming V_S to be a known but unspecified value.

The solution method consists of the following steps:

- Use Kirchhoff's laws and branch relations to write a system of differential equations describing the circuit.

- Determine the dc steady-state circuit response.

- Write assumed solutions in a general form.

- Determine the values of the natural frequencies.

- Determine the coefficients of the exponential terms in the general solution.

- Check the solution by verifying that it satisfies the differential equations, the initial conditions, and the behavior of the circuit as $t \to \infty$.

Applying KCL at the v_{C1} node:

$$\frac{v_{C1} - V_S}{\frac{1}{3}} + (1)\frac{dv_{C1}}{dt} + \frac{v_{C1} - v_{C2}}{\frac{1}{2}} = 0 \tag{3.40}$$

Eq. (3.40) can be rewritten as

$$3(v_{C1} - V_S) + \frac{dv_{C1}}{dt} + 2(v_{C1} - v_{C2}) = 0 \tag{3.41}$$

Solving Eq. (3.41) for the derivative of v_{C1} yields

$$\frac{dv_{C1}}{dt} = -5v_{C1} + 2v_{C2} + 3V_S \tag{3.42}$$

At the v_{C2} node, applying KCL, we obtain

$$\frac{v_{C2} - v_{C1}}{\frac{1}{2}} + (1)\frac{dv_{C2}}{dt} = 0 \tag{3.43}$$

Solving the equation for the derivative of v_{C2}:

$$\frac{dv_{C2}}{dt} = 2v_{C1} - 2v_{C2} \tag{3.44}$$

Expressing Eqs. (3.42) and (3.44) as a system of equations, we have the following:

$$\begin{bmatrix} \dfrac{dv_{C1}}{dt} \\ \dfrac{dv_{C2}}{dt} \end{bmatrix} = \begin{bmatrix} -5 & 2 \\ 2 & -2 \end{bmatrix}\begin{bmatrix} v_{C1} \\ v_{C2} \end{bmatrix} + \begin{bmatrix} 3V_S \\ 0 \end{bmatrix} \tag{3.45}$$

The next step is to determine the dc steady-state voltages V_{C1} and V_{C2}. This is done by replacing the capacitors by their dc steady-state equivalents, open circuits; the result is that $V_{C1} = V_{C2} = V_S$ (see Figure 3.17).

The general solutions are

$$v_{C1}(t) = V_{11}e^{s_1 t} + V_{12}e^{s_2 t} + V_S$$

$$v_{C2}(t) = V_{21}e^{s_1 t} + V_{22}e^{s_2 t} + V_S$$

To find the natural frequencies, we showed [Eq. (3.39)] that we must solve the equation

$$\det[s\mathbf{I} - \mathbf{A}] = 0 \qquad (3.46)$$

where $\mathbf{I}$ is the identity matrix:

$$\mathbf{I} = \begin{bmatrix} 1 & 0 \\ 0 & 1 \end{bmatrix}$$

and $\mathbf{A}$ is the circuit matrix, in this case,

$$\mathbf{A} = \begin{bmatrix} -5 & 2 \\ 2 & -2 \end{bmatrix}$$

Combining:

$$\det\left[s\begin{bmatrix} 1 & 0 \\ 0 & 1 \end{bmatrix} - \begin{bmatrix} -5 & 2 \\ 2 & -2 \end{bmatrix} \right] = 0$$

$$\det\begin{bmatrix} s+5 & -2 \\ -2 & s+2 \end{bmatrix} = 0$$

Evaluating the determinant and solving for the natural frequencies:

$$(s+5)(s+2) - (-2)(-2) = 0$$
$$s^2 + 7s + 6 = 0$$
$$s_1 = -1\ \text{s}^{-1}$$
$$s_2 = -6\ \text{s}^{-1}$$

(*Note:* It is generally more productive to find the natural frequencies by starting from Eq. (3.46), rather than using the quadratic solution expressions given on page 142.)

Substituting the values of the natural frequencies, the general solutions are

$$v_{C1}(t) = V_{11}e^{-t} + V_{12}e^{-6t} + V_S \qquad (3.47)$$

$$v_{C2}(t) = V_{21}e^{-t} + V_{22}e^{-6t} + V_S \qquad (3.48)$$

The next step is to determine the coefficients $V_{11} \ldots V_{22}$. There are a variety of ways of finding them, but they all involve applying the initial conditions to get a system of linear algebraic equations (as opposed to linear differential equations) which yield the values of the constants.

Evaluating Eq. (3.47) and Eq. (3.48) at $t = 0$:

$$v_{C1}(0) = V_{11}e^{-1 \times 0} + V_{12}e^{-6 \times 0} + V_S \Rightarrow V_{11} + V_{12} + V_S = 0$$

$$v_{C2}(0) = V_{21}e^{-1 \times 0} + V_{22}e^{-6 \times 0} + V_S \Rightarrow V_{21} + V_{22} + V_S = 0$$

Remembering that V_S is the *known* source voltage, we can rewrite the above equations as:

$$V_{11} + V_{12} = -V_S \tag{3.49}$$

$$V_{21} + V_{22} = -V_S \tag{3.50}$$

We now have two equations, but we need two more to solve for the four unknowns. To find them, we go back to Eq. (3.45), the system of equations which describe our circuit:

$$\begin{bmatrix} \dfrac{dv_{C1}}{dt} \\ \dfrac{dv_{C2}}{dt} \end{bmatrix} = \begin{bmatrix} -5 & 2 \\ 2 & -2 \end{bmatrix} \begin{bmatrix} v_{C1} \\ v_{C2} \end{bmatrix} + \begin{bmatrix} 3V_S \\ 0 \end{bmatrix}$$

What we want to do is insert the solutions [Eqs. (3.47) and (3.48)] into this equation and evaluate the resulting expressions. As it happens, it is necessary to only evaluate one row of the system; we will choose the second row, since it does not contain a constant term:

$$\frac{dv_{C2}}{dt} = 2v_{C1} - 2v_{C2} \tag{3.51}$$

The left side of Eq. (3.51) is found by differentiating Eq. (3.48):

$$\frac{dv_{C2}}{dt} = \frac{d}{dt}(V_{21}e^{-t} + V_{22}e^{-6t} + V_S)$$

$$= -V_{21}e^{-t} - 6V_{22}e^{-6t}$$

By direct substitution, the right side of Eq. (3.51) is

$$2v_{C1} - 2v_{C2} = 2(V_{11}e^{-t} + V_{12}e^{-6t} + V_S) - 2(V_{21}e^{-t} + V_{22}e^{-6t} + V_S)$$

$$= (2V_{11} - 2V_{21})e^{-t} + (2V_{12} - 2V_{22})e^{-6t}$$

Since both sides must be equal to each other for all values of t, we can equate the coefficients of each exponential term. This results in two equations:

$$-V_{21} = 2V_{11} - 2V_{21}$$

$$-6V_{22} = 2V_{12} - 2V_{22}$$

which simplify to:

$$V_{21} = 2V_{11} \tag{3.52}$$

$$V_{12} = -2V_{22} \tag{3.53}$$

We are now equipped with four equations and four unknowns: Eqs. (3.49), (3.50), (3.52), and (3.53). Simultaneous solution of these four equations yields the four coefficients:

$$V_{11} = -\frac{3}{5}V_S \qquad V_{12} = -\frac{2}{5}V_S$$

$$V_{21} = -\frac{6}{5}V_S \qquad V_{22} = \frac{1}{5}V_S$$

Finally, substitution of the coefficients in Eqs. (3.47) and (3.48) produces the complete solutions:

$$v_{C1}(t) = -\frac{3}{5}V_S e^{-t} - \frac{2}{5}V_S e^{-6t} + V_S \tag{3.54}$$

$$v_{C2}(t) = -\frac{6}{5}V_S e^{-t} + \frac{1}{5}V_S e^{-6t} + V_S \tag{3.55}$$

The last step in finding the solution is to check if our expressions agree with the dc steady state, initial conditions, and the differential equations. We can check the dc steady state by taking the limit as $t \to \infty$:

$$\lim_{t \to \infty} v_{C1} = \lim_{t \to \infty} \left(-\frac{3}{5}e^{-t} - \frac{2}{5}e^{-6t} + 1 \right)V_S = V_S$$

$$\lim_{t \to \infty} v_{C2} = \lim_{t \to \infty} \left(-\frac{6}{5}e^{-t} + \frac{1}{5}e^{-6t} + 1 \right)V_S = V_S$$

This agrees with our previous determination that the capacitor voltages decay toward V_S.

The initial conditions are verified by evaluating the solutions at $t = 0$:

$$v_{C1}(0) = \left(-\frac{3}{5}e^{-0} - \frac{2}{5}e^{-6(0)} + 1 \right)V_S = \left(-\frac{3}{5} - \frac{2}{5} + 1 \right)V_S = 0$$

$$v_{C2}(0) = \left(-\frac{6}{5}e^{-0} + \frac{1}{5}e^{-6(0)} + 1 \right)V_S = \left(-\frac{6}{5} + \frac{1}{5} + 1 \right)V_S = 0$$

So far, so good.

The final check is to substitute the solutions back into the differential equation, Eq. (3.45), and show that they work (i.e., they produce an identity). The first row of Eq. (3.45) is

$$\frac{dv_{C1}}{dt} = -5v_{C1} + 2v_{C2} + 3V_S \tag{3.56}$$

The left side of Eq. (3.56) is found by differentiating the solution for v_{C1}:

$$\frac{dv_{C1}}{dt} = \frac{3}{5}V_S e^{-t} + \frac{12}{5}V_S e^{-6t}$$

By direct substitution, the right side of Eq. (3.56) is:

$$-5v_{C1} + 2v_{C2} + 3V_S = -5\left(-\frac{3}{5}V_S e^{-t} - \frac{2}{5}V_S e^{-6t} + V_S\right)$$

$$+ 2\left(-\frac{6}{5}V_S e^{-t} + \frac{1}{5}V_S e^{-6t} + V_S\right) + 3V_S$$

$$= \frac{3}{5}V_S e^{-t} + \frac{12}{5}V_S e^{-6t}$$

Both sides evaluate to the same expression.

The second row of Eq. (3.45) is

$$\frac{dv_{C2}}{dt} = 2v_{C1} - 2v_{C2} \tag{3.57}$$

Differentiating the solution for v_{C2}:

$$\frac{dv_{C2}}{dt} = \frac{6}{5}V_S e^{-t} - \frac{6}{5}V_S e^{-6t}$$

The righthand side is

$$2v_{C1} - 2v_{C2} = 2\left(-\frac{3}{5}V_S e^{-t} - \frac{2}{5}V_S e^{-6t} + V_S\right) - 2\left(-\frac{6}{5}V_S e^{-t} + \frac{1}{5}V_S e^{-6t} + V_S\right)$$

$$= \frac{6}{5}V_S e^{-t} - \frac{6}{5}V_S e^{-6t}$$

Both sides of Eq. (3.57) are the same. Therefore, we can conclude that the solutions indeed satisfy both of the differential equations.

Note that this last check verifies only that the solutions are consistent with the original differential equations; it does not check whether the differential equations were correctly formulated from the circuit. So a further step in verification is to check the supposed solutions directly with the circuit and make sure they satisfy Kirchhoff's laws and the branch relations.

The total solutions, Eqs. (3.54) and (3.55), are plotted in Figure 3.19, with $V_S = 5$ V.

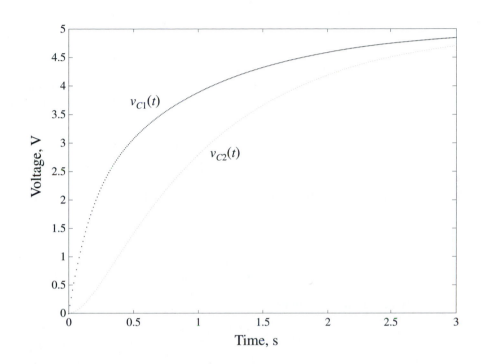

Figure 3.19 Capacitor voltages of Example 3.7, with $V_S = 5$ V.

Example 3.8

Solve for the capacitor voltages in the circuit of Example 3.7 with the following initial conditions:

$$v_{C1}(0) = 1 \text{ V}$$

$$v_{C2}(0) = 2 \text{ V}$$

Changing the initial conditions has no effect on the natural frequencies, since these depend only upon the component values in the circuit. Nor is there any change in the dc steady state. Changing the initial conditions affects only the values of the coefficients V_{11}, V_{12}, V_{21}, and V_{22}.

In the previous example, with the initial capacitor voltages both zero, we derived a set of four algebraic equations to solve for the four coefficients. The change in initial conditions affects two of these equations, those that were derived using the initial conditions. We will now derive the updated pair of equations from the new initial conditions:

$$v_{C1}(0) = V_{11} + V_{12} + V_S = 1$$
$$v_{C2}(0) = V_{21} + V_{22} + V_S = 2$$

which simplify to

$$V_{11} + V_{12} = 1 - V_S$$
$$V_{21} + V_{22} = 2 - V_S$$

The other two equations were found from substituting the assumed solutions into the differential equations. Since we did not use the initial conditions to derive them, they are still valid for this example:

$$V_{21} = 2V_{11}$$
$$V_{12} = -2V_{22}$$

Simultaneous solution of these four equations yields the four coefficients:

$$V_{11} = \frac{5 - 3V_S}{5}, \qquad V_{12} = -\frac{2}{5}V_S$$

$$V_{21} = \frac{10 - 6V_S}{5}, \qquad V_{22} = \frac{1}{5}V_S$$

Hence, the final solutions are:

$$v_{C1}(t) = \frac{5 - 3V_S}{5}e^{-t} - \frac{2}{5}V_S e^{-6t} + V_S$$

$$v_{C2}(t) = \frac{10 - 6V_S}{5}e^{-t} + \frac{1}{5}V_S e^{-6t} + V_S$$

We can check these solutions as we did in the previous example. We first check if they converge to the dc steady-state value of V_S as $t \to \infty$. Since both exponential terms in each solution decay to zero, this criterion is clearly satisfied.

Evaluating the solutions at $t = 0$:

$$v_{C1}(0) = \frac{5 - 3V_S}{5}e^{-0} - \frac{2}{5}V_S e^{-6 \times 0} + V_S = 1$$

$$v_{C2}(0) = \frac{10 - 6V_S}{5}e^{-0} + \frac{1}{5}V_S e^{-6 \times 0} + V_S = 2$$

which agrees with the given initial conditions.

The last check is to verify the solutions satisfy the differential equations. The first row of Eq. (3.45) is

$$\frac{dv_{C1}}{dt} = -5v_{C1} + 2v_{C2} + 3V_S \tag{3.58}$$

The left side of Eq. (3.58) is found by differentiating the solution for v_{C1}:

$$\frac{dv_{C1}}{dt} = \frac{3V_S - 5}{5}e^{-t} + \frac{12}{5}V_S e^{-6t}$$

By direct substitution, the right side of Eq. (3.58) is:

$$-5v_{C1} + 2v_{C2} + 3V_S = -5\left(\frac{5 - 3V_S}{5}e^{-t} - \frac{2}{5}V_S e^{-6t} + V_S\right)$$

$$+ 2\left(\frac{10 - 6V_S}{5}e^{-t} + \frac{1}{5}V_S e^{-6t} + V_S\right) + 3V_S$$

$$= \frac{3V_S - 5}{5}e^{-t} + \frac{12}{5}V_S e^{-6t}$$

Both sides evaluate to the same expression.

The second row of Eq. (3.45) is

$$\frac{dv_{C2}}{dt} = 2v_{C1} - 2v_{C2} \tag{3.59}$$

Differentiating the solution for v_{C2}:

$$\frac{dv_{C2}}{dt} = \frac{6V_S - 10}{5}e^{-t} - \frac{6}{5}V_S e^{-6t}$$

The righthand side is

$$2v_{C1} - 2v_{C2} = 2\left(\frac{5 - 3V_S}{5}e^{-t} - \frac{2}{5}V_S e^{-6t} + V_S\right)$$

$$-2\left(\frac{10 - 6V_S}{5}e^{-t} + \frac{1}{5}V_S e^{-6t} + V_S\right)$$

$$= \frac{6V_S - 10}{5}e^{-t} - \frac{6}{5}V_S e^{-6t}$$

Both sides of Eq. (3.59) are the same. Therefore, we can again conclude that the solutions satisfy the differential equations describing the circuit response.

The total solutions for this example are plotted in Figure 3.20, with $V_S = 5$ V.

3.7 Natural Frequencies and Higher-Order Circuits

There appears to be a pattern forming:

- In the single-lump RC circuit, there was *one* capacitor, *one* exponential term in the solution $(e^{-t/\tau})$, and *one* time constant for the circuit, and *one* initial condition was needed to solve the *one* differential equation describing the circuit.

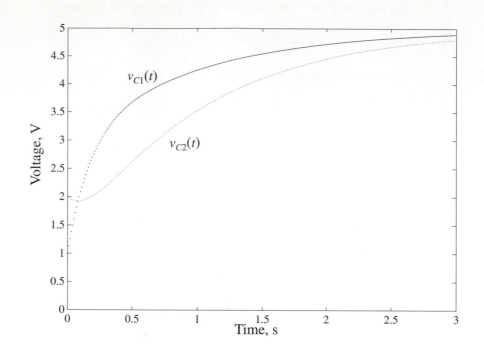

Figure 3.20   Capacitor voltages of Example 3.8, with $V_S = 5$ V.

- In the two-lump *RC* ladder circuit, there are *two* capacitors, *two* exponential terms in the solution, and *two* natural frequencies, and *two* initial conditions are needed to solve the system of *two* differential equations describing the circuit.

This pattern is not a coincidence. In general, if a circuit contains n independent capacitors, it is characterized by n differential equations and has n natural frequencies. In the second-order case two natural frequencies arise, because there are two independent capacitors in the circuit.

(By "independent" we mean that the capacitors cannot be combined into an equivalent capacitance. Capacitors in series or parallel are not independent, nor are capacitors which are "virtually parallel." For example, both of the circuits in Figure 3.21 are second order because each has two independent capacitances.)

If there are n independent capacitors in a circuit, then the total solution contains n transient terms of the form of $e^{s_n t}$. Each of these terms decays to zero as $t \to \infty$. The characteristic time over which the term decays is given by the reciprocal of the natural frequency. This means that when $t = 1/|s_n|$, the nth transient term will have decayed to e^{-1} (approximately 37%) of its initial value. The natural frequencies can therefore be thought of as reciprocal time constants of a sort.

In any circuit with more than one capacitor, there will be several natural frequencies. Often the magnitudes of these natural frequencies will differ by a significant amount. The relative magnitude of each natural frequency, $|s_n|$, defines the time scale over which the

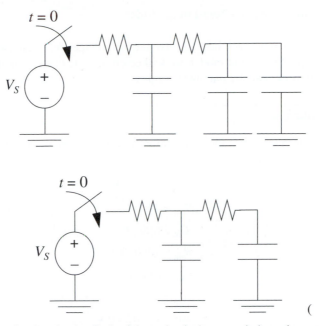

Figure 3.21 Second-order circuits. Both of these circuits have two independent capacitors and two natural frequencies. The two capacitors in parallel in the upper circuit are not independent, since they can be combined into a single equivalent capacitance.

corresponding term is changing. The exponential term with the smallest value of $|s_n|$ will be the last to decay to zero and will thus determine, to a large extent, the overall shape of the response curve. This natural frequency is called the **dominant natural frequency** and can be used to as a rough measure of the circuit speed.

In our example circuit, the two natural frequencies were

$$s_1 = -1 \text{ s}^{-1}$$

$$s_2 = -6 \text{ s}^{-1}$$

The characteristic times associated with the two exponential terms are thus 1.00 s and 0.167 s. The second term has largely died out after $5 \times 0.167 = 0.83$ s; but the first term needs approximately $5 \times 1.00 = 5.0$ s before it has decayed to a negligible value.

Individual natural frequencies almost never characterize individual capacitors. Notice that in the solutions of the second-order examples, both capacitor voltages have terms containing both natural frequencies. In general, if we have n independent capacitors we will have n natural frequencies characterizing every capacitor voltage.

3.8 Timing Delays Using the Two-Lump Model

Now that we have seen how to solve for the voltages in a cascaded inverter pair using the two-lump interconnect model, we will determine the corresponding delay using an example with more realistic parameter values.

Example 3.9

Determine the pull-up transition time for an interconnected pair of inverter gates, using the following parameters:

$$V_S = 5 \text{ V (pull-up case)}$$

$$R_{out} = 1000 \ \Omega$$

$$C_{in} = 1.00 \text{ pF}$$

$$R_{int} = 1000 \ \Omega$$

$$C_{int} = 1.00 \text{ pF}$$

$$V_{th} = 4.0 \text{ V}$$

$$v_{C1}(0) = v_{C2}(0) = 0 \text{ V}$$

The first step is to determine the values of resistance and capacitance in the model of Figure 3.18 appropriate to this problem. Applying the formula of Eqs. (3.9) through (3.12):

$$R_1 = R_{out} + \frac{R_{int}}{2} = 1500 \ \Omega$$

$$R_2 = \frac{R_{int}}{2} = 500 \ \Omega$$

$$C_1 = \frac{C_{int}}{2} = 0.5 \times 10^{-12} \text{ F}$$

$$C_2 = C_{in} + \frac{C_{int}}{2} = 1.5 \times 10^{-12} \text{ F}$$

Inserting these values into Eq. (3.21):

$$\begin{bmatrix} \dfrac{dv_{C1}}{dt} \\ \dfrac{dv_{C2}}{dt} \end{bmatrix} = 10^9 \begin{bmatrix} -5.33 & 4.00 \\ 1.33 & -1.33 \end{bmatrix} \begin{bmatrix} v_{C1} \\ v_{C2} \end{bmatrix} + \begin{bmatrix} 6.66 \times 10^9 \\ 0 \end{bmatrix} \tag{3.60}$$

The natural frequencies are found from

$$\det\left[s\begin{bmatrix} 1 & 0 \\ 0 & 1 \end{bmatrix} - 10^9 \begin{bmatrix} -5.33 & 4.00 \\ 1.33 & -1.33 \end{bmatrix} \right] = 0$$

$$\det\begin{bmatrix} s + (5.33 \times 10^9) & -4.00 \times 10^9 \\ -1.33 \times 10^9 & s + (1.33 \times 10^9) \end{bmatrix} = 0$$

Evaluating the determinant and solving for the natural frequencies:

$$(s + (5.33 \times 10^9))(s + (1.33 \times 10^9)) - (-4.00 \times 10^9)(-1.33 \times 10^9) = 0$$

$$s^2 + (6.66 \times 10^9)s + (1.78 \times 10^{18}) = 0$$

$$s_1 = -278 \times 10^6 \text{ s}^{-1}$$

$$s_2 = -6.38 \times 10^9 \text{ s}^{-1}$$

The solutions to Eq. (3.60) are

$$v_{C1}(t) = V_{11}e^{-278 \times 10^6 t} + V_{12}e^{-6.38 \times 10^9 t} + 5$$

$$v_{C2}(t) = V_{21}e^{-278 \times 10^6 t} + V_{22}e^{-6.38 \times 10^9 t} + 5$$

Applying the initial conditions:

$$v_{C1}(0) = 0 = V_{11} + V_{12} + 5$$

$$v_{C2}(0) = 0 = V_{21} + V_{22} + 5$$

which leads to

$$V_{11} + V_{12} = -5$$

$$V_{21} + V_{22} = -5$$

Evaluating the second row of Eq. (3.60):

$$\frac{dv_{C2}}{dt} = 1.33 \times 10^9 v_{C1} - 1.33 \times 10^9 v_{C2} \qquad (3.61)$$

Differentiation of the general solution gives the left side of Eq. (3.61) as

$$\frac{dv_{C2}}{dt} = -278 \times 10^6 V_{21}e^{-278 \times 10^6 t} - 6.38 \times 10^9 V_{22}e^{-6.38 \times 10^9 t}$$

Substitution of the general solutions gives the right side of Eq. (3.61):

$$1.33 \times 10^9 v_{C1} - 1.33 \times 10^9 v_{C2} = 1.33 \times 10^9 (V_{11} e^{-278 \times 10^6 t} + V_{12} e^{-6.38 \times 10^9 t} + 5)$$

$$-1.33 \times 10^9 (V_{21} e^{-278 \times 10^6 t} + V_{22} e^{-6.38 \times 10^9 t} + 5)$$

Equating coefficients of the like exponential terms in the previous two expressions:

$$-278 \times 10^6 V_{21} = 1.33 \times 10^9 (V_{11} - V_{21})$$

$$-6.38 \times 10^9 V_{22} = 1.33 \times 10^9 (V_{12} - V_{22})$$

Simultaneous solution of the four algebraic equations yields the four constants $V_{11} \dots V_{22}$. Substitution in the general solutions gives the total capacitor voltages:

$$v_{C1}(t) = -4.14 e^{-278 \times 10^6 t} - 0.86 e^{-6.38 \times 10^9 t} + 5$$

$$v_{C2}(t) = -5.23 e^{-278 \times 10^6 t} + 0.23 e^{-6.38 \times 10^9 t} + 5$$

To find the pull-up transition time, we use the fact that the voltage at the input of the load inverter has just reached the high threshold:

$$v_{C2}(t_{\text{low} \to \text{high}}) = V_{\text{th}}$$

$$-5.23 e^{-278 \times 10^6 t_{\text{low} \to \text{high}}} + 0.23 e^{-6.38 \times 10^9 t_{\text{low} \to \text{high}}} + 5 = 4$$

Although this equation cannot be manipulated into an explicit expression for $t_{\text{low} \to \text{high}}$, we can easily find a good numerical approximation by noting that the exponential term containing the dominant natural frequency will overwhelm the other exponential term. Ignoring the second term:

$$-5.23 e^{-278 \times 10^6 t_{\text{low} \to \text{high}}} + 5 = 4$$

$$e^{-278 \times 10^6 t_{\text{low} \to \text{high}}} = \frac{4-5}{-5.23}$$

$$t_{\text{low} \to \text{high}} = 5.95 \times 10^{-9} \text{ s}$$

or 5.95 ns. We need to verify that the second exponential term is indeed negligible in comparison to the dominant term. Substituting the approximate value of $t_{\text{low} \to \text{high}}$:

$$-5.23 e^{-278 \times 10^6 \times 5.95 \times 10^{-9}} + 0.23 e^{-6.38 \times 10^9 \times 5.95 \times 10^{-9}} + 5 = -5.23 e^{-1.654} + 0.23 e^{-37.96} + 5$$

$$= -1.00 + (7.51 \times 10^{-18}) + 5$$

The dominant term is more than seventeen orders of magnitude greater than the other transient term at the point where the voltage is reaching the high threshold; therefore this approximation is quite good. (If the two natural frequencies were close, the transition time could be found by an iterative technique such as Newton's method or successive substitution. See Exercise 3.37.)

3.9 Timing Delays Using Higher-Order Interconnect Models

It follows that an even more accurate approximation can be had by modeling the interconnect with three, four, or more *RC* lumps. Extending the previous analysis is straightforward, but the calculations are best done on a computer.

For the problem described in Example 3.9, a three-lump interconnect model predicts a load inverter input voltage of

$$v_o(t) = -5.30e^{-288 \times 10^6 t} - 0.019e^{-24.9 \times 10^9 t} + 0.319e^{-6.36 \times 10^9 t} + 5$$

which results in a pull-up transition time of 5.78 ns. This represents about a 3% change over the two-lump model. This small increase in accuracy might be worthwhile enough to carry out the needed calculations, but there is less to be gained for each additional *RC* lump added to the circuit.

Note that the dominant natural frequencies in the two-lump model ($s = -278 \times 10^6$ s^{-1}) and the three-lump model ($s = -288 \times 10^6$ s^{-1}) are likewise very close (approximately 3%). Adding the additional lump has not significantly improved our estimate of the speed-limiting transient term. However, the change from the one-lump ($s = -250 \times 10^6$ s^{-1}) to the two-lump model was considerable, more than 11%.

We can conclude that the two-lump interconnect model offers a significant improvement in accuracy over the single *RC* interconnect model, but that extending the model to multiple lumps results in diminishing returns. Indeed, as we have neglected other important circuit features (i.e., parasitic and interconnect fringing capacitances, effects of adjacent interconnects, and inductance) which may have a considerable effect on the system speed, the multiple-lump models may not be appropriate.

3.10 Summary

Interconnects, the wires that connect logic gates in integrated circuits, can introduce significant delays in the propagation of signals. Both the resistance and capacitance of the interconnect contribute to the delay. Metals such as aluminum alloys have a smaller resistivity than polysilicon and are thus preferred for interconnecting gates. Polysilicon can be used for very short interconnect runs without introducing significant delays.

The interconnect resistance and capacitance are **distributed** along the length. Many models of these distributed effects are possible. The most common are multiple lumps of resistance and capacitance arranged as a **ladder circuit**. Modeling the interconnect as a single *RC* lump illustrates the importance of including the effects of interconnect in calculations of gate delay, but it is not sufficiently accurate to provide precise quantitative estimates of circuit speed. An important observation from the single-lump model is that interconnect delay scales as the *square* of the interconnect length; thus as integrated circuits continue to increase in size, the problem of interconnect delay will worsen.

The ladder circuits arising from a two-lump interconnect model illustrate several circuit concepts: **natural frequencies**, initial conditions, interpretation of the dc steady state, and solution methods. The natural frequencies of a circuit depend only on the values of the components (resistors and capacitors, and, as we will see later, inductors) and not on the values of the sources or initial conditions. They describe the rate at which the voltages and currents in the circuit change from their initial values to their final, steady-state values.

The procedure for solving a circuit consists of the following steps:

- Use Kirchhoff's laws and branch relations to write a system of differential equations describing the circuit.

- Determine the dc steady-state circuit response.

- Write assumed solutions in a general form.

- Determine the values of the natural frequencies.

- Determine the coefficients of the exponential terms in the general solution.

- Check the solution by determining if it satisfies the differential equations, the initial conditions, and the behavior of the circuit as $t \to \infty$.

A circuit with n independent capacitors has a solution which can be written as follows:

$$\begin{bmatrix} v_1 \\ v_2 \\ \dots \\ v_n \end{bmatrix} = \begin{bmatrix} V_{11} & V_{12} & \dots & V_{1n} \\ V_{21} & V_{22} & \dots & V_{2n} \\ \dots & \dots & \dots & \dots \\ V_{n1} & V_{n2} & \dots & V_{nn} \end{bmatrix} \begin{bmatrix} e^{s_1 t} \\ e^{s_2 t} \\ \dots \\ e^{s_n t} \end{bmatrix} + \begin{bmatrix} v_1(\infty) \\ v_2(\infty) \\ \dots \\ v_n(\infty) \end{bmatrix}$$

where $v_1 \dots v_n$ are the capacitor voltages and $s_1 \dots s_n$ are the natural frequencies. Often, one of the natural frequencies will have a magnitude much smaller than the rest; it is called the **dominant** natural frequency, since it dominates the behavior of the voltages during the later stages of the transient.

Problems

3.1   (a) Given the interconnect parameters shown below, find the total interconnect resistance and the capacitance between the interconnect and the ground plane.

(b) Model the interconnect using one-, two-, and three-lump approximations.

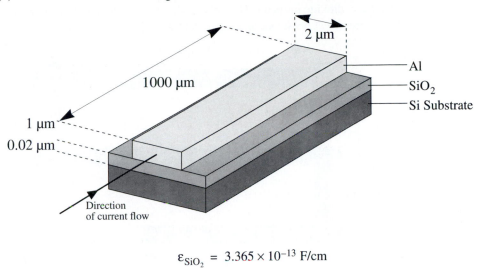

$$\varepsilon_{SiO_2} = 3.365 \times 10^{-13} \text{ F/cm}$$

$$\rho_{Al} = 2.8 \ \mu\Omega\text{-cm}$$

3.2   Repeat Exercise 3.1, assuming the insulating layer is silicon nitride, Si_3N_4, instead of SiO_2.

3.3   Repeat Exercise 3.1, assuming the interconnect material is tungsten instead of aluminum.

3.4   An aluminum integrated-circuit interconnect 1.0 μm thick and 1.5 μm wide connects two gates over a 200-nm SiO_2 dielectric layer. The capacitance of the interconnect is 2.7 pF. Determine the resistance of the interconnect.

3.5   Recently, IC manufacturers have begun using copper instead of aluminum for interconnecting gates. By what percentage would a switch to copper decrease resistance and capacitance for the interconnect of Exercise 3.1?

3.6   The diagram below illustrates a top-down view of a polysilicon interconnect running over a 250-nm-thick silicon dioxide insulator. The poly thickness is 0.5 μm and the width is 1.8 μm. Determine the interconnect resistance and capacitance.

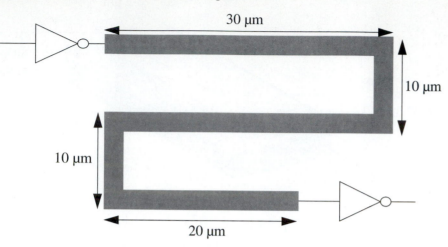

3.7   The diagram below illustrates a top-down view of an polysilicon interconnect running over a 220-nm-thick silicon dioxide insulator. The poly thickness is 0.5 μm and the width is 1.8 μm. The driving inverter, D, is loaded by two inverters A and B. Determine the interconnect resistance and capacitance associated with each of the paths D-A and D-B. Which inverter (A or B) will switch first when D transitions?

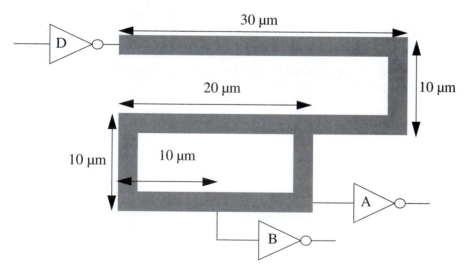

3.8   An IC manufacturer uses the maximum switching speed of a pair of cascaded inverters with a long interconnect as a benchmark for their process. In 1995 they realized that the additional delay due to the interconnect was 50% of the total delay. What percentage of the total delay can be ascribed to the interconnect in the year 2005? Assume that the interconnect length doubles every three years, but the inverter output resistance and input capacitance are reduced by 10% each year.

3.9 When a long interconnect is unavoidable, the delay can sometimes be minimized by the use of a **buffer**. A simple way to make a buffer is to cascade two inverters:

If the distance between the driving and load gates is sufficiently long, then inserting the buffer (which breaks the interconnect into two shorter pieces) can decrease the overall delay time.

In the circuit below, inverter A drives the load inverter B through a long interconnect. The interconnect has a resistance of 0.1 $\Omega/\mu m$ and a capacitance to ground of 1 fF/μm. The inverters making up the buffer are identical to the driving and load inverters.

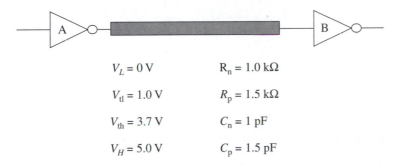

$V_L = 0$ V	$R_n = 1.0$ kΩ
$V_{tl} = 1.0$ V	$R_p = 1.5$ kΩ
$V_{th} = 3.7$ V	$C_n = 1$ pF
$V_H = 5.0$ V	$C_p = 1.5$ pF

(a) Since the buffer itself introduces a delay, this strategy is useful only if the interconnect is sufficiently long. Determine the minimum length of interconnect which will result in an overall decrease in delay time when the buffer is placed at the midpoint of the interconnect. (Use a one-lump model of each interconnect section.)

(b) Suppose the interconnect is 200 mm long. Determine the optimum location of the buffer along the interconnect.

3.10 In Exercise 3.9, determine the minimum interconnect length in which two buffers will result in a lower overall decrease in gate delay. Where should the buffers be placed?

3.11 Instead of directly cascading the two inverters making up the buffer (Exercise 3.9), we could place them at two different locations along the interconnect:

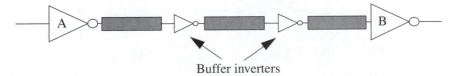

Buffer inverters

Using the parameters of Exercise 3.9, determine the minimum interconnect length which would result in a shorter overall gate delay using this strategy.

3.12 Determine the natural frequencies of the transient response in the circuit below.

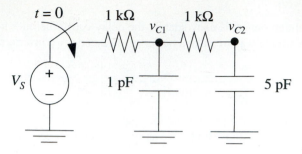

3.13 Determine the natural frequencies of the transient response in the circuit below.

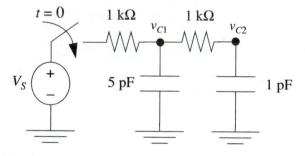

3.14 Determine the natural frequencies of the transient response in the circuit below.

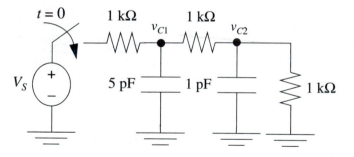

3.15 Determine the natural frequencies of the transient response in the circuit below.

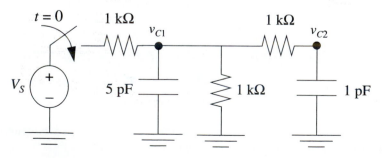

3.16 Determine the natural frequencies of the transient response in the circuit below.

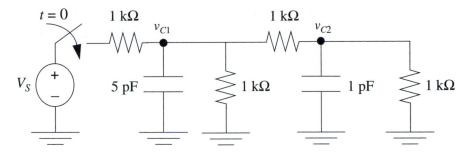

3.17 Determine the dominant natural frequency of the transient response in the circuit below.

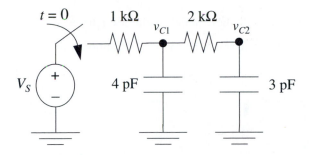

3.18 Find the dc steady-state value of the output voltage V_o.

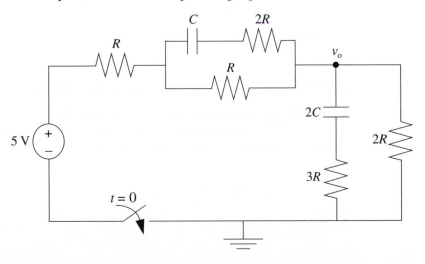

3.19 Show mathematically that the natural frequencies s_1 and s_2 (page 142) for a two-section RC ladder circuit must be real, negative, and distinct.

3.20 (a) Determine the natural frequencies in the circuit below.

   (b) What are the dc steady-state voltages V_1 and V_2?

c) Solve for $v_1(t)$ and $v_2(t)$ for $t \geq 0$, given initial conditions $v_1(0) = 0$ V, $v_2(0) = 0$ V.

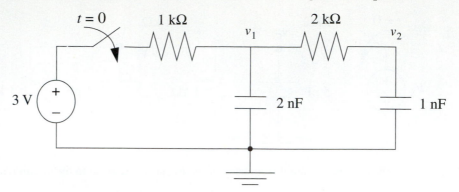

3.21 Repeat Exercise 3.20 with these initial conditions: $v_1(0) = -3$ V, $v_2(0) = 6$ V.

3.22 Solve for $v_1(t)$ and $v_2(t)$ for $t \geq 0$, given initial conditions $v_1(0) = 0$ V, $v_2(0) = 0$ V.

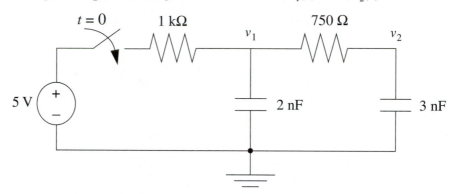

3.23 Repeat Exercise 3.22 with the following initial conditions: $v_1(0) = 1.5$ V, $v_2(0) = 1.2$ V.

3.24 In Section 3.6.4 we used incomplete forms of the transient solution, Eqs. (3.31) and (3.32), to find the natural frequencies. Show that assumption of the complete form of the transient solutions, Eq. (3.38), results in the same values of the natural frequencies.

3.25 Given the following inverter and interconnect parameters, use a two-lump model to solve for $v(t)$ for the case when the output of inverter A is pulling up. Assume that the output of inverter A has been at logic low for a long time before the pull-up occurs. Plot the response $v(t)$.

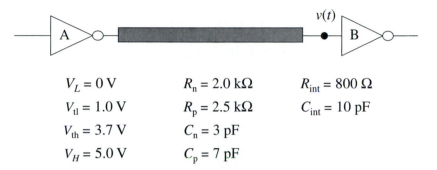

$$V_L = 0 \text{ V} \qquad R_n = 2.0 \text{ k}\Omega \qquad R_{int} = 800 \text{ }\Omega$$

$$V_{tl} = 1.0 \text{ V} \qquad R_p = 2.5 \text{ k}\Omega \qquad C_{int} = 10 \text{ pF}$$

$$V_{th} = 3.7 \text{ V} \qquad C_n = 3 \text{ pF}$$

$$V_H = 5.0 \text{ V} \qquad C_p = 7 \text{ pF}$$

3.26 In Exercise 3.25, determine the pull-up time using one-lump and two-lump models. What is the percentage difference in $t_{low \rightarrow high}$ using the two-lump model?

3.27 Using the circuit of Exercise 3.25, determine the maximum frequency at which this cascaded pair of inverters can be switched continuously, using a one-lump model of the interconnect.

3.28 Calculate the total amount of energy dissipated during a complete pull-down transition in the circuit of Exercise 3.25, using both a one-lump and two-lump model of the interconnect.

3.29 The input of inverter A in the circuit below is driven with a continuous symmetric square wave. The interconnect is modeled with a two-lump model.

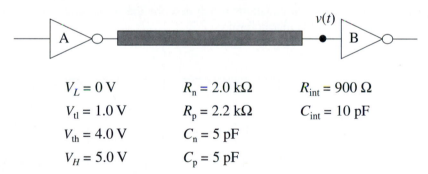

$$V_L = 0 \text{ V} \qquad R_n = 2.0 \text{ k}\Omega \qquad R_{int} = 900 \text{ }\Omega$$

$$V_{tl} = 1.0 \text{ V} \qquad R_p = 2.2 \text{ k}\Omega \qquad C_{int} = 10 \text{ pF}$$

$$V_{th} = 4.0 \text{ V} \qquad C_n = 5 \text{ pF}$$

$$V_H = 5.0 \text{ V} \qquad C_p = 5 \text{ pF}$$

(a) Determine the maximum frequency of the square-wave input signal.

(b) Show that the voltage $v_1(t)$ (across the capacitance $C_{int}/2$) changes slope exactly when the input square wave switches.

(c) Show that the voltage $v(t)$ at the input of the load inverter changes slope *after* the input square wave switches. Calculate this delay for both the pull-up and pull-down transitions. Is it significant?

3.30 Repeat Exercise 3.9 using a two-lump model of each interconnect section.

3.31 For the inverter pairs in Exercise 3.29, determine the maximum aluminum interconnect length which will result in an insignificant ($< 5\%$) increase in gate delay, using a one-lump

RC model. Assume the following interconnect parameters: $w = 1.25$ μm, $h = 0.5$ μm, $d = 0.2$ μm.

3.32 Repeat Exercise 3.31 using a two-*RC*-lump interconnect model. Is the maximum length significantly different?

3.33 Derive a set of differential equations which describe the response of this circuit for $t \geq 0$. Express these equations in matrix form.

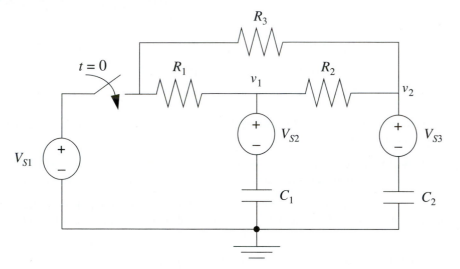

3.34 Derive a set of differential equations which describe the response of this circuit for $t \geq 0$. Express these equations in matrix form.

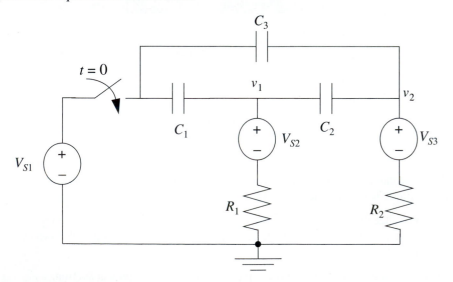

3.35 In Exercise 3.34, is it possible to determine the voltage on the node common to C_1 and C_3 for $t < 0$ from the information given, assuming the switch has been open a long time?

3.36 In the circuit below, assume that the switch has been in the up position for a long time. Solve for v_1 and v_2 for $t \geq 0$.

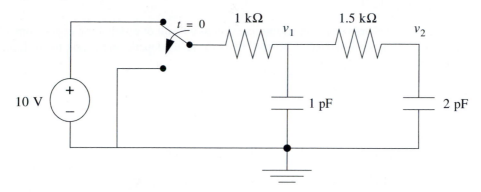

3.37 A two-section *RC* ladder circuit is constructed with these parameters:

$$R_1 = 120 \ \Omega$$
$$R_2 = 2300 \ \Omega$$
$$C_1 = 950 \ \text{fF}$$
$$C_2 = 45 \ \text{fF}$$
$$v_{C1}(0) = 0 \ \text{V}$$
$$v_{C2}(0) = 0 \ \text{V}$$

Assuming the input source is 5 V, determine the time required for v_{C2} to reach 3.8 V.

3.38 Verify that a three-lump interconnect model in the problem of Example 3.9 predicts a load inverter input voltage of

$$v(t) = -5.30e^{-288 \times 10^6 t} - 0.019e^{-24.9 \times 10^9 t} + 0.319e^{-6.36 \times 10^9 t} + 5$$

3.39 In the circuit below, the voltage $v_1(t)$ is given in volts by:

$$v_1(t) = -3.618e^{-3.2725 \times 10^9 t} - 1.382e^{-0.4775 \times 10^9 t} + 5$$

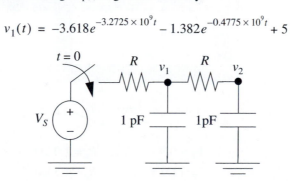

(a) Determine V_S.

(b) Determine $v_1(0)$.

(c) Determine R.

(d) Determine $v_2(t)$.

3.40 Determine the charge stored in an interconnect as a result of a switching event from V_L to V_H using an n-lump RC interconnect model. Does the result depend on the number of lumps used in the model?

3.41 Coupled inverter pairs with

$$R_{out} = 800 \ \Omega$$
$$C_{in} = 500 \ fF$$
$$V_L = 0 \ V$$
$$V_{tl} = 1 \ V$$
$$V_{th} = 4 \ V$$
$$V_H = 5 \ V$$

are cascaded with an interconnect. The interconnect has a resistance of 0.2 Ω/µm and a capacitance to ground of 1 fF/µm. How long does the interconnect need to be before its effect on gate delay becomes significant?

3.42 The diagram below illustrates a top-down view of an polysilicon interconnect running over a 220-nm-thick silicon dioxide insulator. The poly thickness is 0.5 µm and the width is 1.8 µm. Determine the maximum pull-up and pull-down logic transition times for the two inverters A and B, using both one-lump and two-lump interconnect models.

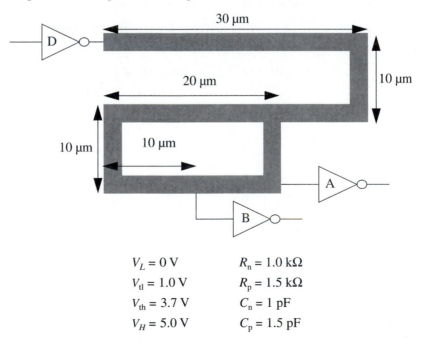

$$V_L = 0 \ V \qquad\qquad R_n = 1.0 \ k\Omega$$
$$V_{tl} = 1.0 \ V \qquad\qquad R_p = 1.5 \ k\Omega$$
$$V_{th} = 3.7 \ V \qquad\qquad C_n = 1 \ pF$$
$$V_H = 5.0 \ V \qquad\qquad C_p = 1.5 \ pF$$

Fanout and Capacitive Coupling

4

4.1 Introduction

In Chapter 2 we learned how to solve for the maximum switching frequency of a coupled inverter pair. Chapter 3 extended this analysis by taking into account effects of the interconnect linking the pair of inverters. In this chapter we further extend this analysis of inverter-pair switching speed. First we investigate how switching speed is affected when a single inverter has to simultaneously drive two or more subsequent inverters, a configuration known as **fanout**. We then discuss aspects of **capacitive coupling**, a more subtle phenomenon which is gaining prominence as interconnect density on integrated circuits increases.

4.2 Fanout

Consider the inverter configuration in Figure 4.1. The connection of several load gates driven by a single gate is called **fanout**. Fanout is quite common in integrated circuits and has significant effects on switching speed. Fanout may involve many different kinds of logic gates, but for simplicity we will examine fanout only with inverters. Also, we will initially ignore the effects of the interconnects by assuming the distance between the driving inverter and each of the fanout inverters is relatively small.

To begin our analysis, we replace the logic elements in Figure 4.1 with their *RC* circuit equivalents, shown in Figure 4.2. Since the *n* capacitors in Figure 4.2 are in parallel, the circuit may be simplified. The result is the first-order *RC* circuit shown in Figure 4.3.

All cases of simple gate fanout may be solved by using the *RC* circuit model of Figure 4.3, whose solution was detailed in Chapter 2. To understand the effects of inverter fanout on maximum switching speed, let us solve an example.

Example 4.1

Compare the maximum switching speeds of an inverter pair and inverters arranged in a fanout configuration.

Consider the inverter configuration shown in Figure 4.4(a), where one inverter drives three inverters. Assume that all four inverters are identical and have input

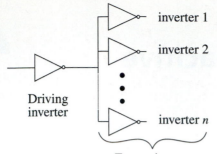

Driving
inverter

inverter 1

inverter 2

inverter *n*

Fanout inverters

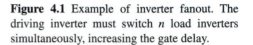

Figure 4.1 Example of inverter fanout. The driving inverter must switch *n* load inverters simultaneously, increasing the gate delay.

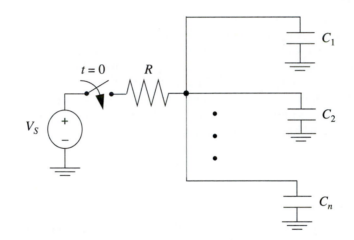

Figure 4.2 *RC* model of an arbitrary inverter fanout circuit. *R* is the output resistance of the driving inverter and C_i is the input capacitance of the *i*th fanout inverter.

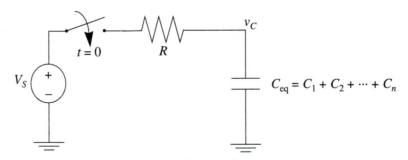

Figure 4.3 Simplification of the circuit of Figure 4.2 with C_{eq} replacing the *n* parallel input capacitances of the load inverters.

capacitance *C* and output resistance *R* (which could be R_p or R_n, depending on whether the load inverters are being pulled up or down). The *RC* circuit model of this configuration is shown in Figure 4.4(b).

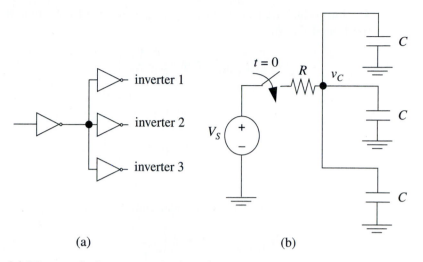

(a)   (b)

Figure 4.4 Diagrams for fanout example of one inverter driving three inverters.

We know that the switching time of a system modeled with a first-order RC circuit can be characterized completely by the time constant. In this case the time constant is:

$$\tau = 3RC \qquad (4.1)$$

This is three times greater than for a circuit with a single load inverter. Consequently, the switching time will also increase by a factor of three.

For example, suppose we have coupled inverters switching between 0 V and 5 V, with:

$$R_n = 1.0 \text{ k}\Omega$$
$$R_p = 1.5 \text{ k}\Omega$$
$$C = 10 \text{ pF}$$
$$V_{tl} = 1.0 \text{ V}$$
$$V_{th} = 3.7 \text{ V}$$

and that the circuit in Figure 4.4(a) has been at rest for a long time. The signal voltage $v_C(t)$ is given by:

$$v_C(t) = (v_{C0} - V_s)e^{-\frac{t}{t}} + V_s \qquad (4.2)$$

As we are interested in transition times, we rearrange Eq. (4.2) to get:

$$t = \tau \ln \left(\frac{v_{C0} - V_s}{v_C(t) - V_s} \right) \qquad (4.3)$$

The low-to-high transition time, $t_{\text{low} \to \text{high}}$, is defined by the time it takes for v_C to rise from 0 V to V_{th}:

$$t_{\text{low} \to \text{high}} = \tau \ln\left(\frac{0 - V_S}{V_{\text{th}} - V_S}\right)$$

$$= 3R_p C_{\text{in}} \ln\left(\frac{0 - V_S}{V_{\text{th}} - V_S}\right)$$

$$= (45 \times 10^{-9}) \ln\left(\frac{0 - 5}{3.7 - 5}\right)$$

$$= 60.6 \text{ ns}$$

The high-to-low transition time is found similarly:

$$t_{\text{high} \to \text{low}} = \tau \ln\left(\frac{5 - V_S}{V_{tl} - V_S}\right)$$

$$= 3R_n C_{\text{in}} \ln\left(\frac{5 - V_S}{V_{tl} - V_S}\right)$$

$$= (30 \times 10^{-9}) \ln\left(\frac{5 - 0}{1 - 0}\right)$$

$$= 48.3 \text{ ns}$$

As the low-to-high transition time is greater than the high-to-low time, the switching speed is limited by the pull-up. The maximum operating frequency is thus:

$$f_{\text{max}} = \frac{1}{2 \times 60.6 \times 10^{-9}} = 8.25 \text{ MHz}$$

We can relate this result to the value of the maximum operating frequency for an inverter *pair* obtained in Chapter 2 using identical inverter parameters. Not surprisingly, the maximum switching frequency for the pair is three times as high, 24.8 MHz. This makes sense, since the time constant τ is linearly proportional to the number of gates being driven.

In general, if we have n identical load gates, the maximum switching speed is:

$$f_{\text{fanout}} = \frac{1}{n} \times f_{\text{pair}} \tag{4.4}$$

If the gates are not identical, then the maximum switching frequency is inversely proportional to the sum of the input capacitances of the driven gates.

4.3 Fanout and Interconnects

In Chapter 3 we saw how interconnects can introduce significant gate delays. If a gate must drive multiple gates through long interconnects, these delays are compounded. In this section we will show how these configurations can be set up. However, we leave the mathematical solution to the reader, as they are simply applications of techniques we learned in Chapters 2 and 3.

4.3.1 One-Lump Model

Consider the configuration shown in Figure 4.5, where two load inverters are driven through an interconnect. The simplest approximation for the interconnect is the one-lump RC model; when inserted in this circuit, the equivalent model of Figure 4.6(a) results. We see that this is again a simple first-order RC circuit, since the inverter output resistance and interconnect resistance can be combined, and the inverter input capacitances can likewise be combined with the interconnect capacitance.

Figure 4.5 Multiple inverters driven through a single interconnect.

4.3.2 Two-Lump model

If we use a two-lump interconnect model for the circuit of Figure 4.5, the resulting circuit is shown in Figure 4.6(b). We recognize this as a two-section RC ladder circuit like those we saw in Chapter 3.

In both of these cases the addition of multiple gates at the end of a single interconnect simply increases the capacitive load. The gate input capacitances are in parallel and can be summed together.

4.3.3 Multiple Interconnects

A more interesting situation results if the load gates are connected through separate interconnects, as shown in Figure 4.7. Here we have used one-lump models for each of the two aluminum lines connecting the driving gate to the fanout inverters. This circuit is different from others we have studied; it cannot be treated as either a one-lump RC or two-section RC ladder circuit.

If we combine the capacitors in parallel and simplify the notation, the circuit in Figure 4.7 can be drawn as shown in Figure 4.8. Applying KCL at the v_1 node:

$$C_1 \frac{dv_1}{dt} = \frac{v_o - v_1}{R_1} \qquad (4.5)$$

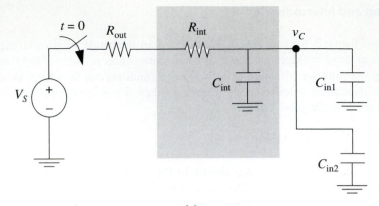

(a)

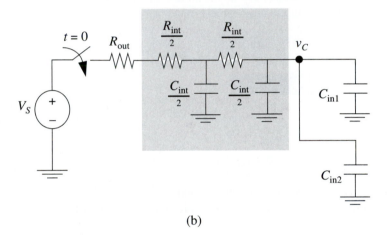

(b)

Figure 4.6 Fanout circuit with interconnect models (shown in shaded rectangles). (a) One-lump interconnect model. (b) Two-lump interconnect model.

and at the v_2 node:

$$C_2\frac{dv_2}{dt} = \frac{v_o - v_2}{R_2} \tag{4.6}$$

The voltage variable v_o can be eliminated by applying KCL at that node:

$$\frac{V_S - v_o}{R_o} = \frac{v_o - v_1}{R_1} + \frac{v_o - v_2}{R_2}$$

$$\frac{V_S}{R_o} + \frac{v_1}{R_1} + \frac{v_2}{R_2} = v_o\left(\frac{1}{R_o} + \frac{1}{R_1} + \frac{1}{R_2}\right)$$

If we let $\beta = R_1R_2 + R_oR_2 + R_oR_1$, then v_o is given by

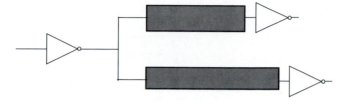

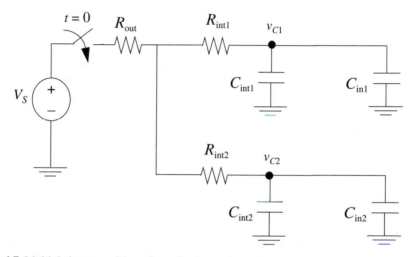

Figure 4.7 Multiple inverters driven through separate interconnects.

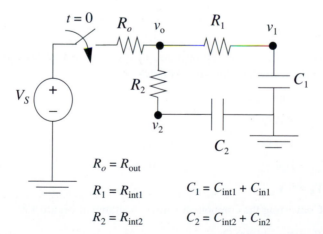

$$R_o = R_{out}$$
$$R_1 = R_{int1} \qquad C_1 = C_{int1} + C_{in1}$$
$$R_2 = R_{int2} \qquad C_2 = C_{int2} + C_{in2}$$

Figure 4.8 Simplified equivalent circuit for multiple inverters driven through separate interconnects.

$$v_o = \frac{\dfrac{V_S}{R_o} + \dfrac{v_1}{R_1} + \dfrac{v_2}{R_2}}{\left(\dfrac{1}{R_o} + \dfrac{1}{R_1} + \dfrac{1}{R_2}\right)}$$

$$v_o = \frac{R_1 R_2 V_S + R_o R_2 v_1 + R_o R_1 v_2}{\beta}$$

Inserting this expression into Eqs. (4.5) and (4.6) results in:

$$\frac{dv_1}{dt} = \frac{1}{R_1 C_1}\left[-\frac{R_1 R_2 + R_o R_1}{\beta}v_1 + \frac{R_o R_1}{\beta}v_2 + \frac{R_1 R_2}{\beta}V_S\right]$$

$$\frac{dv_2}{dt} = \frac{1}{R_2 C_2}\left[\frac{R_o R_2}{\beta}v_1 - \frac{R_1 R_2 + R_o R_2}{\beta}v_2 + \frac{R_1 R_2}{\beta}V_S\right]$$

which can be expressed in matrix form as:

$$\begin{bmatrix} \dfrac{dv_1}{dt} \\[2mm] \dfrac{dv_2}{dt} \end{bmatrix} = \frac{1}{\beta}\begin{bmatrix} -\dfrac{R_2 + R_o}{C_1} & \dfrac{R_o}{C_1} \\[4mm] \dfrac{R_o}{C_2} & -\dfrac{R_1 + R_o}{C_2} \end{bmatrix}\begin{bmatrix} v_1 \\[2mm] v_2 \end{bmatrix} + \frac{1}{\beta}\begin{bmatrix} \dfrac{R_2 V_S}{C_1} \\[4mm] \dfrac{R_1 V_S}{C_2} \end{bmatrix} \qquad (4.7)$$

We recognize Eq. (4.7) as having the same form as the equations for the two-section *RC* ladder circuit, although the individual coefficients are different. We use the same solution method as before, starting with determination of the natural frequencies.

Example 4.2

Determine the pull-up response of inverters driven through two different interconnects, using the following parameters:

$R_{out} = 1000\ \Omega$

$R_{int1} = 200\ \Omega$ $\qquad\qquad C_{int1} = 0.5\ \text{pF}$

$R_{int2} = 400\ \Omega$ $\qquad\qquad C_{int2} = 1.0\ \text{pF}$

$C_{in1} = 1\ \text{pF}$ $\qquad\qquad C_{in2} = 1.0\ \text{pF}$

$V_S = 5\ \text{V}$

Converting the component values as shown in Figure 4.8:

$R_0 = R_{out} = 1000\ \Omega$

$R_1 = R_{int1} = 200\ \Omega$ $\qquad C_1 = C_{int1} + C_{in1} = 1.5\ \text{pF}$

$R_2 = R_{int2} = 400\ \Omega$ $\qquad C_2 = C_{int2} + C_{in2} = 2.0\ \text{pF}$

Substituting these values into Eq. (4.7):

$$\begin{bmatrix} \dfrac{dv_1}{dt} \\[2mm] \dfrac{dv_2}{dt} \end{bmatrix} = 10^9 \begin{bmatrix} -1.37 & 0.980 \\ 0.735 & -0.882 \end{bmatrix} \begin{bmatrix} v_1 \\ v_2 \end{bmatrix} + 10^9 \begin{bmatrix} 1.96 \\ 0.735 \end{bmatrix}$$

The natural frequencies are found using the first coefficient matrix:

$$\det\left[s\mathbf{I} - 10^9 \begin{bmatrix} -1.37 & 0.980 \\ 0.735 & -0.882 \end{bmatrix} \right] = 0$$

$$\det\begin{bmatrix} s + 1.37 \times 10^9 & -0.980 \times 10^9 \\ -0.735 \times 10^9 & s + 0.882 \times 10^9 \end{bmatrix} = 0$$

$$s_1 = -0.244 \times 10^9 \text{ s}^{-1}$$

$$s_2 = -2.01 \times 10^9 \text{ s}^{-1}$$

Using initial capacitor voltages of 0 V, the pull-up responses are

$$v_1(t) = -4.58e^{-0.244 \times 10^9 t} - 0.42e^{-2.01 \times 10^9 t} + 5$$

$$v_2(t) = -5.27e^{-0.244 \times 10^9 t} + 0.27e^{-2.01 \times 10^9 t} + 5$$

4.4 Capacitive Coupling and Crosstalk

Up to this point, we have assumed that interconnect capacitance is exhibited solely between the interconnect and the IC substrate (the ground plane). This is an accurate assumption, provided there are no other nearby interconnects. Generally, though, IC's have a high density of interconnects, both in the same plane and in other, parallel planes. This results in **capacitive coupling** between interconnects. The amount of coupling depends on several factors: the distance between the interconnects; the dielectric thickness; the length of the interconnect overlap; the interconnect dimensions; and the properties of the materials surrounding the interconnects.

Capacitive coupling between interconnects can cause two different kinds of problems. The first is that capacitive coupling represents an additional capacitance, just like another load, which must be charged during a logic transition. This extra load results in additional gate delay and a lowering of the maximum system speed.

The second problem, which is considerably more serious, is that a logic transition in one interconnect can be *coupled* into the other, which may cause a false logic transition. Consider the example illustrated in Figure 4.9. Assume that the input of gate 3 remains constant. This implies that the voltage all along interconnect 2 should be constant; interconnect 2 is a "quiet" line. Now, a logic transition at the input of gate 1 will cause a variation in the voltage along interconnect 1. Ideally, this will not affect gate 4. However,

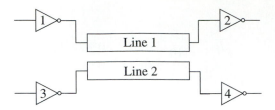

Figure 4.9 A set of logic gates linked by capacitively coupled interconnects. Crosstalk can result if a logic transition on one line is coupled into the other.

if the capacitive coupling between interconnects 1 and 2 is sufficiently strong, the signal may be coupled into gate 4, causing an erroneous transition. This phenomenon is called **crosstalk** and must be avoided.

Capacitive coupling is a major performance–restricting factor in integrated circuits. Clearly, it must be minimized during the IC design phase, otherwise functional errors can be generated when the circuit is operating at high speed.

Figure 4.10 illustrates two configurations of interconnect geometry which exhibit capacitive coupling. In Figure 4.10(a) two interconnects on the same layer run parallel to each other. The heavy parallel lines symbolize the capacitance between the two lines; the geometry is that of two metal plates separated by a dielectric of relative permittivity ε_c. This capacitance is in addition to the usual capacitance to ground, denoted by the capacitor symbol in light lines. (In general, the dielectric separating the interconnects from the substrate is not the same as the "interlevel" dielectric, so we will use ε_i for the permittivity of dielectric directly above the substrate.) Figure 4.10(b) illustrates how two interconnects on different levels can exhibit capacitive coupling. The two lines need not be parallel for there to be coupling.

4.4.1 The Pi-Connection Model

To analyze the effects of coupling capacitance on circuit performance, we need to develop a model of the coupled interconnects. Recall that the single interconnect has a resistance and capacitance distributed along its length, and that many models are possible. The situation for coupled lines is similar, only more complicated, since we now have two lines, both with distributed resistance and capacitance, plus the distributed coupling capacitance.

We will use the *pi-connection model* to analyze coupled interconnects. For example, the model transforms Figure 4.10(a) by lumping the distributed interconnect capacitances into three components: C_c, the coupling capacitance between interconnect #1 and interconnect #2; C_1, the lumped capacitance between interconnect #1 and the ground plane; and C_2, the lumped capacitance between interconnect #2 and the ground plane. These are illustrated in the cross section of Figure 4.11, along with approximate formula for each based on the parallel-plate capacitor relationship.

The pi-connection model, including the interconnect resistances, is shown in Figure 4.12. R_1 and R_2 are the total resistances of the two interconnects. The pi-connection model is so named because it resembles the Greek letter Π, as indicated by the circled region in Figure 4.12(b).

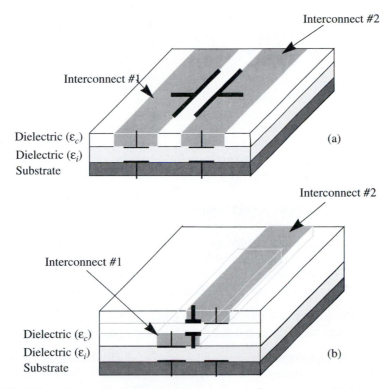

Figure 4.10 (a) Pair of capacitively coupled interconnects on the same layer. Each interconnect has capacitance to ground (denoted by light capacitor symbols) and to the other (capacitor symbol in heavy line). (b) Pair of capacitively coupled interconnects on vertically separated layers. Again, each interconnect has a capacitance to ground and to the other.

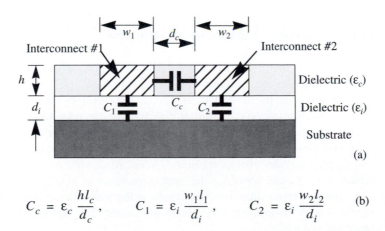

$$C_c = \varepsilon_c \frac{h l_c}{d_c}, \qquad C_1 = \varepsilon_i \frac{w_1 l_1}{d_i}, \qquad C_2 = \varepsilon_i \frac{w_2 l_2}{d_i} \qquad \text{(b)}$$

Figure 4.11 Pi-connection model. (a) Cross–sectional view of parallel interconnects. (b) Approximate formula for each capacitance. The two interconnect lengths are l_1 and l_2; the length over which they overlap is l_c.

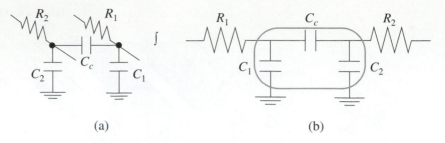

(a)   (b)

Figure 4.12 Circuit representation of the pi-connection model of coupled interconnects. (a) Perspective view. (b) Equivalent circuit.

4.4.2 Adjacent Interconnect Termination

We will analyze in detail three specific cases where capacitive coupling affects system performance. In each case a pair of inverter gates are connected with an interconnect, while an adjacent interconnect is part of the analysis. The circuit will be affected in different ways, depending on what is connected to the adjacent, capacitively coupled interconnect:

- The adjacent interconnect can be grounded at both ends. This is a common situation where the adjacent line is held at a constant potential by a power supply.

- The adjacent interconnect is floating; i.e., it is isolated from other parts of the circuit by open switches. This configuration is common for tristate busses.

- The adjacent interconnect also links a pair of inverter gates.

These three cases are illustrated in Figure 4.13. As we will see, the manner in which the adjacent line is terminated is important in determining the effects of capacitive coupling.

4.5 Capacitive Coupling to a Grounded Adjacent Line

Application of the pi-connection model to the case of a grounded adjacent line, Figure 4.13(a), yields the configuration shown in Figure 4.14.

Assume that the driving inverter on the left (inverter 1) is in the process of pulling up the load inverter (inverter 2) from 0 V to 5 V. We can now add the inverter models to the circuit of Figure 4.14, resulting in the circuit shown in Figure 4.15. Examining this circuit, we notice that v_2 is always zero, since the adjacent line is grounded. This means that R_2 and C_2 play no role whatsoever in the circuit. Also, we can combine the two resistors in series, and the capacitors C_{in2} (the input capacitance of gate 2), C_1, and C_c in parallel.

Making these simplifications results in the first-order circuit of Figure 4.16. We recognize this as the familiar single-lump RC circuit. The equivalent resistance and capacitance are given by

$$R_{eq} = R_{out} + R_1 \tag{4.8}$$

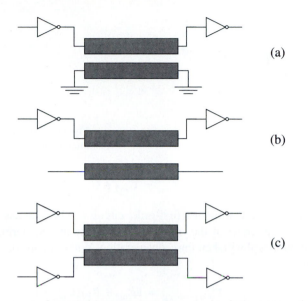

Figure 4.13 Three cases of coupled interconnects. In each circuit the upper pair of inverter gates undergo a transition. (a) Adjacent interconnect grounded at both ends. (b) Adjacent interconnect floating at both ends. (c) Adjacent interconnect (active line) connects two quiet gates.

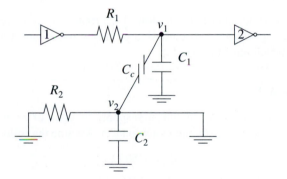

Figure 4.14 Connection of two gates with pi-connection model of coupled interconnects. The adjacent line is grounded at both ends.

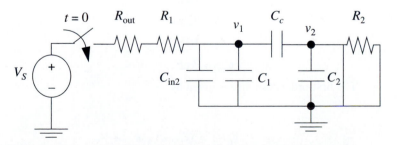

Figure 4.15 Model of coupled interconnects with adjacent grounded line.

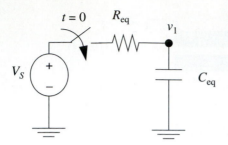

Figure 4.16 Equivalent circuit for coupled interconnects with the adjacent line grounded.

$$C_{eq} = C_{in2} + C_1 + C_c \tag{4.9}$$

The switching time of a first-order circuit can be characterized by its time constant. We can see the effect of the coupling capacitance by comparing the time constants with and without coupling taken into consideration. Without considering the coupling, the time constant is:

$$\tau_{\text{w/o coupling}} = (R_{out} + R_1)(C_{in2} + C_1) \tag{4.10}$$

Including the effect of coupling, the circuit time constant is:

$$\tau_{\text{w/ coupling}} = (R_{out} + R_1)(C_{in2} + C_1 + C_c) \tag{4.11}$$

The grounded adjacent line has slowed the transition by increasing the time constant. In effect, the capacitance of the adjacent grounded interconnect acts as an additional capacitive load which must be charged.

Example 4.3

Calculate and plot the pull-up response of a pair of coupled inverters with and without an adjacent grounded interconnect. Assume the following parameters:

$$R_{out} = 750\ \Omega$$
$$R_1 = 1350\ \Omega$$
$$R_2 = 1500\ \Omega$$
$$C_1 = 20\ \text{pF}$$
$$C_2 = 22\ \text{pF}$$
$$C_c = 10\ \text{pF}$$
$$C_{in2} = 7\ \text{pF}$$

The first-order response is given by

$$v_C(t) = (v_{C0} - V_S)e^{-\frac{t}{\tau}} + V_S \tag{4.12}$$

For a pull-up transition from 0 V to 5 V:

$$v_C(t) = 5\left[1 - e^{-\frac{t}{\tau}}\right] \tag{4.13}$$

Ignoring coupling capacitance, the time constant is

$$\tau_{\text{w/o coupling}} = (R_{\text{out}} + R_1)(C_{\text{in2}} + C_1)$$

$$= (750 + 1350)(7 \times 10^{-12} + 20 \times 10^{-12})$$

$$= 56.7 \times 10^{-9} \text{ s}$$

or 56.7 ns. Including the effect of the adjacent interconnect:

$$\tau_{\text{w/ coupling}} = (R_{\text{out}} + R_1)(C_{\text{in2}} + C_1 + C_c)$$

$$= (750 + 1350)(7 \times 10^{-12} + 20 \times 10^{-12} + 10 \times 10^{-12})$$

$$= 77.7 \times 10^{-9} \text{ s}$$

or 77.7 ns, an increase of 37%. The logic transitions with and without the adjacent interconnect are shown in Figure 4.17. This graph clearly shows how the gate delay increases when coupling capacitance is taken into consideration.

Since we know that the logic transition time is linearly proportional to the time constant, the adjacent grounded interconnect has the effect of increasing the gate delay by 37% in this example. Put another way, the maximum operating frequency has been decreased from 5.5 to 4.1 MHz.

The case of an adjacent grounded interconnect is the worst case of capacitive coupling where the adjacent line is not also switching.[1] (The situation is the same if the adjacent line is held at V_S.) It is the worst case because the additional capacitive load is a maximum; other configurations will add other capacitances in series with C_c, reducing the loading effect of the adjacent line somewhat. This is by no means an uncommon situation, though, since power and ground lines must be run to every logic gate on the integrated circuit.

4.6 Capacitive Coupling to a Floating Adjacent Line

We now turn to the next case: an interconnected pair of inverter gates with an adjacent, floating interconnect. Application of the pi-connection model to the circuit of Figure 4.13(b) yields the circuit shown in Figure 4.18. Adding the inverter models, we have the circuit shown in Figure 4.19.

---

[1] An even worse situation arises if the adjacent line is simultaneously switching in the opposite direction. See Exercise 4.28.

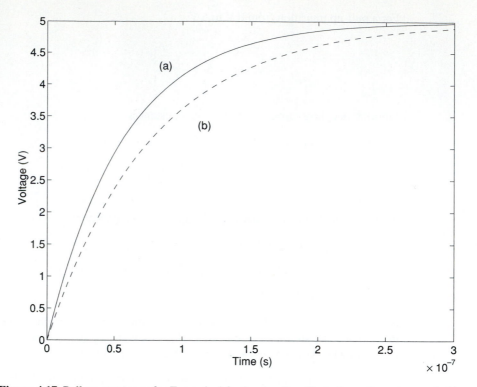

Figure 4.17 Pull-up responses for Example 4.3, showing the effect of an adjacent grounded interconnect. (a) Adjacent interconnect ignored. (b) Coupling of adjacent interconnect included. The presence of an adjacent grounded interconnect significantly slows the logic transition.

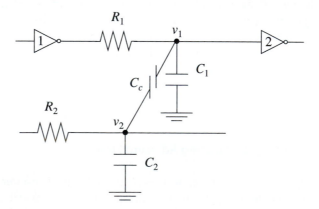

Figure 4.18 Connection of two gates with pi-connection model of coupled interconnects. The adjacent line is floating at both ends.

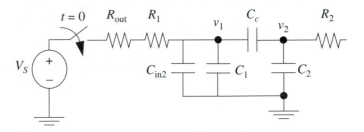

Figure 4.19 Model of coupled inverter gates with an adjacent floating line.

As was the case with the grounded adjacent line, this circuit is simply a first-order *RC* circuit. Although it contains four capacitances, they can all be combined into a single equivalent value. First, we note that C_2 and C_c are in series; they have an equivalent capacitance of $(C_2 C_c)/(C_2 + C_c)$. This series equivalent is in parallel with the remaining two capacitors. Thus the overall equivalent capacitance is given by:

$$C_{eq} = C_{in2} + C_1 + C_c \left(\frac{C_2}{C_2 + C_c} \right) \tag{4.14}$$

The equivalent resistance is equal to the sum of the two resistors in series:

$$R_{eq} = R_{out} + R_1 \tag{4.15}$$

The time constant for this first-order circuit is found by multiplying Eqs. (4.14) and (4.15):

$$\tau_{w/\,coupling} = (R_{out} + R_1) \left(C_{in2} + C_1 + C_c \left(\frac{C_2}{C_2 + C_c} \right) \right) \tag{4.16}$$

Again we see that the time constant has increased compared to the case with no capacitive coupling ($C_c = 0$). However, compared to an adjacent grounded interconnect, Eq. (4.11), the degradation is not as severe. The two expressions for the time constant, Eqs. (4.11) and (4.16), differ by a term multiplying the coupling capacitance. With a grounded adjacent line, the coefficient is unity; with a floating adjacent line, the coefficient is always less than unity. Therefore, a grounded adjacent line will always slow down a logic transition more than a floating adjacent line.

Example 4.4

Calculate the increase in gate delay due to an adjacent floating line, given the parameters in Example 4.3.

Substituting the given parameters in Eq. (4.16), the time constant of the circuit including the effects of interconnect coupling is:

$$\tau_{\text{w/ coupling}} = (R_{\text{out}} + R_1)\left(C_{\text{in2}} + C_1 + C_c\left(\frac{C_2}{C_2 + C_c}\right)\right)$$

$$= (750 + 1350)\left(7 \times 10^{-12} + 20 \times 10^{-12} + 10 \times 10^{-12}\left(\frac{22}{22 + 10}\right)\right)$$

$$= 71.1 \times 10^{-9}\ \text{s}$$

or 71.1 ns. Compare this to the time constant with a grounded adjacent interconnect (77.7 ns) and no adjacent interconnect (56.7 ns). Figure 4.20 illustrates the difference in the logic transition for adjacent grounded and floating interconnects.

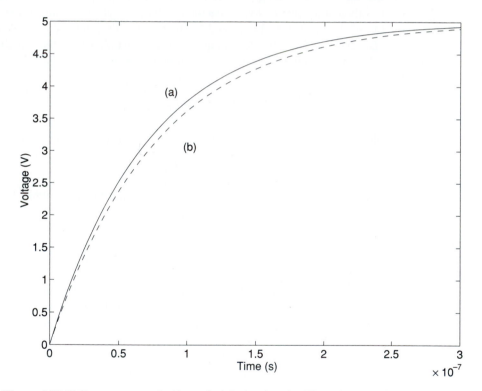

Figure 4.20 Pull-up responses for Example 4.4, showing the effect of (a) an adjacent floating interconnect, and (b) an adjacent grounded interconnect.

This example has verified the observation that the time constant for an interconnect adjacent to a floating line is less than the time constant for an interconnect adjacent to a grounded line. In fact, as far as the effects of coupling capacitance are concerned, the cases of grounded and floating adjacent lines are at opposite ends of a spectrum. The worst case is represented by a grounded adjacent line; this provides the lower bound on the maximum switching speed of the circuit. An adjacent floating line, on the other hand, is the best case of coupling and provides an upper bound on switching speed. The actual speed limitation generally lies between these two extremes.

A floating line will have a signal induced in it by an adjacent noisy line. This is usually not a problem, though, since a floating line, by definition, is not connected to any gates, so spurious logic transitions will not occur.

4.7 Capacitive Coupling to an Adjacent Active Line

If the adjacent interconnect also connects gates, as illustrated in Figure 4.13(c), this second interconnect is an **active line**, i.e., not grounded or floating. We will assume this interconnect is quiet, with no logic transition induced by its own driving gate. As we will see, crosstalk from the other line will be coupled onto the quiet line, making it less quiet.

Application of the pi-connection model to this configuration of gates results in the schematic shown in Figure 4.21. We will assume that inverter 1 is pulling up inverter 2, and that the input to inverter 3 is held constant at logic high. If there were no coupling, holding the inverter 3 input constant would force v_2 to remain at 0 V; however, the induced crosstalk will alter this.

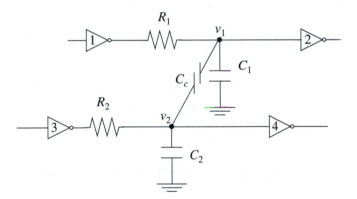

Figure 4.21 Connection of two gates with pi-connection model of coupled interconnects. The adjacent active line also connects two gates. We analyze the problem where the input of inverter 1 falls from logic high to logic low, while the input to inverter 3 remains at logic high.

Inserting the appropriate inverter models, we obtain the circuit shown in Figure 4.22. Combining resistors in series and capacitors in parallel, we can reduce this circuit to that in Figure 4.23. In this circuit:

$$R_a = R_{out1} + R_1$$

$$R_b = R_{out3} + R_2$$

$$C_a = C_{in2} + C_1$$

$$C_b = C_{in4} + C_2$$

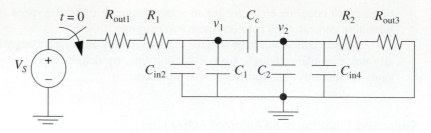

Figure 4.22 Model of coupled inverter gates with an adjacent active line at logic low.

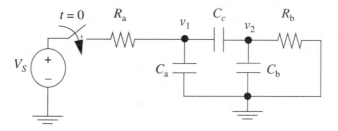

Figure 4.23 Simplified circuit, equivalent to that shown in Figure 4.22.

We analyze the circuit in Figure 4.23 by applying KCL at nodes v_1 and v_2:

$$\frac{V_S - v_1}{R_a} = C_a \frac{d}{dt} v_1 + C_c \frac{d}{dt}((v_1 - v_2)) \tag{4.17}$$

$$C_b \frac{d}{dt} v_2 = C_c \frac{d}{dt}((v_1 - v_2)) - \frac{v_2}{R_b} \tag{4.18}$$

Equations (4.17) and (4.18) can be expressed compactly in matrix form as

$$\begin{bmatrix} C_a + C_c & (-C_c) \\ -C_c & C_b + C_c \end{bmatrix} \begin{bmatrix} \dfrac{dv_1}{dt} \\ \dfrac{dv_2}{dt} \end{bmatrix} = \begin{bmatrix} -\dfrac{1}{R_a} & 0 \\ 0 & \left(-\dfrac{1}{R_b}\right) \end{bmatrix} \begin{bmatrix} v_1 \\ v_2 \end{bmatrix} + \begin{bmatrix} \dfrac{V_s}{R_a} \\ 0 \end{bmatrix} \tag{4.19}$$

It is interesting to note that Eq. (4.19) is the general matrix form of the set of differential equations for each of the three configurations of capacitively coupled interconnects. With an adjacent grounded interconnect, $C_b = 0$ and $v_2 = 0$ is zero. Substituting these values into Eq. (4.19) reduces the system to a first-order RC circuit whose solution is given by Eq. (4.12). With an adjacent floating interconnect, it can also be shown that the system reduces to the correct first-order system. However, with an active adjacent line, the differential equations are coupled and the problem is inherently second order. We can solve this system using the techniques we learned in Chapter 3 with only slight modifications.

Example 4.5

In the logic circuit shown in Figure 4.24, the two interconnects are capacitively coupled. At time $t = 0$, the input to the NOR gate labeled "α" undergoes a high-to-low transition. Before this event, all lines have reached their steady-state values.

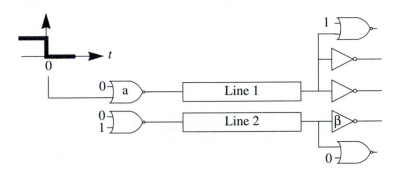

Figure 4.24 Logic circuit for Example 4.5. At $t = 0$, the lower input to the NOR gate "α" drops from logic high to logic low.

Determine the maximum voltage that appears at the input of the inverter labeled "β" and the time at which this occurs. The circuit parameters are:

Gate input capacitances are $C_{in} = 1.8$ pF

Gate pull-up resistances are $R_p = 1.5$ kΩ

Gate pull-down resistances are $R_n = 1.0$ kΩ

Interconnect resistance of Line 1 is $R_1 = 800$ Ω

Interconnect capacitance of Line 1 is $C_1 = 10.0$ pF

Interconnect resistance of Line 2 is $R_2 = 640$ Ω

Interconnect capacitance of Line 2 is $C_2 = 8.0$ pF

Interconnect coupling capacitance is $C_c = 2.0$ pF

Our first task is to combine the CMOS gate models and the interconnect models. Three gates are connected to the output of Line 1 and two gates to the output of Line 2. This results in a capacitance of $3C_{in}$ loading Line 1, and $2C_{in}$ loading Line 2. Application of the pi-connection model of the interconnects to this logic circuit results in the circuit shown in Figure 4.25.

Combining resistances and capacitances where possible, we arrive at the circuit in Figure 4.26. The values in this circuit are:

$$R_a = R_p + R_1 = 1500 + 800 = 2300 \ \Omega$$

$$R_b = R_n + R_2 = 1000 + 640 = 1640 \ \Omega$$

$$C_a = 3C_{in} + C_1 = 3(1.8) + 10 = 15.4 \ \text{pF}$$

$$C_b = 2C_{in} + C_2 = 2(1.8) + 8 = 11.6 \ \text{pF}$$

$$C_c = 2.0 \ \text{pF}$$

$$V_S = 5 \ \text{V}$$

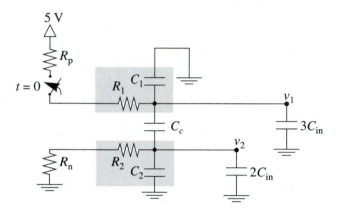

Figure 4.25 Circuit model of the logic circuit in Figure 4.24. The two interconnects, in the shaded rectangles, are coupled through capacitance C_c.

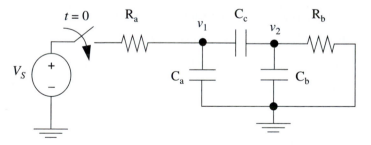

Figure 4.26 Simplified equivalent circuit for Example 4.5.

Substituting these numerical values into Eq. (4.19) gives us the following pair of coupled first-order differential equations:

$$1 \times 10^{-12} \overbrace{\begin{bmatrix} 17.4 & -2 \\ -2 & 13.6 \end{bmatrix}}^{A} \begin{bmatrix} \dfrac{dv_1}{dt} \\ \dfrac{dv_2}{dt} \end{bmatrix} = \begin{bmatrix} -\dfrac{1}{2300} & 0 \\ 0 & -\dfrac{1}{1640} \end{bmatrix} \begin{bmatrix} v_1 \\ v_2 \end{bmatrix} + \begin{bmatrix} \dfrac{5}{2300} \\ 0 \end{bmatrix} \qquad (4.20)$$

We can put this into a more familiar form by multiplying both sides of the equation by the inverse of A. The inverse of a 2×2 matrix is:

$$\begin{bmatrix} a & b \\ c & d \end{bmatrix}^{-1} = \frac{1}{ad - bc}\begin{bmatrix} d & -b \\ -c & a \end{bmatrix}$$

(4.21)

A^{-1} in this example is:

$$\left\{ 1 \times 10^{-12}\begin{bmatrix} 17.4 & -2 \\ -2 & 13.6 \end{bmatrix} \right\}^{-1} = 1 \times 10^{10}\begin{bmatrix} 5.846 & 0.8597 \\ 0.8597 & 7.479 \end{bmatrix}$$

(4.22)

Multiplying Eq. (4.20) by the righthand side of Eq. (4.22), we obtain an expression for the time derivatives of the two node voltages:

$$\begin{bmatrix} \dfrac{dv_1}{dt} \\ \dfrac{dv_2}{dt} \end{bmatrix} = \overbrace{1 \times 10^7 \begin{bmatrix} -2.542 & -0.5242 \\ -0.3738 & -4.561 \end{bmatrix}}^{B}\begin{bmatrix} v_1 \\ v_2 \end{bmatrix} + 1 \times 10^8\begin{bmatrix} 1.271 \\ 0.1869 \end{bmatrix}$$

(4.23)

The natural frequencies of the system are found by setting $\det(s\mathbf{I} - \mathbf{B}) = 0$ as follows:

$$\det\begin{bmatrix} s + 2.542 \times 10^7 & 0.5242 \times 10^7 \\ 0.3738 \times 10^7 & s + 4.561 \times 10^7 \end{bmatrix} = 0$$

(4.24)

Solving for the roots of the resultant quadratic equation derived from Eq. (4.24), we find that

$$s_1 = -2.449 \times 10^7 \text{ s}^{-1}$$
$$s_2 = -4.653 \times 10^7 \text{ s}^{-1}$$

The dc steady-state values can be found by substituting open circuits for each of the capacitors in Figure 4.26. By inspection, we see that

$$v_1(\infty) = V_S = 5$$
$$v_2(\infty) = 0$$

The general solution can thus be written as:

$$v_1(t) = V_{11}e^{s_1 t} + V_{12}e^{s_2 t} + 5$$

(4.25)

$$v_2(t) = V_{21}e^{s_1 t} + V_{22}e^{s_2 t}$$

(4.26)

We now need to determine the initial conditions. From the problem statement, we know that the voltages had reached steady state values before the logic transition at $t = 0$. Given the logic diagram, this means that both $v_1(0)$ and $v_2(0)$ are logic low, or zero volts. Substituting these values in Eqs. (4.25) and (4.26), we get

$$v_1(0) = V_{11} + V_{12} + 5 = 0 \qquad (4.27)$$

$$v_2(0) = V_{21} + V_{22} = 0 \qquad (4.28)$$

We now have two equations in four unknowns. To solve for these unknowns, we will find two more equations using the procedure described in Chapter 3.

The first row in Eq. (4.23) is

$$\frac{dv_1}{dt} = 10^7(-2.542v_1 - 0.5242v_2) + (1.271 \times 10^8) \qquad (4.29)$$

Differentiating Eq. (4.25):

$$\frac{dv_1}{dt} = s_1 V_{11} e^{s_1 t} + s_2 V_{12} e^{s_2 t} \qquad (4.30)$$

Equating Eqs. (4.29) and (4.30), and substituting the general expressions for v_1 and v_2:

$$s_1 V_{11} e^{s_1 t} + s_2 V_{12} e^{s_2 t} = -2.542 \times 10^7 \left(V_{11} e^{s_1 t} + V_{12} e^{s_2 t} + 5 \right)$$

$$-0.5242 \times 10^7 \left(V_{21} e^{s_1 t} + V_{22} e^{s_2 t} \right) \qquad (4.31)$$

$$+1.271 \times 10^8$$

Equating coefficients of each of the exponential terms:

$$s_1 V_{11} = -2.542 \times 10^7 V_{11} - 0.5242 \times 10^7 V_{21}$$

$$s_2 V_{12} = -2.542 \times 10^7 V_{12} - 0.5242 \times 10^7 V_{22} \qquad (4.32)$$

Hence

$$V_{11} = \left[\frac{0.5242 \times 10^7}{s_1 + 2.542 \times 10^7} \right] V_{21} \qquad (4.33)$$

$$V_{12} = \left[\frac{0.5242 \times 10^7}{s_2 + 2.542 \times 10^7} \right] V_{22} \qquad (4.34)$$

We now have four equations in four unknowns. Solving Eqs. (4.27), (4.28), (4.33), and (4.34) simultaneously:

$$V_{11} = -4.789$$

$$V_{12} = -0.211$$

$$V_{21} = 0.8496$$

$$V_{22} = -0.8496$$

Substituting these constants in the general solutions, Eqs. (4.25) and (4.26), yields the expressions for voltage:

$$v_1(t) = -4.789e^{-2.449\times10^7 t} - 0.2110e^{-4.653\times10^7 t} + 5 \tag{4.35}$$

$$v_2(t) = 0.8496e^{-2.449\times10^7 t} - 0.8496e^{-4.653\times10^7 t} \tag{4.36}$$

Before proceeding, we need to check these solutions to be sure they satisfy the original differential equation, Eq. (4.23). Since we used the first row of Eq. (4.23) to find the coefficients, we can use the second row to verify that Eqs. (4.35) and (4.36) are indeed correct. The second differential equation is:

$$\frac{dv_2}{dt} = (1\times10^7)[-0.3738v_1 - 4.561v_2] + (0.1869\times10^8) \tag{4.37}$$

Substituting the solution functions, Eqs. (4.35) and (4.36), into Eq. (4.37) we have:

$$\frac{dv_2}{dt} = -0.3738\times10^7(-4.789e^{-2.449\times10^7 t} - 0.2110e^{-4.653\times10^7 t} + 5)$$

$$-4.561\times10^7(0.8496e^{-2.449\times10^7 t} - 0.8496e^{-4.653\times10^7 t})$$

$$+0.1869\times10^8$$

Evaluating this expression:

$$\frac{dv_2}{dt} = -2.08\times10^7 e^{-2.449\times10^7 t} + 3.95\times10^7 e^{-4.653\times10^7 t} \tag{4.38}$$

This expression must match the derivative of v_2 found by direct differentiation of Eq. (4.36):

$$\frac{dv_2}{dt} = \frac{d}{dt}(0.8496e^{-2.449\times10^7 t} - 0.8496e^{-4.653\times10^7 t})$$

$$= (-2.449\times10^7)0.8496e^{-2.449\times10^7 t} - (-4.653\times10^7)0.8496e^{-4.653\times10^7 t}$$

$$= -2.08\times10^7 e^{-2.449\times10^7 t} + 3.95\times10^7 e^{-4.653\times10^7 t}$$

Since this expression matches Eq. (4.38), our solution checks. The solutions Eq. (4.35) and Eq. (4.36) also check with the initial conditions and dc steady-state behavior of the circuit.

A plot of these two voltages as a function of time is shown in Figure 4.27. Note the crosstalk which appears on the quiet line. In this example, it appears the amount of crosstalk is too small to cause an erroneous logic transition in the gates connected by the adjacent line. We can verify this by determining the maximum voltage which appears at the input of the inverter "β." The maximum is reached when the time derivative of v_2 is zero. From Eq. (4.38):

$$\frac{dv_2}{dt} = -2.08 \times 10^7 e^{-2.449\times10^7 t} + 3.95 \times 10^7 e^{-4.653\times10^7 t} = 0 \qquad (4.39)$$

A few steps of algebra:

$$2.08 \times 10^7 e^{-2.449\times10^7 t} = 3.95 \times 10^7 e^{-4.653\times10^7 t}$$

$$\frac{2.08 \times 10^7}{3.95 \times 10^7} = \frac{e^{-4.653\times10^7 t}}{e^{-2.449\times10^7 t}}$$

$$0.5266 = e^{-2.204\times10^7 t}$$

$$t = 29.1 \times 10^{-9} = 29.1 \text{ ns}$$

$$v_2(29.1 \times 10^{-9}) = 0.197 \text{ V}$$

This maximum value is considerably less that the threshold high voltage for our gates, so the quiet line is indeed safe from spurious switching in this example.

If the integrated-circuit layout were changed so that the coupling capacitance were increased, there would be greater coupling between the lines. Figure 4.28 shows how increases in the coupling capacitance affects both the desired logic transition, $v_1(t)$, and the undesired crosstalk, $v_2(t)$. We see that larger values of coupling capacitance C_c degrade the system speed by slowing down the logic transition, and they make it possible in principle for spurious logic transitions to occur on the quiet line.

4.8 The Capacitance Matrix

Examples 4.3 through 4.5 illustrate how coupling capacitance degrades signal switching speed and can result in crosstalk between supposedly isolated lines. It is thus very important to characterize the capacitance that describes adjacent interconnects, if their coupling capacitance is significant. In a real IC there are many more than two adjacent intercon-

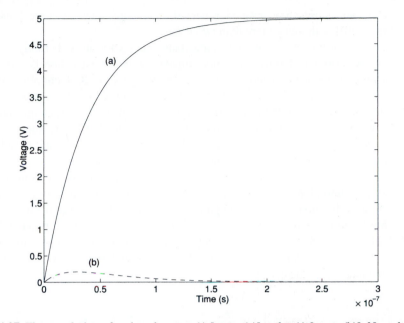

Figure 4.27 Time evolution of node voltages v_1(t) [curve (a)] and v_2(t) [curve (b)]. Note the rise in voltage on the supposedly "quiet" line due to capacitive coupling from the other line. If the coupling capacitance is too large, this crosstalk can cause an erroneous logic transition on the quiet line.

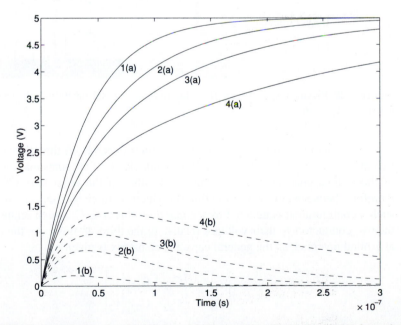

Figure 4.28 Time evolution of node voltages v_1(t) [curves (a)] and v_2(t) [curves (b)] for various values of coupling capacitance. Curves 1 through 4 correspond to C_c = 2, 10, 20, and 50 pF. Note how increasing the coupling capacitance results in both increased crosstalk and increases in the switching time.

nects, and this complexity is reflected in a capacitance matrix [analogous to A in Eq. (4.20)] with many more than two terms.

Unfortunately, a complete capacitance matrix for an IC is rarely available to circuit designers prior to fabrication. Approximate parameter extraction is often done using the two-dimensional interconnect layout geometry plus the IC technology information (i.e., interconnect thicknesses and material properties). In such cases, parallel-plate capacitance formulas plus a correction for the fringing fields near the plate edges are used to obtain approximate values for the pi-connection capacitances C_1, C_2, and C_c. As lithographic dimensions continue to scale downward, the fringing fields account for a larger proportion of the coupling capacitance, so another approach is necessary.

Figure 4.29 shows a cross section of two closely spaced interconnects above the substrate ground plane. The lines represent electric-field lines; they begin and end on charges stored in the interconnects. The parallel-plate formulae, given in Figure 4.11(b), account only for those charges along the bottom and inside edges of the two conductors in Figure 4.29. Clearly, there are many more charges, which are not accounted for, that are responsible for the significant fringing fields. An accurate determination of the capacitance matrix must take this into account.

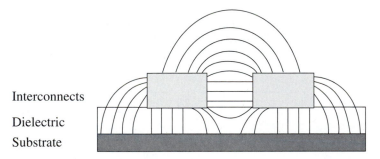

Interconnects

Dielectric

Substrate

Figure 4.29 Illustration of an electric-field configuration set up by closely spaced interconnects above the IC substrate.

The configuration of the electric-field lines depends on the relative voltages on the two interconnects and the substrate. In general, electric field lines may originate from any of these three conductors and terminate at either of the other two. Capacitance exists, therefore, between any two of the three conductors. In circuit analysis the capacitance of such a configuration is denoted by a *capacitance matrix,* **C**, which relates the charges on the two conductors to their voltages relative to the third. In this case, the third electrode is at ground potential, so the general capacitance matrix is given by

$$\begin{bmatrix} q_1 \\ q_2 \end{bmatrix} = \begin{bmatrix} C_{11} & C_{12} \\ C_{12} & C_{22} \end{bmatrix} \begin{bmatrix} v_1 \\ v_2 \end{bmatrix}$$

In standard matrix notation the lower left entry of the 2×2 matrix would be labeled C_{21}. However, we label this entry C_{12} to emphasize the symmetric nature of the capacitance matrix. (This symmetry is an example of a more general circuit property called *reciprocity*.) Note that the pi-connection model we used, even without considering fringing fields, exhibits this symmetry.

In most cases the capacitance matrix can be obtained directly from the interconnect configuration through the use of an automated electrostatic-field solver. These programs compute the capacitance more accurately by taking into consideration properties and conditions peculiar to the given interconnect configuration. Nevertheless, in cases where the capacitance matrix is unavailable, approximate results for the various adjacent interconnect time constants may be obtained from mask geometry and the material's properties or by using the capacitance formulas and fringing corrections.

An in-depth study of the pi-connection model was performed in the previous sections, and the most involved problems resulted in a second-order circuit. Unfortunately, these simple circuit models of capacitively coupled interconnects do not accurately model the true signal response. The reason lies in the fact that we lumped all the coupling capacitance into a single capacitor and treated each interconnect as a one-lump RC circuit.

In reality, the coupling capacitance is distributed, and the interconnects behave as distributed RC lines. Recall that in Chapter 3 we realized a significant increase in accuracy when we treated the interconnects with more than one RC lump. To obtain more realistic approximations of the capacitively coupled interconnect responses, we must do the same. Generalizing from a single-lump pi-connection model to an n-lump model is by no means trivial. (In the end-of-chapter problems, we explore how such an approximation could be set up, but solving the resulting equations is beyond the scope of this text.)

4.9 Summary

The first half of this chapter introduced analysis of inverter fanout. We saw that the greater the number of load gates a driving gate is connected to, the slower the maximum switching frequency. If the gates have identical equivalent resistance and capacitance, the maximum switching frequency is reduced proportionally to the number of fanout gates:

$$f_{\text{fanout}} = \frac{1}{n} \times f_{\text{pair}}$$

where f_{pair} is the maximum operating frequency of an inverter pair, and n is the number of fanout gates.

The second half of this chapter was devoted to an exploration of capacitive coupling, a phenomenon that cannot be excluded in accurate interconnect analysis. Capacitive coupling has two deleterious effects: (1) an increase in gate delay, and (2) crosstalk, the unwanted coupling of a signal from one line to another. Crosstalk is especially noxious, because it can result in functional failure of the logical computation if a gate is falsely triggered.

Capacitive coupling can be modeled using a pi-connection description of the interconnects. This model serves to introduce the concept of a capacitance matrix, which can be extended to higher-order models for more accurate analysis of interconnect coupling.

Problems

4.1 The inverters in the circuit below are all identical. What is the maximum switching speed of this configuration, if two coupled inverters could be driven at 36.6 MHz?

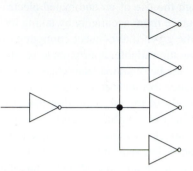

4.2 The coupled inverters in the circuit below can be switched at a maximum frequency of 48.2 MHz:

Determine the maximum switching speed of the following circuit. Assume that the NOR gates use the same kind of NMOS and PMOS transistors as the inverters.

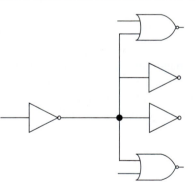

4.3 The coupled inverters in the circuit below can be switched at a maximum frequency of 38.1 MHz:

Determine the maximum switching speed of the following circuit. Assume that the NAND and NOR gates use the same kind of NMOS and PMOS transistors as the inverters.

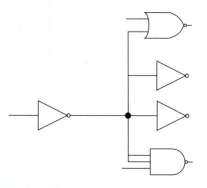

4.4   In the circuit below, the inverters have the following parameters:

$C_{in} = 12.0 \text{ pF}$ $V_L = 0.0 \text{ V}$

$R_p = 1.5 \text{ k}\Omega$ $V_{t1} = 1.0 \text{ V}$

$R_n = 1.0 \text{ k}\Omega.$ $V_{th} = 4.0 \text{ V}$

                                $V_H = 5.0 \text{ V}$

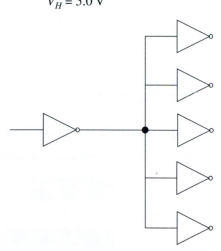

(a) Determine the low-to-high and high-to-low logic transition times.

(b) Determine the maximum switching speed.

4.5   In the circuit below, the inverters have the following parameters:

$C_{in} = 6.0 \text{ pF}$ $V_L = 0.0 \text{ V}$

$R_p = 1.5 \text{ k}\Omega$ $V_{t1} = 1.0 \text{ V}$

$R_n = 1.0 \text{ k}\Omega.$ $V_{th} = 3.7 \text{ V}$

                                $V_H = 5.0 \text{ V}$

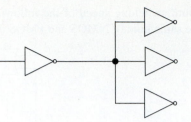

(a) Determine the low-to-high and high-to-low logic transition times.

(b) Determine the maximum switching speed.

4.6   A driving inverter is loaded by two identical inverters through an interconnect with $R_{int} = 1.0\text{ k}\Omega$, $C_{int} = 6$ pF. Determine the maximum switching frequency, modeling the interconnect with a single RC lump. (Use the inverter parameters from Exercise 4.4.)

4.7   Repeat Exercise 4.6, using a two-lump interconnect model. Is the difference in maximum switching speed significant, compared to the single-RC-lump model?

4.8   Just as we might desire a single gate to drive multiple loads, a single load gate could conceivably be driven by multiple gates:

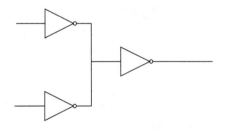

Explain why this circuit configuration might not be such a good idea.

4.9   An inverter drives two identical inverters through separate interconnects, as shown below:

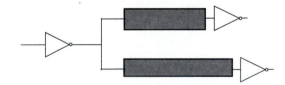

Determine the pull-down transition times of the two load inverters.

$$R_{out} = 1000\ \Omega \qquad\qquad C_{in} = 1\text{ pF}$$
$$R_{int1} = 400\ \Omega \qquad\qquad C_{int1} = 0.5\text{ pF}$$
$$R_{int2} = 200\ \Omega \qquad\qquad C_{int2} = 0.6\text{ pF}$$
$$V_S = 5\text{ V} \qquad\qquad V_{tl} = 1\text{V}$$

4.10 In designing a particular integrated circuit, there is a need to drive two inverters from a single signal. It is possible to use either of the designs (a) or (b) in the diagram below. Which design is preferable? The parameters for the inverters labeled "X" are:

$C_{in} = 2.0$ pF $R_{out} = 1.5$ kΩ

$V_L = 0.0$ V $V_{tl} = 1.0$ V

$V_H = 5.0$ V $V_{th} = 4.0$ V

The inverter labeled "Y" has the same parameters, except for a lower output resistance, which is $R_{out} = 700$ Ω. The interconnect parameters are $R_{int} = 1.0$ kΩ, $C_{int} = 2$ pF. (Assume there is no coupling between the two interconnects in either case.)

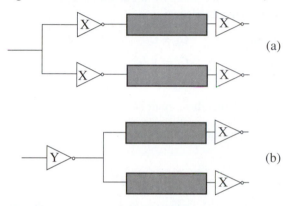

(a)

(b)

4.11 In designing another integrated circuit, there is a need to drive two inverters from a single inverter gate. It is possible to drive the inverters through their own interconnects, each with $R_{int} = 1.0$ kΩ, $C_{int} = 2$ pF [diagram (a)], or to connect them together at the end of a single interconnect which is 10% longer [design (b)]. Which design is preferable? The parameters for the inverters are:

$C_{in} = 2.0$ pF $R_{out} = 1.5$ kΩ

$V_L = 0.0$ V $V_{tl} = 1.0$ V

$V_H = 5.0$ V $V_{th} = 4.0$ V

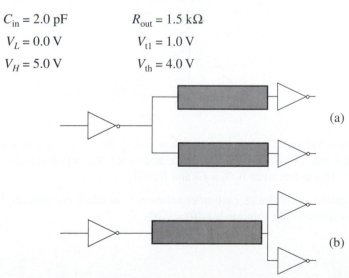

(a)

(b)

4.12 A common subcircuit on nearly all IC's is the **pad driver**. This is a cascade of an even number of inverters of various sizes which optimizes the problem of driving the large capacitance of the bonding pad, C_{pad}.

(a) Consider the circuit shown in (1). The inverter is characterized by input capacitance C and output resistance R. The pad capacitance is given by $M \times C$, where M is a large number. What is the time constant of this circuit when the inverter drives the pad capacitance directly?

(b) Now, consider the circuit shown in (2), where the inverter drives a chain of N inverters ($N = 2$ in this case) interspersed between the drive inverter and the pad. Each inverter is k times larger than the previous inverter, which means that its input capacitance increases by k, while its output resistance decreases by k. For example, if $k = 1.2$, then the three inverters in (2) have the following characteristics:

Leftmost inverter:	$C_{in} = C$	$R_{out} = R$
Center inverter:	$C_{in} = 1.2C$	$R_{out} = R/1.2$
Rightmost inverter:	$C_{in} = 1.44C$	$R_{out} = R/1.44$

Assume $M = 1000$. Show that this circuit has a shorter delay than (1).

(c) Show that the optimum value of the scaling factor k is given by

$$k = M^{\left(\frac{1}{N+1}\right)}$$

(d) A certain IC has $M = 1000$. Determine the optimum number N of inverter stages so as to minimize the delay time. (Remember that N must be an even integer.)

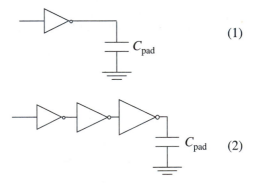

4.13 In Section 4.3.3 we showed that the circuit below can be described by a system of two first-order differential equations, which is to be expected, since the circuit contains two independent capacitors. This is true even if $C_1 = C_2$ and $R_1 = R_2$.

(a) Show that carrying out the second-order solution with this constraint on the component values results in a first-order solution if $v_1(0) = v_2(0)$.

(b) Show how the result can be obtained directly from the circuit, without using the second-order solution method.

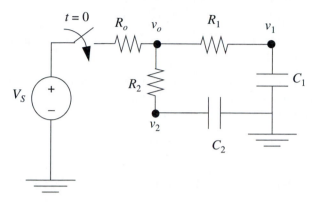

4.14 An inverter drives two gates through separate but capacitively coupled interconnects. Assume that the coupling can be treated as lumped capacitance at the far end of the interconnects:

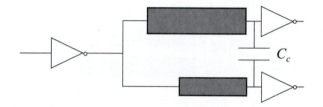

Derive the differential equations which describe this circuit.

4.15 An inverter drives two gates through separate but capacitively coupled interconnects. Assume that the coupling can be treated as a lumped capacitance at the far end of the interconnects, as shown in Exercise 4.14. Determine the pull-up response of the two inverters, given the following parameters:

$$R_{out} = 1000 \ \Omega \qquad\qquad V_S = 5 \ V$$
$$C_{in1} = 1 \ pF \qquad\qquad C_{in2} = 1.2 \ pF$$
$$R_{int1} = 200 \ \Omega \qquad\qquad C_{int1} = 0.5 \ pF$$
$$R_{int2} = 200 \ \Omega \qquad\qquad C_{int2} = 0.4 \ pF$$
$$C_c = 0.3 \ pF$$

4.16 In Exercise 4.15, determine the pull-up transition times (from 0 V to 4 V) of the two load gates with and without the interconnect coupling capacitance.

4.17 An inverter drives two identical inverter gates through separate, identical, but capacitively coupled interconnects. Assume that the coupling can be treated as a lumped capacitance at the far end of the interconnects, as shown in Exercise 4.14. Determine the pull-up response of the two inverters, assuming the system has been quiet for a long time, given the following parameters:

$$R_{out} = 1000 \ \Omega \qquad C_{in} = 1 \ pF \qquad V_S = 5 \ V$$
$$R_{int} = 200 \ \Omega \qquad C_{int} = 0.5 \ pF$$
$$C_c = 0.3 \ pF$$

4.18 In Exercise 4.17, does the response change if the coupling capacitance is larger, say 1.2 pF? Explain.

4.19 The coupled pair of inverter gates in the circuit below are connected with an interconnect parallel to another interconnect grounded at both ends. Determine the low-to-high transition time of the inverter gates. The circuit parameters are:

$$C_{in} = 10.0 \ pF \qquad R_{out} = 1.0 \ k\Omega$$
$$V_L = 0.0 \ V \qquad V_{t1} = 1.0 \ V$$
$$V_H = 5.0 \ V \qquad V_{th} = 4.0 \ V$$
$$C_{int} = 50.0 \ pF \qquad R_{int} = 50 \ \Omega$$

Coupling capacitance = 10 pF.

4.20 In Exercise 4.19, determine the high-to-low transition time.

4.21 The coupled pair of inverter gates in the circuit below are connected with an interconnect parallel to another interconnect floating at both ends. Determine the high-to-low transition time of the inverter gates. The circuit parameters are:

$$C_{in} = 10.0 \ pF \qquad R_{out} = 1.0 \ k\Omega$$
$$V_L = 0.0 \ V \qquad V_{t1} = 1.0 \ V$$
$$V_H = 5.0 \ V \qquad V_{th} = 4.0 \ V$$
$$C_{int} = 50.0 \ pF \qquad R_{int} = 50 \ \Omega$$

Coupling capacitance = 10 pF

4.22 In Exercise 4.21, determine the maximum voltage reached in the adjacent floating line.

4.23 The coupled pair of inverter gates in the circuit below are connected with an interconnect parallel to another interconnect which is grounded at one end and floating at the other. Determine the high-to-low transition time of the inverter gates. The circuit parameters are:

$$C_{in} = 10.0 \text{ pF} \qquad\qquad R_{out} = 1.0 \text{ k}\Omega$$
$$V_L = 0.0 \text{ V} \qquad\qquad\quad V_{t1} = 1.0 \text{ V}$$
$$V_H = 5.0 \text{ V} \qquad\qquad\quad V_{th} = 4.0 \text{ V}$$
$$C_{int} = 50.0 \text{ pF} \qquad\qquad R_{int} = 50 \text{ }\Omega$$
$$\text{Coupling capacitance} = 10 \text{ pF}$$

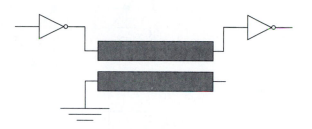

4.24 In Exercise 4.23, determine the maximum voltage reached in the adjacent line.

4.25 In Exercise 4.23, is it relevant which end of the interconnect we consider to be grounded? Explain.

4.26 The coupled pair of inverter gates in the circuit below are connected with an interconnect parallel to another active but quiet interconnect. Determine the high-to-low transition time of the inverter gates, and the maximum crosstalk voltage reached on the quiet line. (Assume the quiet line is initially at logic low.) The circuit parameters are:

$$C_{in} = 10.0 \text{ pF} \qquad\qquad R_{out} = 1.0 \text{ k}\Omega$$
$$V_L = 0.0 \text{ V} \qquad\qquad\quad V_{t1} = 1.0 \text{ V}$$
$$V_H = 5.0 \text{ V} \qquad\qquad\quad V_{th} = 4.0 \text{ V}$$
$$C_{int} = 50.0 \text{ pF} \qquad\qquad R_{int} = 50 \text{ }\Omega$$
$$\text{Coupling capacitance} = 10 \text{ pF}$$

4.27 Repeat Exercise 4.26, but assume that the adjacent active line is at logic high. How does this affect the transition time and crosstalk?

4.28 Repeat Exercise 4.26, assuming that the adjacent active line is not quiet but is undergoing a simultaneous logic transition from low to high.

4.29 A possible extension to the pi-interconnect model, which uses single-lump approximations for each interconnect, is the following two-lump model:

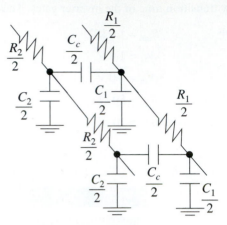

which can be redrawn as:

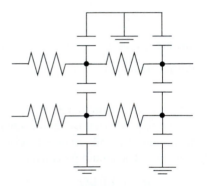

Using this interconnect model, derive the equations describing the system response, assuming a grounded adjacent line.

4.30 Repeat Exercise 4.29 for a floating adjacent line.

4.31 Repeat Exercise 4.29 for an active adjacent line.

4.32 The interconnect model suggested in Exercise 4.29 results in third-order or fourth-order differential equation systems, depending on the adjacent line termination. A more tractable approximation is to use two-lump interconnect models, but lump all of the coupling capacitance together:

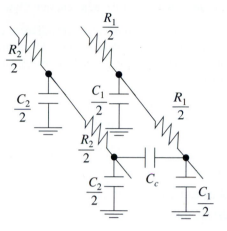

which can be redrawn as:

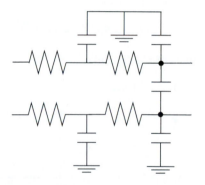

(a) Using this two-lump interconnect model, derive the equations describing the system response, assuming a grounded adjacent line.

(b) Use this model to determine the maximum switching speed for a pair of interconnected inverters, with an adjacent grounded line, with the following parameters:

$C_{in} = 10.0$ pF $\qquad\qquad R_{out} = 1.0$ kΩ

$V_L = 0.0$ V $\qquad\qquad\quad V_{t1} = 1.0$ V

$V_H = 5.0$ V $\qquad\qquad\quad V_{th} = 4.0$ V

$C_{int} = 50.0$ pF $\qquad\qquad R_{int} = 50$ Ω

Coupling capacitance $= 10$ pF

(c) Compare this result to that obtained using the pi-connection model (i.e., Exercise 4.19). Is the difference significant?

4.33 In Example 4.5 we calculated the amount of crosstalk induced in an adjacent active line. Using the same parameter set, determine the minimum value of coupling capacitance C_c which will results in false triggering of the adjacent quiet line.

4.34 A pair of coupled inverter gates has an interconnect adjacent to a very noisy interconnect line which picks up interference from an outside source. Determine the maximum value of voltage spike allowed on the noisy line such that false triggering of the inverter gates will not occur. The circuit parameters are:

$$C_{in} = 10.0 \text{ pF} \qquad R_{out} = 1.0 \text{ k}\Omega$$

$$V_L = 0.0 \text{ V} \qquad V_{tl} = 1.0 \text{ V}$$

$$V_H = 5.0 \text{ V} \qquad V_{th} = 4.0 \text{ V}$$

$$C_{int} = 50.0 \text{ pF} \qquad R_{int} = 50 \text{ }\Omega$$

$$\text{Coupling capacitance} = 10 \text{ pF}$$

4.35 The circuit below has the following responses when the switch is closed:

$$v_1(t) = -5e^{-10^9 t} - 5e^{-2\times10^8 t} + 10$$

$$v_2(t) = -5e^{-10^9 t} + 5e^{-2\times10^8 t}$$

Determine V_S and the values of C_a, C_b, and C_c.

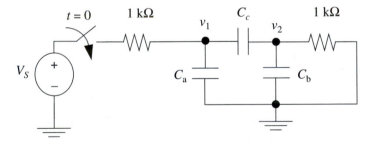

4.36 The circuit below has seven inverters wired with four interconnects, labeled α, β, χ, and δ. Interconnects α and β are capacitively coupled to each other, and both are also coupled to interconnect χ. Using one-lump interconnect models, draw an equivalent circuit which includes the effects of coupling. How many differential equations are needed to describe your equivalent circuit?

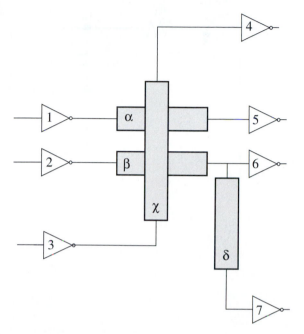

4.37 Write the differential equations in matrix form which describe this circuit.

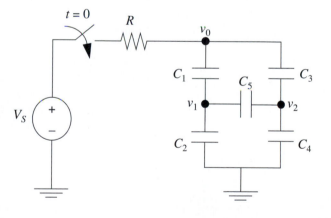

4.38 Write the differential equations in matrix form which describe this circuit.

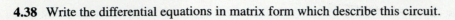
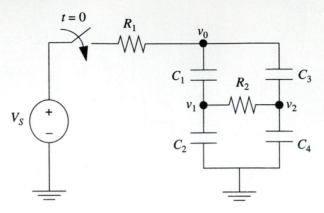

Package Inductance and RLC Circuit Analysis

5

5.1 Introduction

Our models for gates and interconnects contain capacitors and resistors, which affect the switching speed of digital systems. However, these two elements alone are not sufficient to accurately model the actual response. A third fundamental element, the **inductor**, needs to be incorporated into the models. An inductor, which exhibits the property of **inductance, is a device which stores electrical energy in the form of a magnetic field.**

Every IC interconnect has some amount of inductance, but most connections among gates can be modeled using only resistors and capacitors. This approximation is valid because, with one important exception, inductance associated with ICs is negligibly small. It is conceivable that future ICs, with larger areas and thus longer interconnects, will have significant interconnect inductance that must be included for accurate predictions of gate delay.

The one exception, which occurs in every IC, is the inductance associated with the chip package and external wiring. Significant values of inductance arise from the wires connecting the electrical contact pins to the actual integrated circuit, as illustrated in Figure 5.1. The resistance, capacitance, *and* inductance of these wires must be considered to obtain accurate models of the circuit.

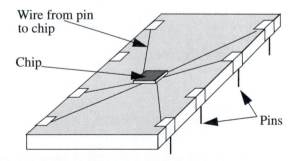

Figure 5.1 View of an IC chip inside a dual-inline package. Signals and power are supplied to the chip by wires much larger than typical IC interconnects.

The resistance of the wires from the pins can often be neglected because they are very thick, compared to the interconnects on the IC itself. For example, commonly used gold bonding wire has a diameter of about 25 µm, which is much greater than a typical IC interconnect, which might have a rectangular cross section of 2 µm × 1 µm. The capacitance to ground is also negligible, since the wire is suspended several mm above the ground plane; this dimension can be compared to typical dielectric thicknesses of less than 1 µm which separate an IC interconnect from the substrate. Since capacitance is inversely proportional to this separation, the capacitance of the bonding wire can also be neglected. However, the inductance of the bonding wire, which scales with both area and length, cannot be neglected. In particular, the inductances associated with power and ground connections are important, because every active logic gate in the IC is connected to the power supply.

In this chapter we will investigate the effects of this inductance. We will see that the combination of capacitance and inductance in a circuit can result in interesting effects, including oscillatory behavior.

5.2 Modelling the Effects of Package Inductance

Inserting the inductance of the wires connecting an inverter to the power and ground pins into our inverter model results in a new equivalent circuit, Figure 5.2.

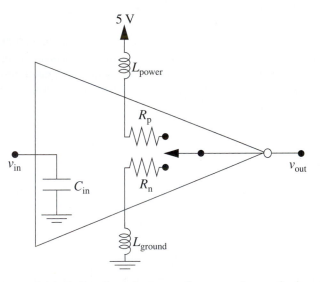

Figure 5.2 Inverter model including the inductance of power and ground wires arising from the packaged device.

Like the capacitor, the inductor has a differential current-voltage relationship. The voltage across the inductor is proportional to the derivative of the current through the inductor:

$$v = L\frac{di}{dt} \tag{5.1}$$

We will make use of the fact that inductor currents are always continuous [since, according to Eq. (5.1), a discontinuity in current results in an infinite, and hence unphysical, voltage drop)].

Another important feature of inductors is their dc steady-state behavior. We previously showed that in steady state a capacitor can be modeled as an open circuit. This is a consequence of the basic relation for capacitor current:

$$i_C = C\frac{dv_C}{dt}$$

"Steady state" means that, in the limit as $t \to \infty$, all time derivatives are zero; hence the steady-state capacitor current is zero. Similar reasoning applies to inductors. Taking the limit as $t \to \infty$ for inductor voltage:

$$\lim_{t \to \infty}[v_L(t)] = \lim_{t \to \infty}\left[L\frac{di_L}{dt}\right] = 0$$

After all transients have decayed away, inductor voltages are zero. This leads to the useful rule: **The dc steady-state equivalent model of an inductor is a short circuit**.

One way to remember the correct dc models for capacitors and inductors is to think about their construction. Capacitors are two conductors separated by an insulator, while inductors can be made by coiling up a length of wire. Clearly, the insulating layer in a capacitor acts more like an open circuit, while a length of wire acts more like a short circuit.

We will determine the response of interconnected gates by solving circuits containing cascaded inverter models like that in Figure 5.2. The resulting circuits contain a resistor, capacitor, and inductor in series and are thus known as series *RLC* circuits. Before we study these circuits, which are second order, we will examine first-order *RL* circuits.

5.3 First-Order *RL* Circuits

Circuits containing only sources, resistors and inductors (and perhaps switches), but no capacitors, are called **RL circuits**. First-order *RL* circuits contain only a single independent inductance (i.e., a single inductor, or multiple inductors that can be reduced to a single equivalent inductance). Methods used for analysis of *RL* circuits are quite similar to those methods used for *RC* circuits.

Consider the *RL* circuit of Figure 5.3. At $t = t_0$, the switch moves downward. Applying KVL for $t > t_0$:

$$-V_S + i_L R + v_L = 0 \tag{5.2}$$

It is convenient in this case to work with inductor current rather than voltage. Substituting the current-voltage relation for the inductor into Eq. (5.2):

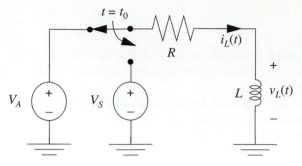

Figure 5.3 A first-order *RL* circuit. At $t = t_0$, the switch moves downward connecting the series *RL* connection to the source V_S.

$$-V_S + i_L R + L\frac{di_L}{dt} = 0 \tag{5.3}$$

Rearranging Eq. (5.3), we obtain:

$$\frac{di_L}{dt} = -\frac{R}{L}i_L + \frac{1}{L}V_S \tag{5.4}$$

Recall that the differential equation describing a series *RC* circuit is given by:

$$\frac{dv_C}{dt} = -\frac{1}{RC}v_C + \frac{1}{RC}V_S \tag{5.5}$$

Note that these two differential equations, characterizing inductor current and capacitor voltage, respectively, have the same form. Therefore, we can apply the general solution, which we developed for the *RC* circuit, to the *RL* circuit, with an appropriate change of variables. The general solution to Eq. (5.5) is:

$$v_C(t) = (v_C(t_0) - V_S)e^{-\frac{t-t_0}{RC}} + V_S$$

Consequently, the solution which satisfies Eq. (5.4) is given by:

$$i_L(t) = \left(i_L(t_0) - \frac{V_S}{R}\right)e^{-\frac{R}{L}(t-t_0)} + \frac{V_S}{R} \tag{5.6}$$

To verify that Eq. (5.6) is the correct solution for the inductor current, we must perform three steps:

- Check the solution for $t = t_0$;

- Check the solution for $t \rightarrow \infty$;

- Check the solution by substituting it back into the original differential equation Eq. (5.4) and determining whether it satisfies the equality.

At $t = t_0$:

$$i_L(t_0) = \left(i_L(t_0) - \frac{V_S}{R}\right)e^{-\frac{R}{L}(t_0 - t_0)} + \frac{V_S}{R}$$

$$= i_L(t_0) - \frac{V_S}{R} + \frac{V_S}{R}$$

$$= i_L(t_0)$$

As $t \to \infty$;

$$i_L(\infty) = \left(i_L(t_0) - \frac{V_S}{R}\right)e^{-\frac{R}{L}(\infty - t_0)} + \frac{V_S}{R}$$

$$= \frac{V_S}{R}$$

This is correct since, in steady state, the inductor is a short circuit. Referring to Figure 5.3, it is easily seen that the steady-state inductor current is given by V_S/R.

Finally, the derivative of i_L is given by:

$$\frac{di_L}{dt} = \left(i_L(t_0) - \frac{V_S}{R}\right)\left(-\frac{R}{L}\right)e^{-\frac{R}{L}(t - t_0)} \tag{5.7}$$

Substitution of Eqs. (5.7) and (5.6) into Eq. (5.4) results in:

$$\left(i_L(t_0) - \frac{V_S}{R}\right)\left(-\frac{R}{L}\right)e^{-\frac{R}{L}(t - t_0)} = -\frac{R}{L}\left[\left(i_L(t_0) - \frac{V_S}{R}\right)e^{-\frac{R}{L}(t - t_0)} + \frac{V_S}{R}\right] + \frac{1}{L}V_S$$

Multiplying through by $-L/R$ yields:

$$\left(i_L(t_0) - \frac{V_S}{R}\right)e^{-\frac{R}{L}(t - t_0)} = \left[\left(i_L(t_0) - \frac{V_S}{R}\right)e^{-\frac{R}{L}(t - t_0)} + \frac{V_S}{R}\right] - \frac{V_S}{R}$$

$$\left(i_L(t_0) - \frac{V_S}{R}\right)e^{-\frac{R}{L}(t - t_0)} = \left(i_L(t_0) - \frac{V_S}{R}\right)e^{-\frac{R}{L}(t - t_0)}$$

which is an identity. Therefore, Eq. (5.6) is the general solution for the inductor current in the series RL circuit shown of Figure 5.3.

An equivalent way of writing the general solution of a first-order RL circuit is

$$i_L(t) = [i_L(t_0) - i_L(\infty)]e^{-\frac{(t - t_0)}{\tau}} + i_L(\infty) \tag{5.8}$$

where $i_L(\infty)$ is the dc steady-state inductor current and the time constant τ is given by

$$\tau = \frac{L}{R} \tag{5.9}$$

We will now use these results to solve a numerical example.

Example 5.1

In the circuit of Figure 5.4, the switch has been in the upper position for a long time. At $t = 0$, it moves down instantaneously. Determine $i_L(t)$.

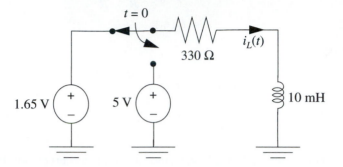

Figure 5.4 Circuit for Example 5.1. The switch in this circuit has been in the upper position for a long time.

We first determine the initial conditions. Since the problem statement specifies that the switch has been idle for a long time preceding $t = 0$, any previous transients have decayed. The circuit is in the dc steady-state condition, and the inductor acts as a short circuit. Therefore, the inductor current for $t = 0^-$ (i.e., the instant before the switch moves) is given by

$$i_L(0^-) = \frac{1.65}{330}$$

$$= 5 \times 10^{-3} A = 5 \text{ mA}$$

Since inductor currents are continuous, $i_L(0^+)$ is also 5 mA.

The dc steady-state value as $t \rightarrow \infty$ is calculated in an analogous fashion:

$$i_L(\infty) = \frac{5.0}{330}$$

$$= 15.2 \times 10^{-3} A = 15.2 \text{ mA}$$

Finally, the time constant is found from:

$$\tau = \frac{L}{R} = \frac{10 \times 10^{-3}}{330}$$

$$= 30.3 \times 10^{-6} \text{ s} = 30.3 \text{ μs}$$

Inserting these values in Eq. (5.8),

$$i_L(t) = (5.0 - 15.2)e^{-\frac{t}{30.3}} + 15.2$$

$$(5.10)$$

$$= -10.2e^{-\frac{t}{30.3}} + 15.2$$

where the inductor current is in mA, and t is in μs. This solution is plotted in Figure 5.5. Note that the waveform, and indeed the entire problem and solution method, is essentially the same as the first-order RC circuit problem.[1]

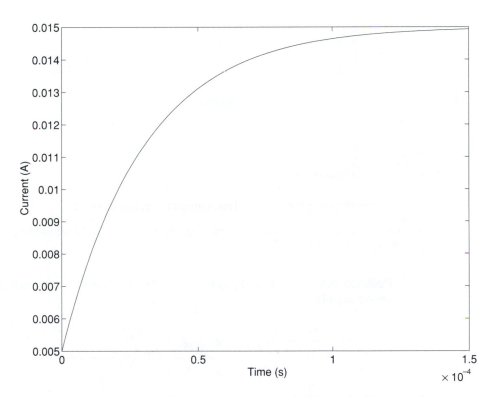

Figure 5.5 Inductor current as a function of time in the series RL circuit of Example 5.1.

The next section introduces the solution of circuits containing a capacitor in series with an inductor and resistor.

---

[1] Two elements which have identical characteristic relations but have their current and voltage variables interchanged are said to be **duals**. Capacitance and inductance are duals, as can be seen from their characteristic equations:

$$i_C = C\frac{dv_C}{dt}, \qquad v_L = L\frac{di_L}{dt}$$

5.4 *RLC* Circuit Model of Coupled Inverter Gates

Figure 5.6 illustrates two cascaded inverter gates, using the gate model of Figure 5.2 including the inductance of the power-supply wires. Using a single-*RC*-lump model for the interconnect, we arrive at the circuit model shown in Figure 5.7. This model is identical to the *RC* model of cascaded gates developed previously, but for the addition of an inductance in series with the switch.

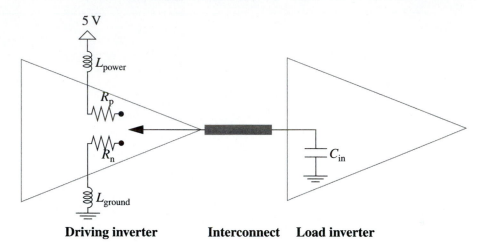

Figure 5.6 Cascaded inverter gates with intervening interconnect, showing the salient parts of the inverter model of Figure 5.2.

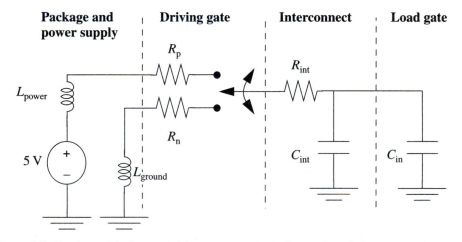

Figure 5.7 Circuit model of cascaded inverter gates, including package inductance, and a single-*RC*-lump interconnect model.

This circuit can be reduced even further to the **series *RLC* circuit** shown in Figure 5.8, with the understanding that:

$$C = C_{in} + C_{int}$$

$$R = R_p + R_{int} \qquad\qquad R = R_p + R_{int}$$

$$L = L_{power} \qquad\qquad L = L_{ground}$$

$$\underbrace{V_S = 5\text{ V}}_{\text{Pull-up}} \qquad\qquad \underbrace{V_S = 0\text{ V}}_{\text{Pull-down}}$$

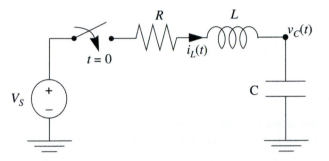

Figure 5.8 Series *RLC* circuit model of cascaded inverter gates.

Like the *RC* circuits discussed in Chapters 2 through 4, equations describing the voltages and currents in *RLC* circuits contain two parts: a transient component which decays to zero, and a time-independent component, the dc steady-state response reached as $t \to \infty$. Symbolically:

$$v_C(t) = v_{Ct}(t) + V_C$$

$$i_L(t) = i_{Lt}(t) + I_L \tag{5.11}$$

To derive the general solution for the *RLC* circuit, we will find the transient and steady-state responses separately. First we discuss the dc steady state.

5.5 dc Steady-State Response of *RLC* Circuits

The dc steady-state circuit response consists of the voltages and currents reached as $t \to \infty$, after the transient portions have decayed away. The general procedure for finding the dc steady-state response of a circuit containing energy storage elements (i.e., capacitors and inductors) is to replace each element by its appropriate steady-state model. Once this is done, the circuit will contain only sources and resistors; conventional dc circuit-analysis techniques will yield the steady-state response. This method is illustrated in the following example.

Example 5.2

Determine the steady-state response of the series *RLC* circuit shown in Figure 5.8.

Replacing the inductor and capacitor with their dc steady state equivalents results in the circuit shown in Figure 5.9.

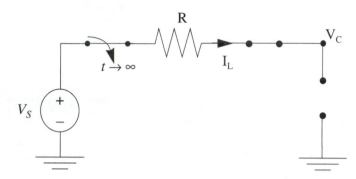

Figure 5.9 Behavior of the series *RLC* circuit as $t \rightarrow \infty$. The inductor and capacitor have been replaced by their dc equivalent models, short and open circuits, respectively.

We observe that I_L is zero, as there is no closed path for current flow. Since there is no current through the resistor, the voltage drop across it is zero, which means that the voltage across the capacitor is equal to V_S. Summarizing, the dc steady-state values for the series *RLC* circuit are:

$$I_L = 0$$
$$V_C = V_S$$

(5.12)

5.6 Series *RLC* Circuit Differential Equations

We solve for the total response of the series *RLC* circuit, Figure 5.10, using the same techniques employed previously. Kirchhoff's laws and appropriate branch relations result in ordinary differential equations governing the voltages and currents. For $t > 0$, application of KCL at the $v_C(t)$ node results in:

$$C \frac{dv_C}{dt} = i_L$$

(5.13)

A second differential equation is found by applying KVL around the series loop. Starting at the bottom lefthand corner and proceeding clockwise:

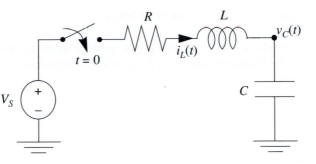

Figure 5.10 Series *RLC* circuit.

$$-V_S + Ri_L + L\frac{di_L}{dt} + v_C = 0 \tag{5.14}$$

Next, we rearrange Eqs. (5.13) and (5.14) so that the derivative terms appear on the left:

$$\frac{dv_C}{dt} = \frac{1}{C}i_L$$

$$\frac{di_L}{dt} = -\frac{1}{L}v_C - \frac{R}{L}i_L + \frac{V_S}{L}$$

Expressed in matrix form:

$$\begin{bmatrix} \dfrac{dv_C}{dt} \\ \dfrac{di_L}{dt} \end{bmatrix} = \begin{bmatrix} 0 & \dfrac{1}{C} \\ -\dfrac{1}{L} & -\dfrac{R}{L} \end{bmatrix} \begin{bmatrix} v_C \\ i_L \end{bmatrix} + \begin{bmatrix} 0 \\ \dfrac{V_S}{L} \end{bmatrix} \tag{5.15}$$

Like the two-section *RC* ladder circuit, the *RLC* circuit is characterized by two coupled first-order differential equations. The reason is that there are two independent energy-storage elements in this circuit, one inductor and one capacitor. In general, if a circuit contains a total of *n* independent inductors and capacitors, it is characterized by at most *n* coupled first-order differential equations.

Our goal is to determine functions $v_C(t)$ and $i_L(t)$ which satisfy Eq. (5.15) [and also satisfy initial conditions $v_C(0)$ and $i_L(0)$, which we assume are known]. The procedure for finding these functions is essentially the same we used for the second-order ladder circuit: we assume transient solutions of the form e^{st}, solve for the natural frequencies *s*, and determine the coefficients of each of the transient terms.

5.7 Natural Frequencies of the Series *RLC* Circuit

The first step is to assume a form for the general solutions. We do this by inserting into Eq. (5.11) the exponential form for the transient solutions and the dc steady-state solutions [Eq. (5.12)]:

$$v_C(t) = v_{Ct}(t) + V_C = Ve^{st} + V_S$$

$$i_L(t) = i_{Lt}(t) + I_L = Ie^{st}$$

$$(5.16)$$

(As before, this approach is correct, but there are actually two exponential terms corresponding to two natural frequencies of the circuit.) Next, we insert these solutions into the differential equation matrix, Eq. (5.15):

$$\begin{bmatrix} \dfrac{d}{dt}(Ve^{st} + V_S) \\[2mm] \dfrac{d}{dt}(Ie^{st}) \end{bmatrix} = \begin{bmatrix} 0 & \dfrac{1}{C} \\[2mm] -\dfrac{1}{L} & -\dfrac{R}{L} \end{bmatrix} \begin{bmatrix} Ve^{st} + V_S \\[2mm] Ie^{st} \end{bmatrix} + \begin{bmatrix} 0 \\[2mm] \dfrac{V_S}{L} \end{bmatrix}$$

Evaluating:

$$\begin{bmatrix} sVe^{st} \\[2mm] sIe^{st} \end{bmatrix} = \begin{bmatrix} 0 & \dfrac{1}{C} \\[2mm] -\dfrac{1}{L} & -\dfrac{R}{L} \end{bmatrix} \begin{bmatrix} Ve^{st} \\[2mm] Ie^{st} \end{bmatrix} + \begin{bmatrix} 0 & \dfrac{1}{C} \\[2mm] -\dfrac{1}{L} & -\dfrac{R}{L} \end{bmatrix} \begin{bmatrix} V_S \\[2mm] 0 \end{bmatrix} + \begin{bmatrix} 0 \\[2mm] \dfrac{V_S}{L} \end{bmatrix}$$

$$e^{st}\begin{bmatrix} sV \\[2mm] sI \end{bmatrix} = e^{st}\begin{bmatrix} 0 & \dfrac{1}{C} \\[2mm] -\dfrac{1}{L} & -\dfrac{R}{L} \end{bmatrix} \begin{bmatrix} V \\[2mm] I \end{bmatrix} + \begin{bmatrix} 0 \\[2mm] \dfrac{V_S}{L} \end{bmatrix} + \begin{bmatrix} 0 \\[2mm] -\dfrac{V_S}{L} \end{bmatrix}$$

$$s\begin{bmatrix} V \\[2mm] I \end{bmatrix} = \begin{bmatrix} 0 & \dfrac{1}{C} \\[2mm] -\dfrac{1}{L} & -\dfrac{R}{L} \end{bmatrix} \begin{bmatrix} V \\[2mm] I \end{bmatrix}$$

$$\begin{bmatrix} s & 0 \\[2mm] 0 & s \end{bmatrix}\begin{bmatrix} V \\[2mm] I \end{bmatrix} - \begin{bmatrix} 0 & \dfrac{1}{C} \\[2mm] -\dfrac{1}{L} & -\dfrac{R}{L} \end{bmatrix} \begin{bmatrix} V \\[2mm] I \end{bmatrix} = \begin{bmatrix} 0 \\[2mm] 0 \end{bmatrix}$$

$$\begin{bmatrix} s & -\dfrac{1}{C} \\ \dfrac{1}{L} & s+\dfrac{R}{L} \end{bmatrix} \begin{bmatrix} V \\ I \end{bmatrix} = \begin{bmatrix} 0 \\ 0 \end{bmatrix} \tag{5.17}$$

We recognize Eq. (5.17) as the familiar eigenvalue equation. Nontrivial solutions for Eq. (5.17) exist only for those values of s_1 which satisfy

$$\det \begin{bmatrix} s & -\dfrac{1}{C} \\ \dfrac{1}{L} & s+\dfrac{R}{L} \end{bmatrix} = 0$$

which, when expanded, yields

$$s\left(s+\frac{R}{L}\right) - \left(-\frac{1}{C}\right)\frac{1}{L} = 0$$

$$s^2 + \frac{R}{L}s + \frac{1}{LC} = 0 \tag{5.18}$$

Equation (5.18) is known as the **characteristic equation** of the series RLC circuit; its roots are the natural frequencies which govern the circuit response. Application of the quadratic formula results in

$$s_1 = -\frac{R}{2L} + \sqrt{\left(\frac{R}{2L}\right)^2 - \frac{1}{LC}} \tag{5.19}$$

$$s_2 = -\frac{R}{2L} - \sqrt{\left(\frac{R}{2L}\right)^2 - \frac{1}{LC}} \tag{5.20}$$

where s_1 and s_2 are the two natural frequencies of the series RLC circuit.

Until this point, the solution of the RLC circuit has been quite similar to that of the two-section RC ladder circuit. In both cases the circuits are described by two coupled first-order differential equations; the response consists of a dc steady-state term plus two transient terms of the form $e^{s_1 t}$ and $e^{s_2 t}$. However, in the case of the ladder circuit, the natural frequencies s_1 and s_2 were always distinct, real, negative quantities. This is not necessarily true for the RLC circuit. The responses of RLC circuits are classified according to the nature of the roots of Eq. (5.18), which can be of four types:

- **Overdamped**: distinct, real and negative roots. Occurs when $\left(\dfrac{R}{2L}\right)^2 > \dfrac{1}{LC}$.

- **Critically damped**: real, negative, repeated roots (i.e., $s_1 = s_2$). Occurs when

$$\left(\frac{R}{2L}\right)^2 = \frac{1}{LC}.$$

- **Underdamped**: complex conjugate roots ($s_1 = s_2^*$). Occurs when $\left(\frac{R}{2L}\right)^2 < \frac{1}{LC}$.

- **Undamped**: imaginary conjugate roots. Occurs as the limiting case of $R \to 0$.

The origin of the terms overdamped, underdamped, etc., will become apparent as we examine each case in detail.

5.8 Series *RLC* Circuit Responses

5.8.1 The Overdamped Response

The overdamped case occurs when the discriminants in Eqs. (5.19) and (5.20) are positive. This will occur for component values such that

$$\left(\frac{R}{2L}\right)^2 > \frac{1}{LC}$$

In this case the two natural frequencies of the series *RLC* circuit, s_1 and s_2, are distinct, negative real numbers. Because there are two natural frequencies, the general solution contains two transient exponential terms:

$$v_C(t) = V_1 e^{s_1 t} + V_2 e^{s_2 t} + V_C \tag{5.21}$$

$$i_L(t) = I_1 e^{s_1 t} + I_2 e^{s_2 t} + I_L \tag{5.22}$$

First we will verify that these expressions satisfy the original differential equations. Inserting the assumed solutions into Eq. (5.15):

$$\begin{bmatrix} V_1 s_1 e^{s_1 t} + V_2 s_2 e^{s_2 t} \\ I_1 s_1 e^{s_1 t} + I_2 s_2 e^{s_2 t} \end{bmatrix} = \begin{bmatrix} 0 & \frac{1}{C} \\ -\frac{1}{L} & -\frac{R}{L} \end{bmatrix} \begin{bmatrix} V_1 e^{s_1 t} + V_2 e^{s_2 t} + V_C \\ I_1 e^{s_1 t} + I_2 e^{s_2 t} + I_L \end{bmatrix} + \begin{bmatrix} 0 \\ \frac{V_s}{L} \end{bmatrix}$$

Expanding:

$$\begin{bmatrix} V_1 s_1 e^{s_1 t} + V_2 s_2 e^{s_2 t} \\ I_1 s_1 e^{s_1 t} + I_2 s_2 e^{s_2 t} \end{bmatrix} = \begin{bmatrix} \frac{I_1}{C} e^{s_1 t} + \frac{I_2}{C} e^{s_2 t} + \frac{I_L}{C} \\ \left(-\frac{V_1}{L} - \frac{R}{L} I_1\right) e^{s_1 t} + \left(-\frac{V_2}{L} - \frac{R}{L} I_2\right) e^{s_2 t} - \frac{V_C}{L} - \frac{R}{L} I_L + \frac{V_s}{L} \end{bmatrix}$$

Inserting the known values for the steady-state capacitor voltage (V_S) and inductor current (zero), this simplifies to:

$$\begin{bmatrix} V_1 s_1 e^{s_1 t} + V_2 s_2 e^{s_2 t} \\ I_1 s_1 e^{s_1 t} + I_2 s_2 e^{s_2 t} \end{bmatrix} = \begin{bmatrix} \dfrac{I_1}{C} e^{s_1 t} + \dfrac{I_2}{C} e^{s_2 t} \\ \left(-\dfrac{V_1}{L} - \dfrac{R}{L} I_1 \right) e^{s_1 t} + \left(-\dfrac{V_2}{L} - \dfrac{R}{L} I_2 \right) e^{s_2 t} \end{bmatrix} \tag{5.23}$$

Since the solution must be valid for all values of time t, the coefficients of each exponential term must match. For the $e^{s_1 t}$ terms, we have:

$$V_1 s_1 = \frac{1}{C} I_1$$

$$I_1 s_1 = -\frac{1}{L} V_1 - \frac{R}{L} I_1$$

Rearranging:

$$I_1 = V_1 s_1 C$$

$$I_1 \left(s_1 + \frac{R}{L} \right) + \frac{1}{L} V_1 = 0$$

Combining these two by substituting the top expression for I_1 into the second equation yields:

$$V_1 s_1 C \left(s_1 + \frac{R}{L} \right) + \frac{1}{L} V_1 = 0$$

which, after rearranging, becomes

$$V_1 \left(s_1^2 + \frac{R}{L} s_1 + \frac{1}{LC} \right) = 0$$

The term in parentheses is equal to zero, by Eq. (5.18); therefore, this expression is an equality.

Turning to the $e^{s_2 t}$ terms in Eq. (5.23), equating coefficients results in:

$$V_2 s_2 = \frac{1}{C} I_2$$

$$I_2 s_2 = -\frac{1}{L} V_2 - \frac{R}{L} I_2$$

Solving the top equation for I_2 and substituting the result in the bottom equation gives:

$$V_2 \left(s_2^2 + \frac{R}{L} s_2 + \frac{1}{LC} \right) = 0$$

which again is an equality, since the term in parentheses is zero. Since substitution of the assumed solutions into the original differential equations results in equalities, the solutions are valid.

We now consider a numerical example of an overdamped series *RLC* circuit.

Example 5.3

Determine the response of the *RLC* circuit shown in Figure 5.11.

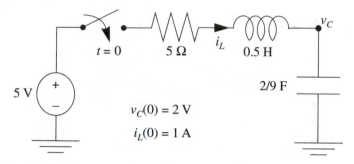

Figure 5.11 Overdamped series *RLC* circuit for Example 5.3.

The component values in this and subsequent examples are unrealistic in that you are highly unlikely to encounter them in a real circuit; they have been chosen to illustrate the solution method with simple arithmetic results. After we have examined the *RLC* circuit solutions in detail, we will look at examples with more realistic parameters.

The dc steady-state solutions are apparent from inspection of the circuit:

$$I_L = 0 \text{ A}$$
$$V_C = 5 \text{ V}$$

The two natural frequencies are found from Eq. (5.19) and Eq. (5.20):

$$s_1 = -\frac{R}{2L} + \sqrt{\left(\frac{R}{2L}\right)^2 - \frac{1}{LC}}$$

$$= -\frac{5}{2 \times 0.5} + \sqrt{\left(\frac{5}{2 \times 0.5}\right)^2 - \frac{1}{0.5 \times {}^2\!/_9}}$$

$$= -1 \text{ s}^{-1}$$

$$s_2 = -\frac{R}{2L} - \sqrt{\left(\frac{R}{2L}\right)^2 - \frac{1}{LC}}$$

$$= -\frac{5}{2 \times 0.5} - \sqrt{\left(\frac{5}{2 \times 0.5}\right)^2 - \frac{1}{0.5 \times \frac{2}{9}}}$$

$$= -9 \text{ s}^{-1}$$

Since the values of the natural frequencies in this circuit are distinct real negative numbers, we know that the response is overdamped and is given by the general solutions of Eqs. (5.21) and (5.22). Substituting the natural frequency and steady-state values into the general solution:

$$v_C(t) = V_1 e^{-t} + V_2 e^{-9t} + 5 \tag{5.24}$$

$$i_L(t) = I_1 e^{-t} + I_2 e^{-9t} \tag{5.25}$$

To complete the solution, we use the same techniques developed in Chapter 3 to solve for the values of the four unknown coefficients V_1, V_2, I_1, and I_2. We begin by inserting the known value of capacitor voltage at $t = 0$:

$$v_C(0) = 2 = V_1 e^{-1(0)} + V_2 e^{-9(0)} + 5$$

which simplifies to:

$$V_1 + V_2 = -3 \tag{5.26}$$

Similarly for the initial inductor current:

$$i_L(t) = 1 = I_1 e^{-1(0)} + I_2 e^{-9(0)}$$

which can be written more simply as

$$I_1 + I_2 = 1 \tag{5.27}$$

We now have two equations for the four constants V_1, V_2, I_1, and I_2 in Eqs. (5.24) and (5.25). To obtain the other two needed equations, we substitute the assumed solutions into either of the original differential equations and equate the coefficients of like exponential terms. Either of the two original differential equations can be used; we will use Eq. (5.13), since it is simpler.

Substituting Eqs. (5.24) and (5.25) into Eq. (5.13):

$$C \frac{dv_C}{dt} = i_L$$

$$\frac{2}{9} \frac{d}{dt}(V_1 e^{-t} + V_2 e^{-9t} + 5) = I_1 e^{-t} + I_2 e^{-9t}$$

$$-\frac{2}{9} V_1 e^{-t} - \frac{2}{9} 9 V_2 e^{-9t} = I_1 e^{-t} + I_2 e^{-9t}$$

Equating the coefficients of the e^{-t} terms:

$$-V_1 = \frac{9}{2} I_1 \qquad\qquad (5.28)$$

and the e^{-9t} terms:

$$-2V_2 = I_2 \qquad\qquad (5.29)$$

Equations (5.26) through (5.29) constitute a set of four algebraic equations which completely determine the four constants in the circuit solution. These can be solved using any standard algebraic technique. We proceed by substituting Eqs. (5.26) and (5.27) into Eq. (5.29):

$$-2(-3 - V_1) = 1 - I_1 \qquad\qquad (5.30)$$

Substituting the expression for V_1 [given by Eq. (5.28)] into Eq. (5.30) and solving for I_1:

$$-2\left(-3 + \frac{9}{2} I_1\right) = 1 - I_1$$

$$6 - 9 I_1 = 1 - I_1$$

$$5 = 8 I_1$$

$$I_1 = \frac{5}{8}$$

From Eq. (5.27),

$$I_1 + I_2 = 1$$

Therefore:

$$I_2 = 1 - I_1 = \frac{3}{8}$$

Similarly, from Eqs. (5.28) and (5.29) we have

$$V_1 = -\frac{9}{2} I_1 = -\frac{9}{2} \times \frac{5}{8} = -\frac{45}{16}$$

and

$$V_2 = -\frac{1}{2}I_2 = -\frac{1}{2} \times \frac{3}{8} = -\frac{3}{16}$$

Finally, substituting the values of the four constants into the solutions gives the complete expressions for the capacitor voltage and inductor current as a function of time:

$$v_C(t) = -\frac{45}{16}e^{-t} - \frac{3}{16}e^{-9t} + 5 \qquad (5.31)$$

$$i_L(t) = \frac{5}{8}e^{-t} + \frac{3}{8}e^{-9t} \qquad (5.32)$$

Once the final solutions are obtained, it is always a good idea to verify them by checking them against the original differential equations. Since we used Eq. (5.13) to derive two of the expressions for the constants, we will use the other differential equation, Eq. (5.14), to check the results:

$$-V_S + Ri_L + L\frac{di_L}{dt} + v_C = 0$$

$$-5 + 5\left(\frac{5}{8}e^{-t} + \frac{3}{8}e^{-9t}\right) + \frac{1}{2}\frac{d}{dt}\left(\frac{5}{8}e^{-t} + \frac{3}{8}e^{-9t}\right) + \left(-\frac{45}{16}e^{-t} - \frac{3}{16}e^{-9t} + 5\right) = 0$$

Evaluating:

$$-5 + 5 + e^{-t}\left[\frac{25}{8} + \frac{1}{2}\left(-\frac{5}{8}\right) - \frac{45}{16}\right] + e^{-9t}\left[\frac{15}{8} + \frac{1}{2}\left(-\frac{27}{8}\right) - \frac{3}{16}\right] = 0$$

$$e^{-t}\left[\frac{50 - 5 - 45}{16}\right] + e^{-9t}\left[\frac{30 - 27 - 3}{16}\right] = 0$$

which is an equality. Therefore, the solutions in Eqs. (5.31) and (5.32) are seen to satisfy both differential equations and are likely to be correct.

The capacitor voltage and inductor current are plotted as a function of time in Figures 5.12 and 5.13.

5.8.2 The Critically Damped Response

The critically damped case occurs when the value of the discriminants in Eqs. (5.19) and (5.20) are zero. This will happen for component values such that

$$\left(\frac{R}{2L}\right)^2 = \frac{1}{LC}$$

Although it may seem unlikely that any particular circuit would have this exact combination of resistance, inductance, and capacitance, critical damping is in fact quite frequently

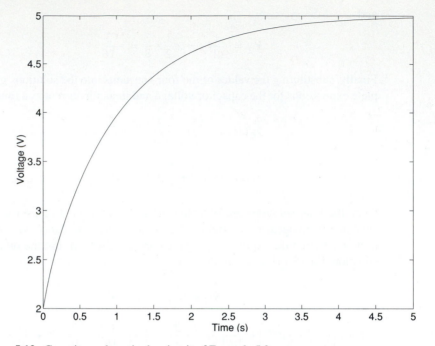

Figure 5.12   Capacitor voltage in the circuit of Example 5.3.

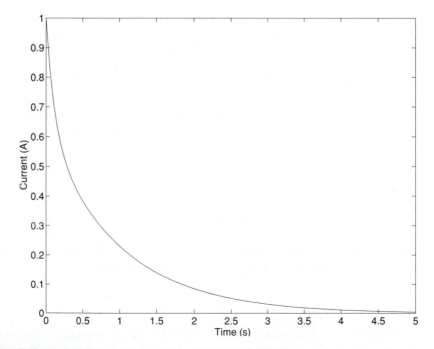

Figure 5.13   Inductor current in the circuit of Example 5.3.

designed into circuits. This is because critically damped responses represent the dividing line between behaviors that are overdamped (with slow transient responses) and under-damped (with faster but oscillatory transient behavior).

The natural frequencies of the *RLC* circuit are, in general,

$$s = -\frac{R}{2L} \pm \sqrt{\left(\frac{R}{2L}\right)^2 - \frac{1}{LC}}$$

When the term under the radical is zero, the roots are thus

$$s = -\frac{R}{2L}$$

which we write as

$$s_1 = s_2 = -\alpha$$

For the critically damped case, the two natural frequencies are real, negative, and equal. The parameter α, which in the case of a series *RLC* circuit is equal to

$$\frac{R}{2L},$$

is known as the **damping factor**. The origin of this nomenclature will become apparent when we examine the underdamped case in the next section.

Because the two natural frequencies are equal, the general solutions are slightly different than for the overdamped case. The solutions to the differential equations are given by:

$$v_C(t) = V_1 e^{-\alpha t} + V_2 t e^{-\alpha t} + V_C \tag{5.33}$$

$$i_L(t) = I_1 e^{-\alpha t} + I_2 t e^{-\alpha t} + I_L \tag{5.34}$$

These differ from the form of the overdamped solutions in the second term, where there is a *t* term multiplying the exponential term. (A method of deriving these solutions is described in Exercise 5.24.)

First we will verify these expressions by inserting them into the original coupled differential equations:

$$\begin{bmatrix} \frac{d}{dt}(V_1 e^{-\alpha t} + V_2 t e^{-\alpha t} + V_C) \\ \frac{d}{dt}(I_1 e^{-\alpha t} + I_2 t e^{-\alpha t} + I_L) \end{bmatrix} = \begin{bmatrix} 0 & \frac{1}{C} \\ -\frac{1}{L} & -\frac{R}{L} \end{bmatrix} \begin{bmatrix} V_1 e^{-\alpha t} + V_2 t e^{-\alpha t} + V_C \\ I_1 e^{-\alpha t} + I_2 t e^{-\alpha t} + I_L \end{bmatrix} + \begin{bmatrix} 0 \\ \frac{V_S}{L} \end{bmatrix}$$

Expanding this expression, and using the fact that the dc steady-state values are $V_C = V_S$ and $I_L = 0$,

$$
\begin{bmatrix} -\alpha V_1 e^{-\alpha t} - \alpha V_2 t e^{-\alpha t} + V_2 e^{-\alpha t} \\ -\alpha I_1 e^{-\alpha t} - \alpha I_2 t e^{-\alpha t} + I_2 e^{-\alpha t} \end{bmatrix} = \begin{bmatrix} 0 & \dfrac{1}{C} \\ -\dfrac{1}{L} & -\dfrac{R}{L} \end{bmatrix} \begin{bmatrix} V_1 e^{-\alpha t} + V_2 t e^{-\alpha t} + V_S \\ I_1 e^{-\alpha t} + I_2 t e^{-\alpha t} \end{bmatrix} + \begin{bmatrix} 0 \\ \dfrac{V_S}{L} \end{bmatrix}
$$

$$
= \begin{bmatrix} \dfrac{I_1}{C} e^{-\alpha t} + \dfrac{I_2}{C} t e^{-\alpha t} \\ \left(-\dfrac{V_1}{L} - \dfrac{R}{L} I_1 \right) e^{-\alpha t} + \left(-\dfrac{V_2}{L} - \dfrac{R}{L} I_2 \right) t e^{-\alpha t} + \left(-\dfrac{1}{L} + \dfrac{1}{L} \right) V_S \end{bmatrix}
$$

Factoring out the common $e^{-\alpha t}$ term,

$$
\begin{bmatrix} -\alpha V_1 + V_2 - \alpha V_2 t \\ -\alpha I_1 + I_2 - \alpha I_2 t \end{bmatrix} = \begin{bmatrix} \dfrac{I_1}{C} + \dfrac{I_2}{C} t \\ \left(-\dfrac{V_1}{L} - \dfrac{R}{L} I_1 \right) + \left(-\dfrac{V_2}{L} - \dfrac{R}{L} I_2 \right) t \end{bmatrix}
$$

Since the solutions must be valid for all values of $t > 0$, equating the coefficients of like terms results in the following four expressions, all of which must hold:

$$
-\alpha V_1 + V_2 = \frac{I_1}{C} \tag{5.35}
$$

$$
-\alpha I_1 + I_2 = -\frac{V_1}{L} - \frac{R}{L} I_1 \tag{5.36}
$$

$$
-\alpha V_2 = \frac{I_2}{C} \tag{5.37}
$$

$$
-\alpha I_2 = -\frac{V_2}{L} - \frac{R}{L} I_2 \tag{5.38}
$$

To show that these represent a true, consistent statement about the circuit, we begin by solving Eq. (5.37) for I_2 $(= -\alpha C V_2)$ and substituting the result in Eq. (5.38):

$$
-\alpha(-\alpha C V_2) = -\frac{V_2}{L} - \frac{R}{L}(-\alpha C V_2)
$$

$$
\alpha^2 C V_2 = \left(-\frac{1}{L} + \frac{\alpha C R}{L} \right) V_2
$$

Remembering that $\alpha = \dfrac{R}{2L}$, this becomes

$$
\alpha^2 = -\frac{1}{LC} + \frac{\alpha R}{L}
$$

$$\left(\frac{R}{2L}\right)^2 = -\frac{1}{LC} + \frac{R^2}{2L^2}$$

$$\frac{1}{LC} = \frac{R^2}{2L^2} - \frac{R^2}{4L^2} = \frac{R^2}{4L^2}$$

which is true, since it is the defining criterion for the underdamped case.

The remaining equations, (5.35) and (5.36), can be combined by elimination of the variable I_1. The result is:

$$V_1\left[\alpha^2 LC + 1 - \alpha RC\right] + LI_2 - \alpha LCV_2 = -RCV_2$$

The term in square brackets is recognized as the left side of the characteristic equation Eq. (5.18), and is thus identically zero. Making the substitution $I_2 = -\alpha CV_2$ [Eq. (5.37)], this becomes

$$- \alpha CV_2 L - \alpha LCV_2 = -RCV_2$$

$$\alpha L + \alpha L = R$$

which is an identity. Hence, we have shown that the general solutions for the critically damped case, given by Eqs. (5.21) and (5.22), satisfy the first order coupled differential equations for the series RLC circuit.

We now consider a numerical example of a critically damped *RLC* circuit.

Example 5.4

Consider the *RLC* circuit shown in Figure 5.14. Solve for the voltage and current responses.

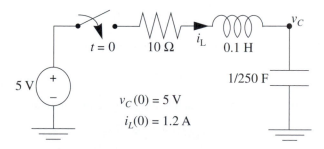

Figure 5.14 Critically damped series *RLC* circuit for Example 5.4.

The dc steady-state solutions are again apparent from inspection of the circuit:

$$I_L = 0 \text{ A}$$

$$V_C = 5 \text{ V}$$

The natural frequencies are found from Eqs. (5.19) and (5.20):

$$s = -\frac{R}{2L} \pm \sqrt{\left(\frac{R}{2L}\right)^2 - \frac{1}{LC}}$$

$$= -\frac{10}{2 \times 0.1} \pm \sqrt{\left(\frac{10}{2 \times 0.1}\right)^2 - \frac{1}{0.1 \times \dfrac{1}{250}}}$$

$$= -50 \pm \sqrt{(50)^2 - 2500} = -50 \pm \sqrt{0}$$

$$= -50 \text{ s}^{-1}$$

We see that there is a single natural frequency, so the response is critically damped. The capacitor voltage and inductor current can thus be written as

$$v_C(t) = V_1 e^{-50t} + V_2 t e^{-50t} + 5 \qquad (5.39)$$

and

$$i_L(t) = I_1 e^{-50t} + I_2 t e^{-50t} \qquad (5.40)$$

The procedure for calculating the unknown coefficients for the critically damped case is much like that used in the overdamped case. We determine the values of the unknown constant coefficients by using the initial conditions and one of the differential equations describing the circuit response. This will result in a set of four algebraic equations in four unknowns.

Applying the initial condition for capacitor voltage to Eq. (5.39):

$$v_C(0) = V_1 + 5 = 5$$

which means that

$$V_1 = 0 \qquad (5.41)$$

Similarly, using the initial inductor current gives:

$$i_L(0) = I_1 = 1.2 \qquad (5.42)$$

Substituting these coefficients into the general solutions:

$$v_C(t) = V_2 t e^{-50t} + 5$$

$$i_L(t) = 1.2 e^{-50t} + I_2 t e^{-50t}$$

These expressions are substituted into Eq. (5.13), the branch relation for capacitor current:

$$C\frac{dv_C}{dt} = i_L$$

$$\frac{1}{250}\frac{d}{dt}(V_2 te^{-50t} + 5) = 1.2e^{-50t} + I_2 te^{-50t}$$

Evaluating,

$$V_2 e^{-50t} - 50V_2 te^{-50t} = 300e^{-50t} + 250I_2 te^{-50t}$$

Equating coefficients of like terms results in

$$V_2 = 300$$

and

$$I_2 = \frac{-50V_2}{250} = -60$$

The circuit response is therefore:

$$v_C(t) = 300te^{-50t} + 5 \tag{5.43}$$

$$i_L(t) = 1.2e^{-50t} - 60te^{-50t} \tag{5.44}$$

Equations (5.43) and (5.44) completely characterize the response of the capacitor voltage and inductor current.

Again, it is always a good idea to verify the numerical solutions by inserting them into the original differential equations. We will use Eq. (5.14) as a check, since Eq. (5.13) was used in the derivation. The KVL loop equations is

$$Ri_L + L\frac{di_L}{dt} + v_C = V_S$$

Inserting circuit parameters and the solutions:

$$10(1.2e^{-50t} - 60te^{-50t}) + 0.1\frac{d}{dt}[1.2e^{-50t} - 60te^{-50t}] + 300te^{-50t} + 5 = 5$$

Evaluating:

$$e^{-50t}[12 - 6 - 6] + te^{-50t}[-600 + 300 + 300] = 0$$

The coefficients of each exponential term are identically zero, so the solution checks.

The critically damped solutions to Example 5.4 are plotted in Figures 5.15 and 5.16. Note that the capacitor voltage begins at its initial value of 5 V and then increases in an approximately linear fashion for a certain period after the switch closure. This qualitative behavior, characteristic of critically damped systems, can be understood by examining the Taylor expansion of the exponential function. One term in the critically damped solution has the form $te^{-\alpha t}$, which can be expanded as

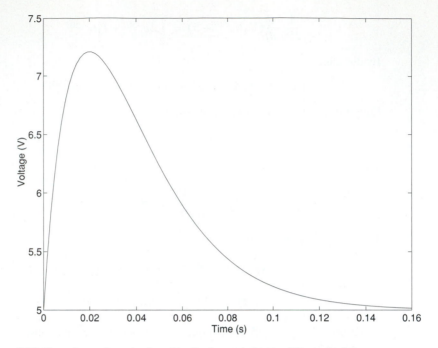

Figure 5.15 Capacitor voltage in the critically damped circuit of Example 5.4.

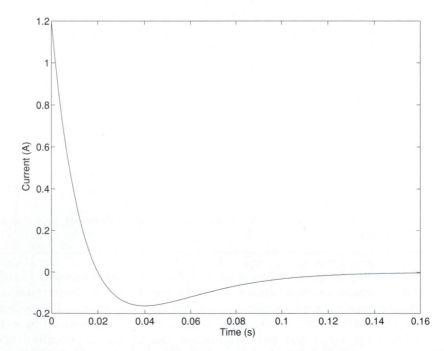

Figure 5.16 Inductor current in the critically damped circuit of Example 5.4.

$$te^{-\alpha t} = t\left[1 - \alpha t + \frac{1}{2}(\alpha t)^2 - \cdots\right]$$

$$= t - \alpha t^2 + \frac{1}{2}\alpha^2 t^3 - \cdots$$

For small values of t this expansion is dominated by the leading linear term. The same behavior is seen in the graph of inductor current.

We now consider the underdamped case.

5.8.3 The Underdamped Response

An underdamped response results when the circuit components satisfy the inequality

$$\left(\frac{R}{2L}\right)^2 < \left(\frac{1}{LC}\right)$$

This causes the argument of the square-root term in the natural frequency expression to be negative, with the result that the two natural frequencies are *complex* numbers.

To treat the underdamped case, we will find it useful to introduce some additional notation. For the series *RLC* circuit, we define the **undamped natural frequency** ω_0 by

$$\omega_0 \equiv \sqrt{\frac{1}{LC}}$$

and the **damped natural frequency** ω_d by

$$\omega_d \equiv \left|\sqrt{\omega_0^2 - \alpha^2}\right|$$

where, as before, the **damping factor** α is given by

$$\alpha = \frac{R}{2L}$$

The natural frequencies of the series *RLC* circuit are found from

$$s = -\frac{R}{2L} \pm \sqrt{\left(\frac{R}{2L}\right)^2 - \frac{1}{LC}}$$

which can be recast in the new symbols as

$$s = -\alpha \pm \sqrt{\alpha^2 - \omega_0^2}$$

By definition, the underdamped case has $\alpha^2 < \omega_0^2$, so this expression can be written as[2]

---

[2] By long-standing electrical engineering convention the symbol j is used for the imaginary number $\sqrt{-1}$, since i is reserved for current.

$$s = -\alpha \pm \sqrt{(-1)(\omega_0^2 - \alpha^2)}$$

$$= -\alpha \pm \sqrt{j^2 \omega_d^2}$$

$$= -\alpha \pm j\omega_d$$

The general solutions for the underdamped case are:

$$v_C(t) = \Phi_1 e^{(-\alpha + j\omega_d)t} + \Phi_2 e^{(-\alpha - j\omega_d)t} + V_C \qquad (5.45)$$

$$i_L(t) = \Gamma_1 e^{(-\alpha + j\omega_d)t} + \Gamma_2 e^{(-\alpha - j\omega_d)t} + I_L \qquad (5.46)$$

These solutions can be verified by substituting them back into the original differential equations. Since they are the same form as for the overdamped case, they have already been verified. (Note that the verification procedure described in Section 5.8.1 does not depend on the form of the natural frequencies, but only on the fact that the natural frequencies satisfy the characteristic equation, Eq. (5.18). Thus it is valid for both the underdamped and overdamped cases.)

The coefficients of the exponential terms in Eqs. (5.45) and (5.46) have been written as $\Phi_1 \ldots \Gamma_2$, rather than the more familiar $V_1 \ldots I_2$ to emphasize that they are complex numbers. Although it is possible to solve circuit problems using these forms of the solutions, it is more useful to recast them into expressions which contain only real numbers. We will show how to accomplish this using the expression for capacitor voltage, Eq. (5.45); the same procedure applies to Eq. (5.46).

We begin with **Euler's identity**:

$$e^{j\theta} = \cos\theta + j\sin\theta \qquad (5.47)$$

Applying Euler's Identity to the two exponential terms, we can write

$$e^{j\omega_d t} = \cos\omega_d t + j\sin\omega_d t$$

$$e^{-j\omega_d t} = \cos\omega_d t - j\sin\omega_d t$$

where we have made use of the fact that the cosine is an even function. We can now write the capacitor voltage as

$$v_C(t) = e^{-\alpha t}[\Phi_1(\cos\omega_d t + j\sin\omega_d t) + \Phi_2(\cos\omega_d t - j\sin\omega_d t)] + V_C$$

Collecting terms,

$$v_C(t) = e^{-\alpha t}[(\Phi_1 + \Phi_2)\cos\omega_d t + j(\Phi_1 - \Phi_2)\sin\omega_d t] + V_C$$

Since $v_C(t)$ represents a voltage signal, it must be real valued. Consequently, the constants Φ_1 and Φ_2 are constrained such that

$$Im[\Phi_1 + \Phi_2] = 0$$

$$Im[j(\Phi_1 - \Phi_2)] = 0$$

If we write the coefficient Φ_1 as

$$\Phi_1 = \frac{1}{2}(V_1 - jV_2)$$

then the above constraints force Φ_2 to be the complex conjugate of Φ_1:

$$\Phi_2 = \overline{\Phi_1} = \frac{1}{2}(V_1 + jV_2)$$

Substituting these forms into values into the capacitor voltage expression results in:

$$v_C(t) = e^{-\alpha t}\left[\frac{1}{2}(V_1 - jV_2 + V_1 + jV_2)\cos\omega_d t\right]$$

$$+ e^{-\alpha t}\left[\frac{j}{2}(V_1 - jV_2 - (V_1 + jV_2))\sin\omega_d t\right] + V_C$$

which after simplification becomes:

$$v_C(t) = e^{-\alpha t}(V_1\cos\omega_d t + V_2\sin\omega_d t) + V_C \qquad (5.48)$$

Similarly, for the inductor current,

$$i_L(t) = e^{-\alpha t}(I_1\cos\omega_d t + I_2\sin\omega_d t) + I_L \qquad (5.49)$$

Note that when the signal responses are written in this form, they are clearly real-valued functions.

We see from Eqs. (5.48) and (5.49) that the solutions of an underdamped circuit are **exponentially damped sinusoids**. The voltage and current waveforms oscillate about their dc steady-state values. The amplitude of these oscillations decreases with time, owing to the decaying exponential prefactor $e^{-\alpha t}$. This oscillatory nature does not occur for overdamped or critically damped circuits; their natural frequencies are purely real, and consequently their waveforms do not have sinusoidal components.

The underdamped solutions also provide a context for our use of the term **natural frequency** for those values of s which satisfy the characteristic equation of a circuit containing capacitors and inductors. We see that the magnitude of the imaginary part of s, given by ω_d, describes the angular frequency of the sinusoidal terms. Hence, it is appropriate to express ω_d (and also ω_0, when we encounter it later) in units of radians per second.

Let us now solve a numerical example of a series *RLC* circuit whose response is underdamped.

Example 5.5

Solve for the voltage and current responses in the RLC circuit shown in Figure 5.17.

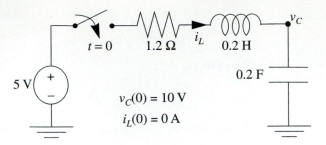

Figure 5.17 Underdamped series *RLC* circuit for Example 5.5

The dc steady-state solutions are self-evident from inspection of the circuit:

$$I_L = 0 \text{ A}$$

$$V_C = 5 \text{ V}$$

The natural frequencies are found from Eqs. (5.19) and (5.20):

$$s = -\frac{R}{2L} \pm \sqrt{\left(\frac{R}{2L}\right)^2 - \frac{1}{LC}}$$

$$= -\frac{1.2}{2 \times 0.2} \pm \sqrt{\left(\frac{1.2}{2 \times 0.2}\right)^2 - \frac{1}{0.2 \times 0.2}}$$

$$= -3 \pm \sqrt{(3)^2 - 25} = -3 \pm j4$$

Inspection of the natural frequencies shows that the response of this circuit is underdamped. The damping factor and damped natural frequency are

$$\alpha = 3 \text{ s}^{-1}$$

$$\omega_d = 4 \text{ rad/s}$$

Substituting these values into the general solutions, Eqs. (5.48) and (5.49):

$$v_C(t) = e^{-3t}(V_1\cos 4t + V_2\sin 4t) + 5$$

$$i_L(t) = e^{-3t}(I_1\cos 4t + I_2\sin 4t)$$

Application of the given initial conditions yields

$$v_C(0) = 10 = e^{-3(0)}(V_1\cos 4(0) + V_2\sin 4(0)) + 5$$

$$V_1 = 5$$

$$i_L(0) = 0 = e^{-3(0)}(I_1\cos 4(0) + I_2\sin 4(0))$$

$$I_1 = 0$$

As before, we determine the remaining two coefficients by substituting the voltage and current expressions into one of the original differential equations, Eq. (5.13):

$$C\frac{dv_C}{dt} = i_L$$

$$0.2\frac{d}{dt}[e^{-3t}(5\cos4t + V_2\sin4t) + 5] = I_2e^{-3t}\sin4t$$

$$0.2e^{-3t}[(-20\sin4t + 4V_2\cos4t) - 3(5\cos4t + V_2\sin4t)] = I_2e^{-3t}\sin4t$$

Equating coefficients of like terms:

$$0.8\dot{V}_2 - 3 = 0$$

$$V_2 = 3.75$$

$$I_2 = -4 - 0.6V_2 = -6.25$$

Therefore the complete solutions are

$$v_C(t) = e^{-3t}(5\cos4t + 3.75\sin4t) + 5 \tag{5.50}$$

$$i_L(t) = -6.25e^{-3t}\sin4t \tag{5.51}$$

As a check, we verify these solutions by substituting them back into the other original differential equation, Eq. (5.14):

$$Ri_L + L\frac{di_L}{dt} + v_C = V_S$$

$$1.2(-6.25e^{-3t}\sin4t) + 0.2\frac{d}{dt}(-6.25e^{-3t}\sin4t)$$

$$+ (e^{-3t}(5\cos4t + 3.75\sin4t) + 5) = 5$$

$$-7.5e^{-3t}\sin4t - 1.25e^{-3t}(-3\sin4t + 4\cos4t) + e^{-3t}(5\cos4t + 3.75\sin4t) = 0$$

$$(-7.5 + 3.75 + 3.75)e^{-3t}\sin4t - (-5 + 5)e^{-3t}\cos4t = 0$$

which is valid for all values of t.

The underdamped responses in this example are shown in Figures 5.18 and 5.19.

Examining Figure 5.18, we see that the capacitor voltage begins at its initial value, 10 V, and asymptotically approaches the dc steady-state value of 5 V via the transient response. The voltage signal "overshoots" the final steady-state value, and in fact oscillates around it with an ever-decreasing amplitude. The inductor current, Figure 5.19, behaves in the same fashion.

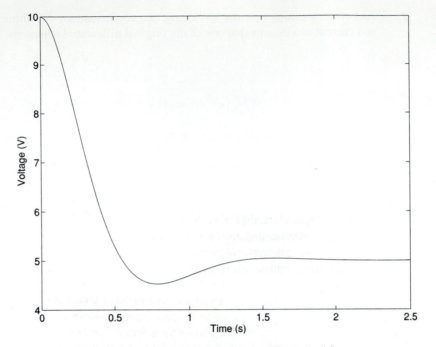

Figure 5.18 Capacitor voltage in the underdamped circuit of Example 5.5.

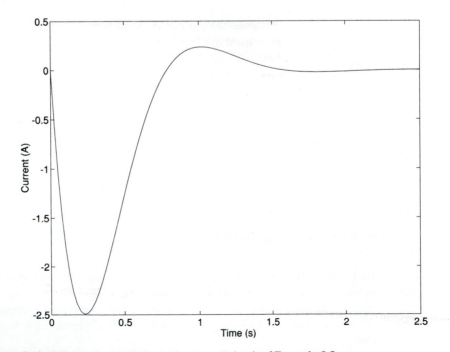

Figure 5.19 Inductor current in the underdamped circuit of Example 5.5.

5.8.4 The Undamped Response

The underdamped response is characterized by exponentially decaying sinusoidal terms of the form

$$e^{-\alpha t}\cos\omega_d t$$

$$e^{-\alpha t}\sin\omega_d t$$

Each of these consists of a sinusoidal term, bounded by ± 1, multiplied by an "envelope function," $e^{-\alpha t}$. For example, the $e^{-\alpha t}\cos\omega_d t$ term, graphed in Figure 5.20(a) for $\alpha = 1$ s^{-1} and $\omega_d = 10$ rad/s, shows how the signal oscillates inside an envelope defined by $\pm e^{-\alpha t}$. In other words, the oscillations are "damped out" by the decaying exponential; the rate of decay is characterized by α, hence the term "damping factor." (When several oscillations occur without significant damping, the response is said to "ring"; the acoustic signal produced by a bell has a similar waveform.) An "underdamped" response is characterized by oscillatory behavior; there is insufficient damping to prevent the signal from oscillating. "Overdamped" circuits have more than enough damping to prevent oscillations. The boundary between these two behaviors, where there is just enough damping to prevent oscillation, is "critical damping."

In an *RLC* circuit, the oscillations are manifestations of an exchange of energy between the capacitor and inductor. The capacitor stores energy in the form of an electric field, while the inductor stores energy as a magnetic field. As the energy is shuttled between the two components, the exchange is mediated by the resistor, which converts part of the energy to heat on each cycle. As a result, the energy available in the circuit decreases, and the oscillations are damped. The rate of damping, given by α, is directly proportional to the resistance in a series *RLC* circuit. Therefore, if the resistance in the circuit is decreased, the damping factor also decreases, which lengthens the period over which the signal oscillations have a significant amplitude. This effect is illustrated in Figure 5.20 (a) through (c), where the damping factor α has been decreased (while holding the oscillation frequency ω_d constant). We see that a lower resistance corresponds to a longer transient response where the oscillations are significant.

The limiting case, shown in Figure 5.20 (d), occurs when the resistance is zero. This undamped response, consisting of a pure sinusoid, is the underdamped response in the limit where $\alpha \to 0$. For a series *RLC* circuit this occurs when $R \to 0$. Taking this limit in the expression for natural frequency,

$$s = \lim_{R \to 0}\left(-\frac{R}{2L} \pm \sqrt{\left(\frac{R}{2L}\right)^2 - \frac{1}{LC}}\right)$$

$$s = \pm\sqrt{-\frac{1}{LC}} = \pm j\omega_0 \qquad (5.52)$$

The general solutions for the capacitor voltage and inductor current are similar to that of the underdamped case, but contain no decaying exponential term:

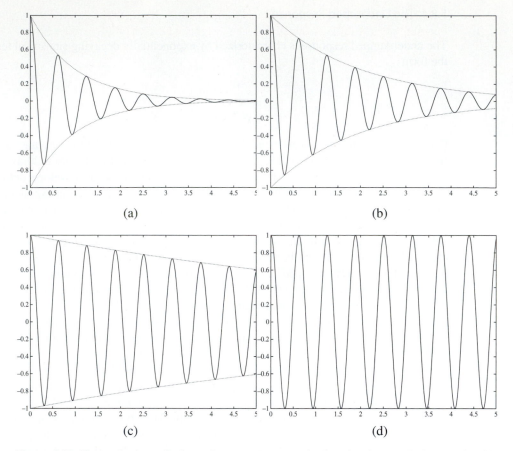

Figure 5.20 Change in the underdamped response term as the damping factor α is decreased (with ω_d held constant at 10 rad/s). The solid lines are plots of $e^{-\alpha t}\cos\omega_d t$ vs. t; the dotted lines show the envelope, $\pm e^{-\alpha t}$, which bounds the response. (a) $\alpha = 1\ \text{s}^{-1}$; (b) $\alpha = 0.5\ \text{s}^{-1}$; (c) $\alpha = 0.1\ \text{s}^{-1}$; (d) limiting undamped case with $\alpha = 0$.

$$v_C(t) = V_1\cos\omega_0 t + V_2\sin\omega_0 t + V_C \qquad (5.53)$$

$$i_L(t) = I_1\cos\omega_0 t + I_2\sin\omega_0 t + I_L \qquad (5.54)$$

Equations (5.53) and (5.54) can be verified by following the same technique used to check the general solutions for the underdamped case. Note that the angular frequency of the oscillations in the undamped case is given by ω_0, the **undamped natural frequency**, which depends only on L and C.

The undamped response is similar to the underdamped response, but the lack of any damping means that the "transient" part of the response lasts an infinitely long time. Because the circuit without resistance never reaches steady state, the usual meanings of V_C and I_L must be interpreted. V_C and I_L are the **average values** of the capacitor voltage and inductor current. This can be seen by considering the definition of the average value, denoted $\overline{f(t)}$, of a periodic signal $f(t)$ (i.e., a signal where $f(t + T) = f(t)$):

$$\overline{f(t)} = \frac{1}{T} \int_{t}^{t+T} f(t)dt$$

For the periodic capacitor voltage, Eq. (5.53), applying this expression gives an expression for the average voltage $v_C(t)$:

$$\overline{v_C(t)} = \frac{1}{T} \int_{t}^{t+T} (V_1 \cos \omega_0 t + V_2 \sin \omega_0 t + V_C)dt$$

$$= \frac{1}{T} \int_{t}^{t+T} \left(V_1 \cos \frac{2\pi}{T} t + V_2 \sin \frac{2\pi}{T} t + V_C \right)dt$$

$$= \frac{1}{T} \left(\frac{V_1 T}{2\pi} \sin \frac{2\pi}{T} t - \frac{V_2 T}{2\pi} \cos \frac{2\pi}{T} t + V_C t \right)\Bigg|_{t}^{t+T}$$

$$= \frac{V_1}{2\pi} \left(\sin \frac{2\pi}{T}(t + T) - \sin \frac{2\pi}{T} t \right)$$

$$- \frac{V_2}{2\pi} \left(\cos \frac{2\pi}{T}(t + T) - \cos \frac{2\pi}{T} t \right) + \frac{V_C}{T}(t + T - t)$$

Since both the sine and cosine functions have period T, the first two terms are zero. This leaves

$$\overline{v_C(t)} = V_C$$

and similarly for the average inductor current.

V_C and I_L are found in the same manner as for other circuits: the dc models (i.e., open and short circuits for capacitors and inductors, respectively) are substituted and the circuit analyzed. One way to think of the steady-state values in an undamped circuit is to consider the circuit as the limiting case of an underdamped circuit. The circuit will reach the dc steady state only after an infinitely long transient.

Because every real circuit contains resistance, the undamped response is an abstraction. However, it is a useful approximation in many instances, especially for order-of-magnitude calculations of oscillation frequency.

Example 5.6

Determine the capacitor voltage and inductor current in the undamped circuit of Figure 5.21.

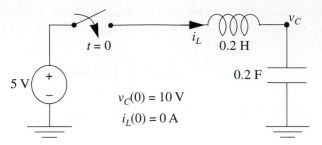

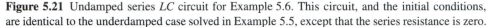

Figure 5.21 Undamped series *LC* circuit for Example 5.6. This circuit, and the initial conditions, are identical to the underdamped case solved in Example 5.5, except that the series resistance is zero.

We will first explicitly determine the "steady state" (i.e., average) values V_C and I_L. Substituting the dc models for the capacitor and inductor results in the circuit of Figure 5.22. By inspection, we see that the average values are

$$I_L = 0 \text{ A}$$
$$V_C = 5 \text{ V}$$

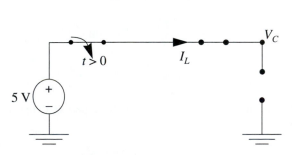

Figure 5.22 Circuit model for determining V_C and I_L. The capacitor and inductor in Figure 5.21 are replaced with their respective dc models, an open circuit and a short circuit. The resulting values are the average capacitor voltage and inductor current in the undamped circuit.

Next, we find the natural frequencies from Eq. (5.52):

$$s = \pm\sqrt{-\frac{1}{LC}} = \pm j\omega_0$$

$$= \pm j\sqrt{\frac{1}{0.2 \times 0.2}} = \pm j5$$

The general solutions are:

$$v_C(t) = V_1 \cos 5t + V_2 \sin 5t + 5 \tag{5.55}$$

and

$$i_L(t) = I_1 \cos 5t + I_2 \sin 5t \tag{5.56}$$

Applying the initial conditions yields:

$$v_C(0) = 10 = V_1 \cos 5(0) + V_2 \sin 5(0) + 5$$
$$V_1 = 5$$

and

$$i_L(0) = 0 = I_1 \cos 5(0) + I_2 \sin 5(0)$$
$$I_1 = 0$$

The inductor current is given by the differential equation

$$i_L = C \frac{dv_C}{dt}$$

$$I_2 \sin 5t = 0.2 \frac{d}{dt}((5 \cos 5t + V_2 \sin 5t + 5))$$

$$I_2 \sin 5t = -5 \sin 5t + V_2 \cos 5t$$

Equating coefficients of like terms results in:

$$V_2 = 0$$
$$I_2 = -5$$

Hence:

$$v_C(t) = 5 \cos 5t + 5$$

and

$$i_L(t) = -5 \sin 5t$$

These solutions can be verified by substituting them into the differential equation resulting from application of KVL to the circuit in Figure 5.21 [i.e., Eq. (5.14) with $R = 0$]:

$$L \frac{di_L}{dt} + v_C = 5$$

$$0.2 \frac{d}{dt}(-5 \sin 5t) + 5 \cos 5t + 5 = 5$$

$$-5 \cos 5t + 5 \cos 5t = 0$$

which is an equality.

The undamped responses are illustrated in Figures 5.23 and 5.24. Note that the oscillations of the voltage and current are purely sinusoidal and never decay. Also note that the signals oscillate about the average values, 5 V and 0 A, found from the dc equivalent circuit of Figure 5.22.

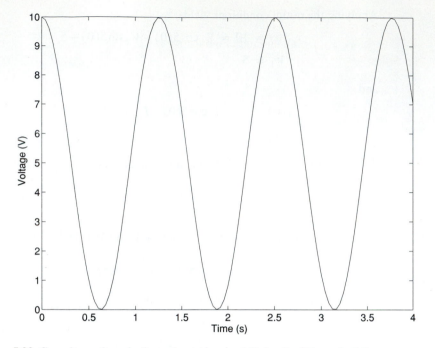

Figure 5.23 Capacitor voltage in the undamped series *LC* circuit of Example 5.6.

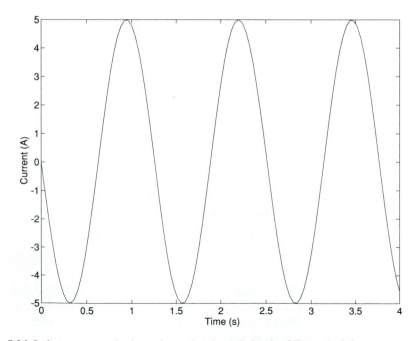

Figure 5.24 Inductor current in the undamped series *LC* circuit of Example 5.6.

5.9 Application to Digital-System Switching Speed

All wires have a certain amount of inductance. In an IC, the inductance associated with the on-chip conductors is generally negligible, but the external power and ground wires can introduce enough inductance to measurably affect switching speed. We will calculate the change in switching speed engendered by this effect.

Example 5.7

Determine the effect on the pulldown transition time of an interconnected pair of inverter gates ($C_{in} = 1.0$ pF, $R_{out} = 100 \ \Omega$) by the presence of a series inductance of $L = 10$ nH. Assume the system switches between 0 V and 5 V, with logic low and high thresholds of 1.3 V and 4.0 V.

Neglecting the interconnect, a first-order RC model results in a voltage at the input of the second inverter given by

$$v_C(t) = (v_C(0) - V_S)e^{-\frac{t}{RC}} + V_S$$

Assuming the first inverter output has been at logic high for a long time, this results in

$$v_C(t) = 5e^{-\frac{t}{0.1}} \tag{5.57}$$

where t is in ns. The pull-down transition time is defined by the interval required for $v_C(t)$ to reach 1.3 V, which is given by:

$$t_{hl} = 0.1 \ln\left(\frac{5}{1.3}\right) = 0.135 \text{ ns}$$

To determine the voltage at the input of the load inverter, including the series inductance, we first calculate the values of α and ω_0 from the given parameters:

$$\alpha = \frac{R}{2L} = \frac{100}{2 \times 10 \times 10^{-9}}$$

$$= 5 \times 10^9 \text{ s}^{-1}$$

$$\omega_0 = \sqrt{\frac{1}{LC}} = \sqrt{\frac{1}{10 \times 10^{-9} \times 1 \times 10^{-12}}}$$

$$= 10 \times 10^9 \text{ rad/s}$$

Since $\alpha^2 < \omega_0^2$, the response is underdamped. The damped natural frequency is found from

$$\omega_d = \sqrt{\omega_0^2 - \alpha^2} = \sqrt{(10 \times 10^9)^2 - (5 \times 10^9)^2}$$

$$= 8.66 \times 10^9 \text{ rad/s}$$

Substituting these values into the general underdamped solutions, Eqs. (5.48) and (5.49):

$$v_C(t) = e^{-5 \times 10^9 t}(V_1 \cos 8.66 \times 10^9 t + V_2 \sin 8.66 \times 10^9 t)$$

$$i_L(t) = e^{-5 \times 10^9 t}(I_1 \cos 8.66 \times 10^9 t + I_2 \sin 8.66 \times 10^9 t)$$

Application of the given initial conditions yields

$$v_C(0) = 5 = e^0(V_1 \cos(0) + V_2 \sin(0))$$
$$V_1 = 5$$

$$i_L(0) = 0 = e^0(I_1 \cos(0) + I_2 \sin(0))$$
$$I_1 = 0$$

The remaining two coefficients are found by substituting the voltage and current expressions into the original differential equations, Eq. (5.13), and equating like coefficients of the resulting expression:

$$C \frac{dv_C}{dt} = i_L$$

$$10^{-12} \frac{d}{dt}[e^{-5 \times 10^9 t}(5 \cos(8.66 \times 10^9 t) + V_2 \sin(8.66 \times 10^9 t))]$$

$$= I_2 e^{-5 \times 10^9 t}(\sin 8.66 \times 10^9 t)$$

$$10^{-12} e^{-5 \times 10^9 t}\left[-43.3 \times 10^9 \sin(8.66 \times 10^9 t) + 8.66 \times 10^9 V_2 \cos(8.66 \times 10^9 t)\right.$$

$$\left. -5 \times 10^9(5 \cos(8.66 \times 10^9 t) + V_2 \sin(8.66 \times 10^9 t))\right]$$

$$= I_2 e^{-5 \times 10^9 t} \sin 8.66 \times 10^9 t$$

After performing the algebra:

$$V_2 = 2.89$$
$$I_2 = -0.0577$$

Therefore the complete solutions are

$$v_C(t) = e^{-5t}(5\cos(8.66t) + 2.89\sin(8.66t)) \tag{5.58}$$

$$i_L(t) = -0.0577e^{-5t}\sin(8.66t) \tag{5.59}$$

where t is in ns. The pull-down transition time t_{hl} is found by setting $v_C(t_{hl}) = 1.3$ V in Eq. (5.58); by trial and error the pull-down time is

$$t_{hl} = 0.176 \text{ ns}$$

which is 30% longer than the pull-down time calculated from the RC model.

The results of the RC and RLC models, Eqs. (5.57) and (5.58), are shown in Figure 5.25. Although inclusion of the inductance has increased the estimate of switching time in this case, this is by no means a general feature of RLC models. We see that the underdamped response has the inverter signal actually going negative, below the ground level of the circuit.

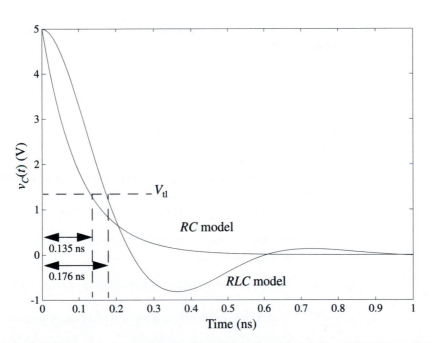

Figure 5.25 Voltage at the input of the load inverter in Example 5.7. The RC model neglects the effect of inductance and predicts a pull-down transition time of 0.135 ns. The RLC model illustrates the underdamped response which results when the power-supply wire inductance is included. The pull-down time with the RLC model is considerably longer in this case, 0.176 ns.

It is instructive to consider what happens in this example if the system is switched again.

Example 5.8

Suppose the coupled pair of inverter gates in Example 5.7 are switched again at $t = 0.3$ ns. Determine the time when the load inverter reaches the high threshold voltage (4.0 V) using each circuit model.

Using the solution for the first-order RC model, the capacitor voltage at 0.3 ns is given by

$$v_C(0.3) = 5e^{-\frac{0.3}{0.1}}$$

$$= 0.249 \text{ V}$$

Substituting this value into the general first-order solution (with $V_S = 5$):

$$v_C(t) = (0.249 - 5)e^{-\frac{t-0.3}{0.1}} + 5$$

$$= 5 - 4.75e^{-\frac{t-0.3}{0.1}}$$

Setting the lefthand side of this equation to $v_{C(t_{lh})} = 4$ and solving for the transition time results in:

$$t_{lh} = 0.3 + 0.1 \ln\frac{-4.75}{4-5}$$

$$= 0.456 \text{ ns}$$

To use the second-order model, we first use Eqs. (5.58) and (5.59) to determine the initial conditions $v_C(0.3)$ and $i_L(0.3)$:

$$v_C(0.3) = e^{-5(0.3)}(5\cos(8.66(0.3)) + 2.89\sin(8.66(0.3))) = -0.621 \text{ V}$$

$$i_L(0.3) = -0.0577e^{-5(0.3)}\sin(8.66(0.3)) = -6.66 \text{ mA}$$

Following the procedure we used previously, the capacitor voltage and inductor current for $t > 0.3$ ns are given by:

$$v_C(t) = e^{-5(t-0.3)}(-5.62\cos(8.66(t-0.3))-4.02\sin(8.66(t-0.3)))$$

$$i_L(t) = e^{-5(t-0.3)}(-0.00666\cos(8.66(t-0.3)) + 0.0611\sin(8.66(t-0.3)))$$

The expression for $v_C(t)$ can be solved iteratively to find the pull-up time; the solution is

$$t_{lh} = 0.505 \text{ ns}$$

The first- and second-order solutions for both switching events are plotted in Figure 5.26. One striking difference between the *RC* and *RLC* models is the behavior at the pull-up switching event, $t = 0.3$ ns. In the *RC* model the slope of the capacitor voltage (which is proportional to the current in the circuit) changes abruptly. However, in the *RLC* model the slope does not change abruptly; in fact, the voltage continues to *decrease* for a short time. This behavior is caused by the inductor. Recall that inductor current must be continuous; therefore, since the derivative of the capacitor voltage is linearly related to the inductor current, the slope of $v_C(t)$ must also be continuous.

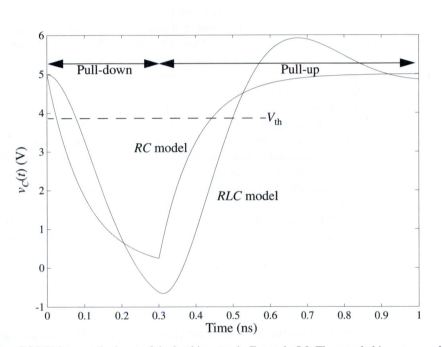

Figure 5.26 Voltage at the input of the load inverter in Example 5.8. The coupled inverters undergo a pull-down transition at $t = 0$ and are subsequently pulled up at $t = 0.3$ ns. Note how the voltage in the *RLC* model continues to fall for a period after the second switching event; this is due to the presence of the inductor, which attempts to force the current to flow in the same direction.

5.10 Gate Conductance and *RLGC* Circuits

In the previous sections we have studied the series *RLC* circuit in detail. The analysis methods developed can be used for any other second-order circuit with inductance and capacitance, regardless of the configuration. In particular, the classification of a circuit as overdamped, etc., depending on the form of the natural frequencies, is valid for all such circuits. Indeed, higher-order circuits with more than two independent energy-storage elements are also treated in the same manner, although the algebraic solution methods may become more involved.

As a means of demonstrating that other circuit topologies are solved in the same way, we will examine the *RLGC* circuit. This combination of elements is found in our interconnected gate example when the finite conductivity to ground of the interconnect and load gate is included in the model. The *RLGC* circuit is also used as a model for "lossy waveguides," such as long lengths of coaxial cable found in cable TV installations. It can be shown (Exercise 5.35) that the series *RLC* is a special case of the *RLGC* circuit.

5.10.1 Gate Conductance

In Chapter 2 we described an elementary model of an MOS transistor, consisting of an input capacitor, an output resistor, and a switch. We have used this simple model throughout to predict the performance of interconnected gates composed of these devices. Although this model is quite useful, it does not capture all of the subtleties of real transistors.

One improvement to this model is to include the finite resistance to ground of the transistor gate. Generally, this resistance is very high[3] and need not be included for accurate modeling of interconnect delay. (See Exercise 5.43.) However, no insulator is perfect, so there will always be a finite leakage current through the gate terminal whenever a voltage is present. This leakage current is significant in certain kinds of devices, especially dynamic random-access memories (DRAM's), which constitute the majority of fast memory in modern computers. DRAM's store a 0 or 1 as the absence or presence of a bundle of gate charge; as the charge leaks away, the data can change. Thus it is important to understand the effects of gate conductance in calculating the possible data storage time in semiconductor memories, and hence the rates at which they must be refreshed.

Figure 5.27 updates our model of the CMOS inverter gate to include this resistance, labeled R_g. Using this model, we can approximate the behavior of interconnected inverters using the circuit of Figure 5.28. In this diagram we have specified the gate leakage resistance as $1/G$ instead of R_g; the two are related by:

$$R_g = \frac{1}{G}$$

G is known as a **conductance**; its units are siemens (symbol: S) which are equal to A/V or reciprocal ohms. We use the leakage conductance, rather than resistance, in this problem because the resulting equations are simpler. The other component values in this diagram are the same as defined previously (page 221).

(The combination of *RLGC* elements in Figure 5.28 also represents a highly accurate model for a section of interconnect. In Chapter 3 we approximated interconnects with a finite number of series R and parallel C elements. A more accurate approximation includes the series inductance and parallel conductance. However, since the leakage conductances of IC interconnects are vanishingly small, this model is too sophisticated to be useful. In some cases, the series inductance will have measurable effects on circuit delay. This case is covered in a later chapter. The complete *RLGC* model of a transmission line is useful in other applications, such as coaxial cables used for radio-frequency and microwave communication systems, and is known as the "lossy waveguide" model.)

[3.] The resistance is high because the gate electrode in most MOSFETS is isolated from ground by a dielectric layer of silicon dioxide, a very good insulator.

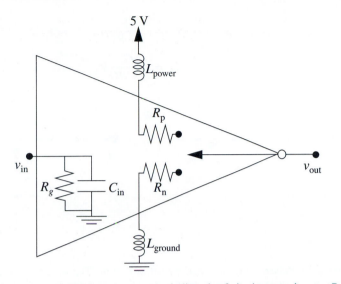

Figure 5.27 Model of a CMOS inverter gate, including the finite input resistance R_g, representing the resistance to ground responsible for gate leakage current.

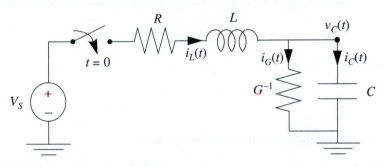

Figure 5.28 RLGC circuit model of interconnected inverter gates. The reciprocal of the leakage resistance R_g is known as a *conductance*.

5.10.2 Analysis of the *RLGC* Circuit

The *RLGC* circuit in Figure 5.28 has two independent energy-storage elements; hence it is second order. We will analyze this circuit as before: solve for the dc steady state; use Kirchhoff's laws to write a system of differential equations describing the circuit response; find the natural frequencies; assume general solutions; and determine the unknown constants in the general expressions.

To find the dc steady-state response, we replace the inductor and capacitor with their dc equivalents, as shown in Figure 5.29. Straightforward analysis shows that the dc steady-state capacitor voltage and inductor current are given by

$$V_C = \frac{1}{1 + RG} V_S$$

$$I_L = \frac{G}{1 + RG} V_S$$

(5.60)

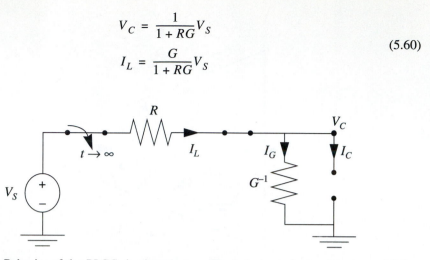

Figure 5.29 Behavior of the *RLGC* circuit as $t \to \infty$. The inductor and capacitor are modeled as their dc steady-state equivalents, short and open circuits.

Next, we apply Kirchhoff's voltage law around the outer loop of the circuit, yielding:

$$v_C + L \frac{d}{dt}(i) + iR = V_S$$

which can be written as

$$\frac{di}{dt} = -\frac{1}{L} v_C - \frac{R}{L} i + \frac{V_S}{L}$$

(5.61)

Using Kirchhoff's current law at the $v_C(t)$ node:

$$i_G + i_C = i_L$$

Substituting the appropriate branch relations, this expression becomes

$$G v_C + C \frac{dv_C}{dt} = i_L$$

which can be transposed to

$$\frac{dv_C}{dt} = -\frac{G}{C} v_C + \frac{1}{C} i_L$$

(5.62)

Writing Eqs. (5.61) and (5.62) in matrix form:

$$\begin{bmatrix} \dfrac{dv_C}{dt} \\[2mm] \dfrac{di_L}{dt} \end{bmatrix} = \begin{bmatrix} -\dfrac{G}{C} & \dfrac{1}{C} \\[2mm] -\dfrac{1}{L} & -\dfrac{R}{L} \end{bmatrix} \begin{bmatrix} v_C \\[2mm] i_L \end{bmatrix} + \begin{bmatrix} 0 \\[2mm] \dfrac{V_S}{L} \end{bmatrix} \qquad (5.63)$$

Equation (5.63) governs the response of the *RLGC* circuit. It is essentially similar to Eq. (5.15) (for the series *RLC* circuit), except for the additional $-G/C$ term in the matrix, which accounts for the leakage conductance.

Following the procedure described in Section 5.7, the natural frequencies of the *RLGC* circuit are found from the coefficient matrix in Eq. (5.63) using the eigenvalue equation:

$$\det \begin{bmatrix} s\mathbf{I} - \begin{bmatrix} -\dfrac{G}{C} & \dfrac{1}{C} \\[2mm] -\dfrac{1}{L} & -\dfrac{R}{L} \end{bmatrix} \end{bmatrix} = 0$$

$$\det \begin{bmatrix} s + \dfrac{G}{C} & -\dfrac{1}{C} \\[2mm] \dfrac{1}{L} & s + \dfrac{R}{L} \end{bmatrix} = 0$$

Expanding,

$$\left(s + \frac{G}{C} \right)\left(s + \frac{R}{L} \right) - \left(-\frac{1}{C} \right)\frac{1}{L} = 0$$

$$s^2 + \frac{GL + RC}{LC}s + \frac{1 + GR}{LC} = 0 \qquad (5.64)$$

Equation (5.64) is the characteristic equation for the *RLGC* circuit.

It is again useful to substitute different symbols for the various coefficients in the characteristic equation. For this circuit, we have

$$\omega_0 \equiv \sqrt{\frac{1 + GR}{LC}}$$

$$\alpha = \frac{1}{2}\frac{GL + RC}{LC}$$

$$\omega_d \equiv \left| \sqrt{\omega_0^2 - \alpha^2} \right|$$

The natural frequencies of the series *RLGC* circuit are thus

$$s = -\alpha \pm \sqrt{\alpha^2 - \omega_0^2}$$

As before, there are four cases of the *RLGC* circuit, depending on the relative values of α and ω_0. The solutions for each case are completely analogous to the corresponding cases of the series *RLC* circuit.

5.11 Neglecting Unimportant Components in Circuit Analysis

A recurring theme in this text is that a reasonable speed estimate of an enormously complicated system, such as a digital computer, can be found from analyzing a model consisting of only a few components. The skills required to perform this kind of approximate calculation are among the most valuable possessed by a practicing engineer. The main objective in an approximate analysis is to determine which effects are the most important, thereby including only those components which contribute to these effects. In practice, this amounts to simplifying a circuit by *discarding* the unimportant components.

Given a particular circuit, how can one tell which parts to neglect? There is no easy, general answer, as it depends highly on quantifying terms such as "reasonable," "approximate," and "unimportant." However, there are many guidelines which are useful in selecting an accurate circuit model. One of these is the concept of relevant time scales: the approximate solution should include effects on the time scale of interest, whether that be picoseconds, microseconds, or longer. Effects which are apparent only on other, vastly different time scales can often be safely neglected. We will illustrate this concept by looking in detail at an approximation we implicitly made earlier, the modeling of a series *RLC* circuit by an *RC* circuit.

5.11.1 Current Discontinuity in an *RC* Circuit

Consider the simple *RC* circuit in Figure 5.30. This is a first-order approximation of a pull-down transition of two interconnected inverter gates. The capacitor current in this circuit is given by

$$i_C(t) = \begin{cases} 0, & t < 0 \\ -\dfrac{v_C(0)}{R}e^{-\frac{t}{RC}}, & t > 0 \end{cases} \tag{5.65}$$

Of particular note in this circuit is the discontinuity of the current when the switch is closed. The sudden jump in current does not violate any physical law (capacitor *voltage* must be continuous), but it does seem vaguely "unphysical."

In fact, this behavior *is* unphysical, but not because of any error in the above analysis. The problem is that the circuit in Figure 5.30 cannot actually be built with real components. We know that the wires in the circuit contain a small but finite amount of inductance. Therefore, a real circuit built with real components is more closely approximated with the *RLC* model of Figure 5.31. In this circuit, the current is continuous, even if the inductance is immeasurably tiny.

Although the more accurate *RLC* model correctly describes an aspect of the real circuit which is missed by the *RC* model (i.e., the fact that the current does not undergo a dis-

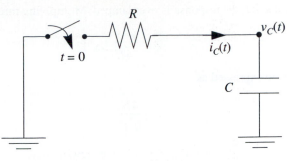

Figure 5.30 *RC* circuit model of a pull-down transition.

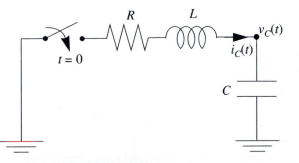

Figure 5.31 More accurate *RLC* model of a pull-down transition, including the small but always present inductance of the connecting wires.

continuous jump), this gain in accuracy is offset by the complexity of the *RLC* analysis. Perhaps we should employ an even more complicated model which incorporates the capacitor leakage or other effects. How does one know which model to use?

The problem can be posed as follows: under what conditions is the *RC* model an appropriate approximation to the *RLC* circuit? To answer this question, we will examine the natural frequencies of the *RLC* circuit where the inductive component is small, and quantify the meaning of "small."

5.11.2 Natural Frequencies in a Highly Overdamped *RLC*

The natural frequencies in the series *RLC* circuit are given by

$$s = -\frac{R}{2L} \pm \sqrt{\left(\frac{R}{2L}\right)^2 - \frac{1}{LC}} \tag{5.66}$$

Of the two terms under the square root, the first is a stronger function of *L* (inverse square versus reciprocal). Therefore, for sufficiently small values of *L*, we will have:

$$\left(\frac{R}{2L}\right)^2 \gg \frac{1}{LC} \tag{5.67}$$

which means that the response is overdamped. Multiplying through by LC:

$$\frac{R^2 C}{4L} \gg 1$$

which can be expressed as

$$\frac{4L}{R^2 C} \ll 1 \tag{5.68}$$

Factoring out the $\frac{R}{2L}$ term, Eq. (5.66) may be rewritten as

$$s = \left(\frac{R}{2L}\right)\left[-1 \pm \sqrt{1 - \frac{4L}{R^2 C}}\right] \tag{5.69}$$

From Eq. (5.68), the second term under the square root is much less than the first; consequently, we can use the binomial expansion[4] to approximate Eq. (5.69) as:

$$s \approx \left(\frac{R}{2L}\right)\left[-1 \pm \left(1 - \frac{1}{2}\frac{4L}{R^2 C}\right)\right] \tag{5.70}$$

Taking the lower sign in Eq. (5.70), and remembering that $\frac{4L}{R^2 C} \ll 1$:

$$s_1 \approx \left(\frac{R}{2L}\right)\left[-1 - \left(1 - \frac{1}{2}\frac{4L}{R^2 C}\right)\right]$$

$$s_1 \approx \left(\frac{R}{2L}\right)(-2) = -\frac{R}{L}$$

Taking the upper sign in Eq. (5.70):

$$s_2 \approx \left(\frac{R}{2L}\right)\left[-1 + \left(1 - \frac{1}{2}\frac{4L}{R^2 C}\right)\right]$$

$$s_2 \approx \left(\frac{R}{2L}\right)\left(-\frac{1}{2}\frac{4L}{R^2 C}\right) = -\frac{1}{RC}$$

Note that Eq. (5.68), which expresses the regime of inductance where this approximation is valid, can be written as

$$4 \times \frac{1}{RC} \times \frac{L}{R} \ll 1$$

---

[4] $(1 + x)^n = 1 + nx + \frac{n(n-1)}{2}x^2 + \dots$, for $x \ll 1$.

$$4|s_2|\frac{1}{|s_1|} << 1$$

which means, equivalently, that $|s_1| >> |s_2|$.

Therefore, for sufficiently small inductance, the current in the series RLC circuit (for $t > 0$) is given **approximately** by

$$i_C(t) \approx I_1 e^{-\frac{R}{L}t} + I_2 e^{-\frac{1}{RC}t} \tag{5.71}$$

Equation (5.71) consists of two terms, each of which are the solution to a first-order approximation of the RLC circuit. Referring to Figure 5.32, we see there are two possible first-order approximations. If we examine the behavior of the circuit for very small values of t (i.e., where $t < 5L/R$), then the voltage across the capacitor is essentially constant. Therefore, we can approximate the RLC circuit with an RL circuit, as shown in the top diagram of Figure 5.32. For values of time greater than this, $t > 5L/R$, the first term in Eq. (5.71) is negligibly small; therefore the response is dominated by the second term. This term corresponds to the response of an RC circuit, which results from neglecting the inductive component in the original RLC circuit.

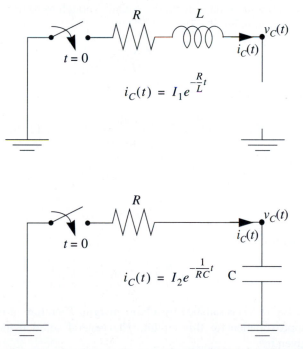

Figure 5.32 First-order approximations to the second-order RLC circuit. *Top: RL* approximation, neglecting the capacitor. This approximation is appropriate for small values of t, of the order of L/R. *Bottom: RC* approximation, neglecting the inductor. This approximation is valid for larger values of t.

The discontinuity in current which is apparent in the *RC* model at $t = 0$ is actually a very fast-changing transient term in a real circuit. If we are only concerned with calculating the logic transition time, it is sufficient to disregard this fast transient and use the *RC* model. On the other hand, there may be occasions where we do want to know the details of the transition at the early stages. For example, it might be necessary to accurately determine the peak value of current through the circuit. (This information is often needed so that long-term failure rates of the circuit may be estimated. Accurate values are important because certain kinds of failures are exponentially related to the current density.) Since the current peak occurs shortly after the switching event, examination of the response at short times is needed; thus the effect of inductance must be included.

In the following example, we use typical values for an interconnected pair of on-chip inverters, including the interconnect inductance rather than the external power-supply wire inductance.

Example 5.9

An interconnected pair of inverters is modeled in pull-down using these circuit parameters: $R = 100 \ \Omega$, $C = 100$ fF, $L = 15$ fH. Determine and justify an appropriate model for this circuit for the purpose of calculating the logic transition time.

The criteria for an inductance "small" enough to neglect is given by the inequality expressed in Eq. (5.67):

$$\left(\frac{R}{2L}\right)^2 >> \frac{1}{LC}$$

Inserting values:

$$\left(\frac{R}{2L}\right)^2 = \left(\frac{100}{2 \times 15 \times 10^{-15}}\right)^2 = 11.1 \times 10^{30}$$

and

$$\frac{1}{LC} = \frac{1}{15 \times 10^{-15} \times 100 \times 10^{-15}} = 667 \times 10^{24}$$

Therefore,

$$\left(\frac{R}{2L}\right)^2 = (16.7 \times 10^3)\frac{1}{LC}$$

so Eq. (5.67) is satisfied by a large margin. Therefore, it is appropriate to use an *RC* approximation for this circuit. The general solution including the inductance is given by

$$i_C(t) \approx I_1 e^{-\frac{R}{L}t} + I_2 e^{-\frac{1}{RC}t}$$

Inserting numerical values, the complete solution for current, including the effects of the inductor, is thus

$$i_{RLC}(t) \; = \; 0.05e^{-6.67\times10^{15}t} - 0.05e^{-1.00\times10^{11}t} \tag{5.72}$$

Ignoring the inductor results in a first-order RC circuit, whose current is given by:

$$i_{RC}(t) \; = \; -0.05e^{-1.00\times10^{11}t} \tag{5.73}$$

The two expressions for current, Eqs. (5.72) and (5.73) are plotted in Figure 5.33. We see that they are essentially identical, differing only by a short transient term just after the switching event. This plot illustrates why neglecting the inductance in this case makes no difference in determining the behavior at long times, where the logic transition will occur.

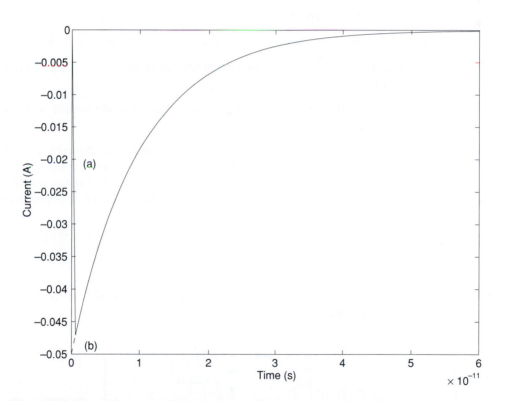

Figure 5.33 Current in the pull-down transition of Example 5.9. (a) When the inductor is included, the current is continuous. (b) The RC approximation, resulting from neglecting the inductor, shows a current discontinuity at the switching event. Except for this short transient, the two expressions for current are identical.

5.12 Summary

Interconnect inductance does not yet play a significant role in limiting the switching speed of present-day integrated circuits. Inductances associated with external wiring, especially chip packaging, printed circuit boards, and cable assemblies, however, are sufficiently large to impact circuit performance. As ICs continue to scale, interconnect lengths increase, so on-chip inductance may yet become an important limiting factor.

Inductance is described by the branch relation

$$v = L\frac{di}{dt}.$$

In the dc steady-state, an inductor is equivalent to a short circuit. Circuits containing resistors, capacitors, and inductors, known as *RLC* circuits, have four fundamental types of responses, which are classified according to the roots of the characteristic equation (i.e., the natural frequencies). These responses are:

- Overdamped, where the natural frequencies are real, negative, and distinct;

- Critically damped, where the natural frequencies are repeated, real, negative numbers;

- Underdamped, where the natural frequencies have nonzero real and imaginary components;

- Undamped, the limiting case of underdamped, where the natural frequencies are purely imaginary.

Complex circuits containing many inductors and capacitors can often be approximated by simpler circuits. A useful guideline in determining whether or not a component is significant is to calculate the first-order time constant associated with that component; if the time scales of each energy-storage component are vastly different, then a first-order approximation is valid.

Problems

5.1   In the *RL* circuit below, determine $v_L(t)$ for $t > 0$. Assume the switch has been in the upper position for a long time.

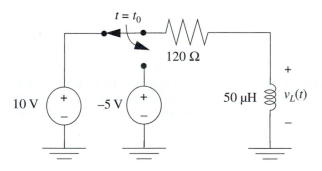

5.2   In the *RL* circuit below, determine $v_L(t)$ for $t > 0$. Assume the switch has been in the upper position for a long time.

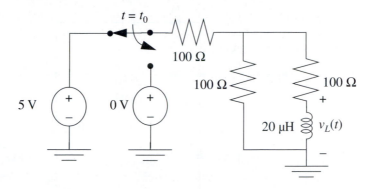

5.3   In the circuit below, the switch has been closed for a long time. At $t = 0$, the switch opens. Determine $v_1(t)$ and $v_2(t)$.

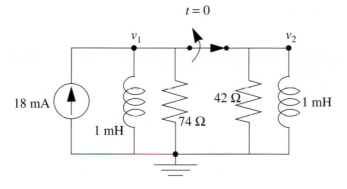

5.4   For the circuit below:

(a) Draw the dc steady-state equivalent circuit.

(b) Determine the dc steady-state voltage at point *A*.

(c) Determine the dc steady-state currents I_1 through I_5.

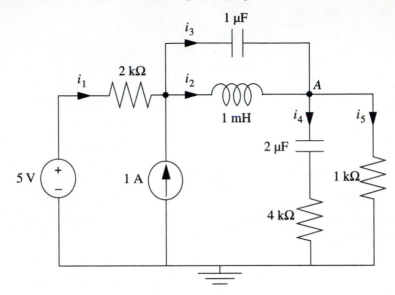

5.5 Derive the differential equations which describe the response of the circuit below. Express your solution in matrix form.

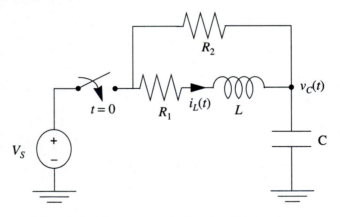

5.6 Determine ranges of C such that the response in the circuit below is:

(a) Overdamped.

(b) Critically damped.

(c) Underdamped.

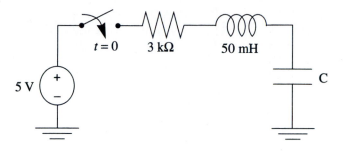

5.7 Determine ranges of L such that the response in the circuit below is:

(a) Overdamped.

(b) Critically damped.

(c) Underdamped.

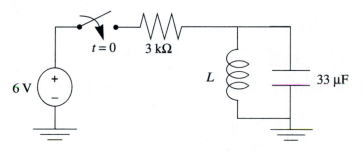

5.8 Find the natural frequencies of the circuit below, and determine the nature of the response (i.e., overdamped, underdamped, critically damped, or undamped).

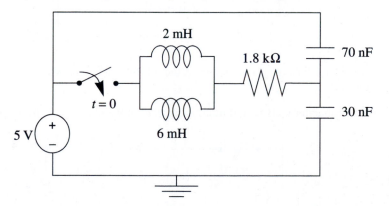

5.9   Solve for the capacitor voltage and inductor current in the circuit below.

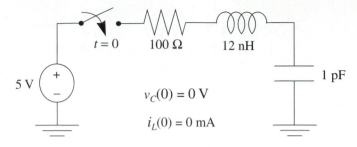

5.10 Solve for the capacitor voltage and inductor current in the circuit below.

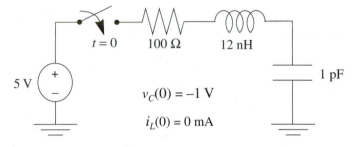

5.11 Solve for the capacitor voltage and inductor current in the circuit below.

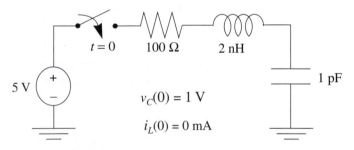

5.12 In Exercise 5.11, what is the average value of the capacitor voltage over the interval $0 < t < 100$ ps?

5.13 Solve for the capacitor voltage and inductor current in the undamped circuit below.

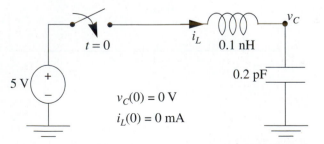

5.14 Repeat Exercise 5.13 assuming an initial capacitor voltage $v_C(0) = 2.5$ V.

5.15 Solve for the capacitor voltage and inductor current in the undamped circuit below.

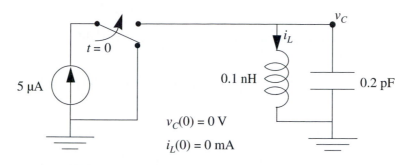

$v_C(0) = 0$ V

$i_L(0) = 0$ mA

5.16 Repeat Exercise 5.15 assuming an initial capacitor voltage $v_C(0) = 2.5$ V.

5.17 An interconnected pair of inverter gates, each with an input capacitance of 1.0 pF and an output resistance of 200 Ω, are placed in a circuit with an equivalent series inductance of $L = 1$ nH. Assume the system switches between 0 V and 5 V, with logic low and high thresholds of 1.3 V and 4.0 V. Determine the effect of the series inductance on the high-to-low switching time.

5.18 An interconnected pair of inverter gates, each with an input capacitance of 1.0 pF and an output resistance of 200 Ω, are placed in a circuit with an equivalent series inductance of $L = 10$ nH. Assume the system switches between 0 V and 5 V, with logic low and high thresholds of 1.3 V and 4.0 V. Determine the effect of the series inductance on the high-to-low switching time.

5.19 An interconnected pair of inverter gates, each with an input capacitance of 1.0 pF and an output resistance of 150 Ω, are placed in a circuit with an equivalent series inductance of $L = 10$ nH. Assume the system switches between 0 V and 5 V, with logic low and high thresholds of 1.3 V and 4.0 V. Determine the effect of the series inductance on the high-to-low switching time.

5.20 Consider the interconnected inverter gates of Exercise 5.19. The input to the first inverter gate is switched from logic high to logic low, and then back to logic high 0.45 ns later. Calculate and plot the transient response.

5.21 Each of an interconnected pair of inverter gates has an input capacitance of 2.0 pF and an output resistance of 100 Ω. The system switches between 0 V and 5 V, with logic low and high thresholds of 1.3 V and 4.0 V. The inverters can be placed on the chip in one of two configurations. In the first alternative, the series inductance is 8 nH and the interconnect resistance is 100 Ω. The second alternative results in a higher series inductance, 10 nH, but a lower interconnect resistance, 50 Ω. Determine which of the two configurations is more desirable.

5.22 Each of two interconnected pairs of inverter gates has an input capacitance of 2.0 pF and an output resistance of 50 Ω. The circuit has an equivalent series inductance of 2 nH. The system switches between 0 V and 5 V, with logic low and high thresholds of 1.3 V and 4.0 V.

(a) The inverters are connected with a very short wire, with negligible resistance and capacitance. Determine the nature of the transient response.

(b) Suppose the inverters are connected with an aluminum interconnect 1.25 μm wide and 0.5 μm thick, which runs over a 0.2-μm layer of SiO_2. Determine the ranges of interconnect length which result in underdamped, critically damped, and overdamped transient responses.

5.23 Derive the symbolic solution for an overdamped series *RLC* circuit with $v_C(0) = v_{C0}$ and $i_L(0) = i_{L0}$:

$$v_C(t) = \frac{1}{s_2 - s_1}\left[s_2(v_{C0} - V_S) - \frac{1}{C}(i_{L0} - I_L)\right]e^{s_1 t}$$

$$+ \frac{1}{s_2 - s_1}\left[-s_1(v_{C0} - V_S) + \frac{1}{C}(i_{L0} - I_L)\right]e^{s_2 t} + V_S$$

$$i_L(t) = \frac{1}{s_2 - s_1}\left[-s_1(i_{L0} - I_L) + \frac{1}{L}(v_{C0} - V_S)\right]e^{s_1 t}$$

$$+ \frac{1}{s_2 - s_1}\left[s_2(i_{L0} - I_L) - \frac{1}{L}(v_{C0} - V_S)\right]e^{s_2 t}$$

5.24 Show that the general solutions for the critically damped case of the series RLC circuit, given by Eqs. (5.33) and (5.34), follow from the overdamped solutions (Exercise 5.23) where $s_1 = s_2$. (*Hint:* Let the two overdamped roots be given by $s_1 = s + \varepsilon$ and $s_2 = s - \varepsilon$; the critically damped case corresponds to the limit as $\varepsilon \to 0$.)

5.25 Derive the symbolic solution for a critically damped series *RLC* circuit with $v_C(0) = v_{C0}$ and $i_L(0) = i_{L0}$:

$$v_C(t) = (v_{C0} - V_S)e^{st} + \left[-s_2(v_{C0} - V_S) + \frac{1}{C}(i_{L0} - I_L)\right]te^{st} + V_S$$

$$i_L(t) = (i_{L0} - I_L)e^{st} + \left[s_2(i_{L0} - I_L) - \frac{1}{L}(v_{C0} - V_S)\right]te^{st}$$

5.26 Derive the symbolic solution for an underdamped series *RLC* circuit with $v_C(0) = v_{C0}$ and $i_L(0) = i_{L0}$:

$$v_C(t) = (v_{C0} - V_S)e^{-\alpha t}\cos(\omega_d t)$$

$$+ \left[\frac{\alpha C(v_{C0} - V_S) + (i_{L0} - I_L)}{\omega_d C}\right]e^{-\alpha t}\sin(\omega_d t) + V_S$$

$$i_L(t) = (i_{L0} - I_L)e^{-\alpha t}\cos(\omega_d t)$$

$$+ \left[\frac{\alpha L i_{L0} - (v_{C0} - V_S)}{\omega_d L}\right]e^{-\alpha t}\sin(\omega_d t)$$

5.27 Derive the symbolic solution for an undamped series RLC circuit with $v_C(0) = v_{C0}$ and $i_L(0) = i_{L0}$:

$$v_C(t) = (v_{C0} - V_S)\cos(\omega_0 t) + \left[\frac{(i_{L0} - I_L)}{\omega_0 C}\right]\sin(\omega_0 t) + V_S$$

$$i_L(t) = (i_{L0} - I_L)\cos(\omega_0 t) + \left[-\frac{(v_{C0} - V_S)}{\omega_0 L}\right]\sin(\omega_0 t)$$

5.28 The circuit below is an example of a parallel *GLC* circuit. Determine the capacitor voltage and inductor current for $t > 0$. Assume: $G = 30$ mS, $L = 2$ mH, $C = 1$ μF.

$v_C(0) = 2$ V

$i_L(0) = 5$ mA

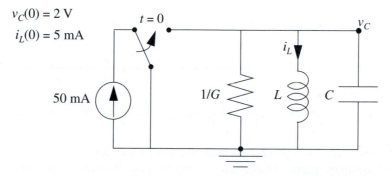

5.29 Derive the symbolic solution for the general *GLC* circuit below. Assume the response is underdamped.

$v_C(0) = v_{C0}$

$i_L(0) = i_{L0}$

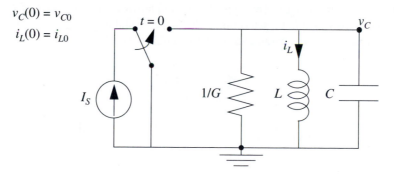

5.30 In the circuit below, the component values are: $R = 200\ \Omega$, $L = 1$ mH, $C = 2\ \mu$F. Determine the maximum current which flows through the voltage source, and the time at which it occurs.

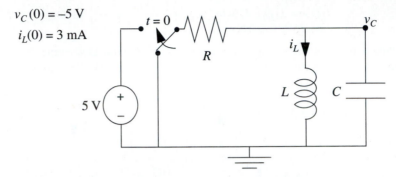

$$v_C(0) = -5\ \text{V}$$
$$i_L(0) = 3\ \text{mA}$$

5.31 (a) Solve for the natural frequencies s_1 and s_2 of the circuit below, as a function of the resistance R.

(b) Assume that $R = 1000\ \Omega$. Determine the natural frequencies of the circuit and plot them on the complex plane. Is the circuit overdamped, underdamped, critically damped, or undamped?

(c) Now assume that R varies over the range of $0\ \Omega$ to 10 kΩ. Plot the positions of the natural frequencies as R changes. This type of plot is known as a **root locus** of the system. Root-locus plots are valuable in determining the range of stability of systems (especially systems with feedback) as some parameter, such as amplifier gain, is varied.

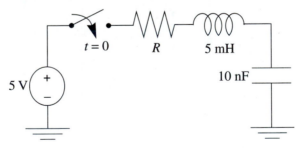

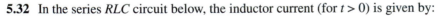

5.32 In the series RLC circuit below, the inductor current (for $t > 0$) is given by:

$$i_L(t) = 0.0048 e^{-(5.0 \times 10^8)t} \sin(8.66 \times 10^8)t$$

Determine R, C, and $v_C(t)$.

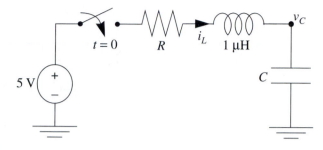

5.33 Solve for $v_C(t)$ and $i_L(t)$ in the *RLGC* circuit below.

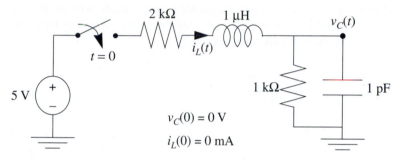

5.34 Derive the general symbolic solution of an underdamped *RLGC* circuit like that of Figure 5.28.

5.35 Show that the natural frequencies in an *RLGC* circuit, given by the roots of Eq. (5.64), include the series *RLC* and parallel *GLC* circuits as special cases.

5.36 The damping factor, α, is a measure of how effectively the natural oscillations of a circuit are suppressed. Damping is the physical dissipation of energy; a higher damping factor corresponds to more effective conversion of stored energy into heat. Explain physically why α is proportional to the resistance in a series *RLC* circuit but is inversely proportional to resistance in a parallel *RLC* circuit.

5.37 In our discussion of the underdamped transient response (Example 5.5), we observed that the capacitor voltage "overshoots" the dc steady-state value. Show that overshoot can also occur for critically damped and overdamped responses.

5.38 Consider the transient response of a series *RLC* circuit, where the resistance R can be varied over a large range. As R is increased, the transient response varies from underdamped, through critically damped, to overdamped. Suppose a considerable amount of overshoot occurs even when the circuit is slightly overdamped. Show that this overshoot can always be eliminated with a sufficiently high damping (i.e., at high values of R).

5.39 A low-to-high transient logic response is seen to overshoot the final dc steady-state value, and subsequently it undershoots the dc steady-state value. Is it possible to conclude that the transient response is underdamped?

5.40 Derive the differential equations (in matrix form) that describe the response of this circuit.

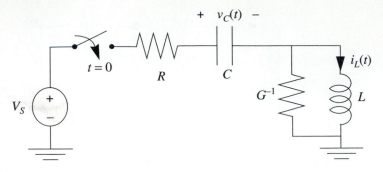

5.41 Instead of using a single-RC-lump interconnect approximation in our interconnected inverter model, in we could use a two-section RC ladder circuit. Including inductance, the resulting third-order model would look like the following:

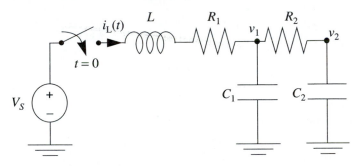

(a) Derive the differential equations (in matrix form) describing the response of this circuit.

(b) Assume the following circuit parameters: $L = 10\ \mu\text{H}$, $R_1 = R_2 = 1\ \text{k}\Omega$, $C_1 = C_1 = 1\ \text{pF}$. Show that one of the natural frequencies is equal to

$$s_1 = -1.9737 \times 10^9\ \text{s}^{-1}$$

and determine the other two natural frequencies.

(c) Suppose the second-order RLC model were used to approximate the behavior (i.e., the one-RC-lump interconnect model). Determine the natural frequencies, using the appropriate parameter values.

(d) For these parameter values, is a first-order RC model (i.e., neglecting the inductance) an accurate approximation?

(e) Comparing the values of natural frequencies obtained for the first-, second-, and third-order models, justify an appropriate choice of model.

5.42 A digital system operating between logic levels of 0 V and 5 V has logic threshold values of $V_{tl} = 0.5\ \text{V}$ and $V_{th} = 4.5\ \text{V}$. A pair of interconnected inverter gates in this system is well characterized by a first-order RC model with $R_{out} = 1\ \text{k}\Omega$ and $C_{in} = 0.2\ \text{pF}$; the interconnect itself is very short and has negligible resistance and capacitance.

(a) Assume the gates have been quiescent for a long time, such that the input of the load gate is at logic low. At $t = 0$, the input of the driving-gate goes low; at $t = 1.0$ ns, the driving gate input goes high. Calculate the values of t where the load gate transitions into logic high and low.

(b) Suppose a 75-nH inductance is inserted in series with the interconnect. Calculate the resulting transition times including this inductance. (Such a deliberately inserted inductance is known as a "speed-up inductor." Because inductors take up a lot of space on an IC, this technique is rarely employed in VLSI designs.)

5.43 An interconnected inverter pair can be accurately modeled with an *RLGC* circuit (Figure 5.28) with the following parameters: $R = 1$ kΩ, $L = 1$ nH, $C = 0.1$ pF, $G = 2$ nS. Determine which model is appropriate in calculating logic transition times: the full *RLGC* model, a series *RLC* model neglecting the gate conductance, or a parallel *GLC* model neglecting the interconnect and output resistance.

Printed-Circuit Interconnects and Lossless Transmission Lines

6

6.1 Introduction

Thus far, we have studied single-lump RC circuits, RC ladder circuits, and RLC circuits. These models are useful to approximate inverter-pair switching characteristics, interconnect limitations, and power-line influence on inverter switching, respectively.

We now turn to another circuit configuration used to model the behavior of interconnects, the **lossless transmission line** (also known as an LC **line** and **delay line**). Such lines are found linking ICs in multichip modules (MCMs) and printed circuit boards. They are relatively thick and composed of low-resistivity materials, hence their resistance may often be neglected. Unlike on-chip interconnects, though, lossless transmission lines possess significant amounts of inductance. Also, the capacitance of the line to ground is not negligible.

In this chapter we will develop a continuum model for lossless transmission lines and show that the voltage and current signals are given by wave solutions. Practical problems involving transmission lines will be solved by using simple equivalent-circuit models. We will see that circuits containing transmission lines can have several interesting behaviors, such as time delay and signal reflection.

6.2 *n*-lump Transmission-Line Model

Consider a long, low-resistance conductor running a short distance over the ground plane. We will assume the line resistance is negligible (i.e., it is much less than the output resistance of the driving gate). The total inductance of the line is given by L, and the total capacitance to ground is C. Such a conductor could be a metal trace on a circuit board or MCM carrying signals between two chips, as shown in Figure 6.1.

Ideally, if chip 1 sends a logic signal to chip 2, that signal will arrive instantaneously at chip 2 and be identical in form to the transmitted signal. However, the signal propagating down the transmission line is an **electromagnetic wave**, which has a high but finite speed. Therefore the signal takes a finite period of time to travel the length of the line. Also, we will see that the shape of the signal when it arrives at its destination is usually altered.

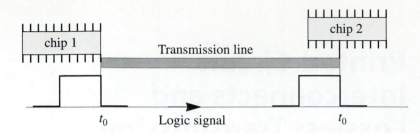

Figure 6.1 Top view of a transmission line connecting two chips in an MCM. The metal line runs over a conducting ground plane parallel to the plane of the page. Ideally, the signal at the output of chip 1 appears instantaneously at the input of chip 2, with no change in pulse shape. Unfortunately, neither of these ideal behaviors is ever encountered.

To calculate the inevitable time delay, and pulse distortion, we will use the n-lump LC transmission-line model. In Chapter 3 we illustrated how the distributed resistance and capacitance of an RC interconnect can be approximated with 1-lump, 2-lump, and n-lump models. The same holds for a transmission line. The inductance and capacitance are distributed uniformly along the length of the line, which can be thought of using the combination symbol shown in Figure 6.2(a). For analysis purposes, the distributed line is broken into discrete lumps. The n-lump model of a lossless LC transmission line is shown in Figure 6.2(b).

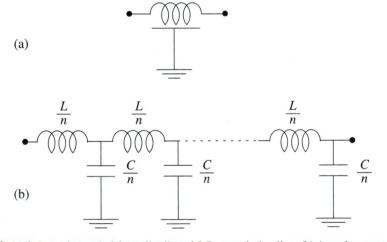

Figure 6.2 (a) Schematic symbol for a distributed LC transmission line. (b) An n-lump model of the lossless line. The total inductance L and capacitance C of the line is broken up into n equal lumps.

Unlike the RC line, the lossless LC line can be easily modeled when the number of lumps approaches infinity. Therefore, instead of using a small number of LC lumps, which will give only approximate results, we will derive an exact model, using an infinite number of infinitesimal lumps connected in a ladder configuration.

6.3 Analysis of the Lossless Transmission-Line Model

6.3.1 Derivation of the Telegrapher's Equations

Consider a lossless LC line of length l (cm) with total inductance L and capacitance C. For a uniform line; these parameters are distributed equally along the line, thus we can speak of the inductance per unit length,

$$\frac{L}{l} \tag{6.1}$$

in H/cm, and the capacitance per unit length,

$$\frac{C}{l} \tag{6.2}$$

in F/cm.

Figure 6.3(a) illustrates the physical configuration of the line. We specify the position along the line with the variable x which can take on values from zero to l, the total length of the line. Figure 6.3(b) illustrates the circuit symbol we will use to represent the transmission line in schematic diagrams.

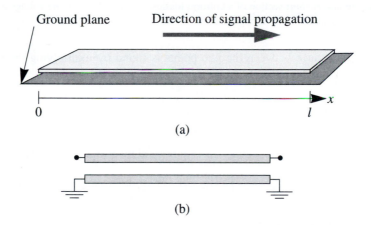

(a)

(b)

Figure 6.3 (a) Physical layout of an LC line of length l, showing the x-axis along the direction of signal propagation. (b) Schematic symbol for the transmission line.

To analyze the transmission line, we focus on a short section extending from x to $x + \Delta x$ anywhere along the line. The circuit model for this short section is shown in Figure 6.4. The currents flowing **into** each of the two marked nodes through the inductances are given by

$$i(x, t) \qquad \text{and} \qquad i(x + \Delta x, t)$$

and the voltages (with respect to ground) at the two nodes are

$$v(x, t) \qquad \text{and} \qquad v(x + \Delta x, t)$$

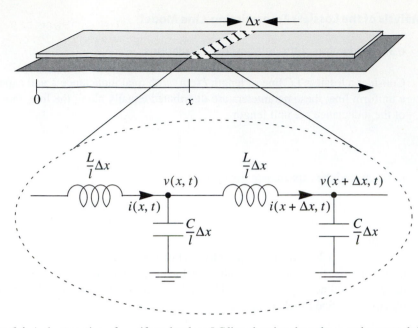

Figure 6.4 A short section of a uniform lossless LC line showing the voltage and current definitions at points x and $x + \Delta x$ along the line.

The inductance between the two points is found by multiplying the inductance per unit length, given by Eq. (6.1), by the length Δx; the result is

$$\frac{L}{l}\Delta x$$

and similarly for the capacitance.

Applying KVL around the loop formed by ground and the two nodes, we obtain

$$v(x, t) - v(x + \Delta x, t) - \left(\frac{L}{l}\Delta x\right)\frac{d}{dt}(i(x + \Delta x, t)) = 0$$

which can be rearranged as

$$\frac{v(x + \Delta x, t) - v(x, t)}{\Delta x} = -\frac{L}{l}\frac{d}{dt}i(x + \Delta x, t)$$

Applying KCL at the $v(x, t)$ node,

$$i(x, t) = \left(\frac{C}{l}\Delta x\right)\frac{d}{dt}(v(x, t)) + i(x + \Delta x, t)$$

which when rearranged can be written as

$$\frac{i(x + \Delta x, t) - i(x, t)}{\Delta x} = -\frac{C}{l}\frac{d}{dt}(v(x, t))$$

Now, suppose we shrink the length Δx; this corresponds to increasing the number of lumps in the model. Taking the limit of the two Kirchhoff law expressions as $\Delta x \to 0$ yields

$$\frac{\partial v}{\partial x} = -\frac{L}{l}\frac{\partial i}{\partial t} \tag{6.3}$$

$$\frac{\partial i}{\partial x} = -\frac{C}{l}\frac{\partial v}{\partial t} \tag{6.4}$$

where

$$v \equiv v(x, t)$$

$$i \equiv i(x, t)$$

Since the voltage and the current are functions of two variables (x, the position along the *LC* line, and t, time), we use **partial derivatives** to describe how each changes with respect to one variable as the other is kept constant.

The coupled partial differential equations, Eqs. (6.3) and (6.4), are known as the **telegrapher's equations** for a lossless transmission line. Transmission-line analysis first arose in the nineteenth century in the study of wires carrying telegraph signals that represented Morse code. Signal distortion was a problem then, too, even though there were only two kinds of signals: long pulses ("dashes") and short pulses ("dots"). Care had to be taken in transmission, else erroneous dashes or dots would appear or correct ones could disappear entirely.

In general, solving partial differential equations is considerably more difficult than solving the ordinary differential equations we encountered previously. Fortunately, this pair of equations are an exception to the general rule. The functions $i(x, t)$ and $v(x, t)$ which satisfy Eqs. (6.3) and (6.4) are:

$$i(x, t) = f_+\left(t - \frac{x}{v_0}\right) - f_-\left(t - \frac{l-x}{v_0}\right) \tag{6.5}$$

$$v(x, t) = Z_0\left[f_+\left(t - \frac{x}{v_0}\right) + f_-\left(t - \frac{l-x}{v_0}\right)\right] \tag{6.6}$$

The functions

$$f_+\left(t - \frac{x}{v_0}\right) \text{ and } f_-\left(t - \frac{l-x}{v_0}\right)$$

are traveling waves. We provide a detailed interpretation of these functions in the next section.

The constant v_0 is given by

$$v_0 = \frac{1}{\sqrt{\left(\frac{L}{l}\right)\left(\frac{C}{l}\right)}} = \frac{l}{\sqrt{LC}} \tag{6.7}$$

We call v_0 the **characteristic velocity**. Recall from Chapter 5 that the combination $1/\sqrt{LC}$ has units of s^{-1}; therefore v_0 has units of cm/s, which is a velocity unit. We will see that v_0 is the speed at which waves propagate along the transmission line. (Note that the symbol v_0 does NOT designate the value of a voltage at $t = 0$!)

The constant Z_0 is given by

$$Z_0 = \sqrt{\frac{L}{C}} \tag{6.8}$$

and is called the **characteristic impedance** of the line. Z_0 has units of ohms (Ω), just like resistance. This can be easily seen from the fundamental definitions of inductance and capacitance:

$$L = \frac{v}{di/dt}$$

$$C = \frac{i}{dv/dt}$$

Taking the ratio of these expressions and expressing the right side in terms of the appropriate units:

$$\sqrt{\frac{L}{C}} = \sqrt{\frac{V}{A/s} \div \frac{A}{V/s}} = \sqrt{\left(\frac{V}{A}\right)^2} = \Omega$$

(In fact, we will see in later chapters that impedance is a more general form of resistance.)

Another useful parameter is the **delay time** T_D:

$$T_D = \frac{l}{v_0} = \sqrt{LC} \tag{6.9}$$

T_D is the time taken by the wave, traveling at v_0 cm/s, to propagate the length l of the line.

The characteristic velocity depends strongly on the dielectric material separating the two conductors of the transmission line (i.e., the metal trace and the ground plane). If the two conductors are separated by a vacuum, then the characteristic velocity is equal to the speed of light. This is an upper limit on the propagation speed of signals in an electronic system; generally, the speed will be much less. To see this, we begin with the expression for inductance developed in Chapter 1. The transmission line can be thought of as a single-turn solenoid, which means the inductance is given by

$$L = \mu\frac{ld}{w} \tag{6.10}$$

where w is the linewidth, d is the dielectric thickness, and μ is the dielectric permeability. The capacitance of the line is

$$C = \varepsilon\frac{lw}{d} \tag{6.11}$$

where ε is the dielectric permittivity. Substituting Eqs. (6.10) and (6.11) into Eq. (6.7), we obtain

$$v_0 = \frac{l}{\sqrt{LC}} = \frac{l}{\sqrt{\mu \dfrac{ld}{w} \varepsilon \dfrac{lw}{d}}} = \frac{1}{\sqrt{\varepsilon\mu}} \qquad (6.12)$$

Minimum values of ε and μ are the free-space values, ε_0 and μ_0. When these are inserted in Eq. (6.12), the result is $v_0 = c$, the speed of light. Generally, the insulating materials in transmission lines have $\mu \approx \mu_0$, but the permittivity is higher than the free-space permittivity. For example, the dielectric constant of materials used as printed circuit-board substrates ranges from 2.1 to 2.6, so the wave speed is at most $0.7c$.

6.3.2 Verification of the Wave Solutions

We first want to verify that the expressions in Eqs. (6.5) and (6.6) indeed satisfy the partial differential equations describing the signals on the transmission line. The salient fact about these equations is the form of the functional argument. Because the two functions have different arguments, we will find it convenient to define two argument variables:

$$\xi_+ = t - \frac{x}{v_0} \qquad (6.13)$$

$$\xi_- = t - \frac{l-x}{v_0} \qquad (6.14)$$

Using these variables, the current and voltage solutions can be expressed as

$$i(x, t) = f_+(\xi_+) - f_-(\xi_-)$$

$$v(x, t) = Z_0[f_+(\xi_+) + f_-(\xi_-)]$$

This notation is useful because we must evaluate the various partial derivatives in Eqs. (6.3) and (6.4). For example, the partial of current with respect to x is found using the chain rule, as follows:

$$\frac{\partial i}{\partial x} = \frac{\partial}{\partial x}[f_+(\xi_+) + f_-(\xi_-)]$$

$$= \frac{\partial f_+(\xi_+)}{\partial \xi_+}\frac{\partial \xi_+}{\partial x} - \frac{\partial f_-(\xi_-)}{\partial \xi_-}\frac{\partial \xi_-}{\partial x}$$

$$= -\frac{1}{v_0}\frac{\partial f_+(\xi_+)}{\partial \xi_+} - \frac{1}{v_0}\frac{\partial f_-(\xi_-)}{\partial \xi_-}$$

The other partial derivatives are derived similarly:

$$\frac{\partial v}{\partial x} = -\frac{Z_0}{v_0}\left(\frac{\partial f_+(\xi_+)}{\partial \xi_+} - \frac{\partial f_-(\xi_-)}{\partial \xi_-}\right)$$

$$\frac{\partial i}{\partial t} = \frac{\partial f_+(\xi_+)}{\partial \xi_+} - \frac{\partial f_-(\xi_-)}{\partial \xi_-}$$

$$\frac{\partial v}{\partial t} = Z_0\left(\frac{\partial f_+(\xi_+)}{\partial \xi_+} + \frac{\partial f_-(\xi_-)}{\partial \xi_-}\right)$$

Inserting the appropriate relations into Eq. (6.3):

$$\frac{\partial v}{\partial x} = -\frac{L}{l}\frac{\partial}{\partial t}(i)$$

$$-\frac{Z_0}{v_0}\left(\frac{\partial f_+(\xi_+)}{\partial \xi_+} - \frac{\partial f_-(\xi_-)}{\partial \xi_-}\right) = -\frac{L}{l}\left(\frac{\partial f_+(\xi_+)}{\partial \xi_+} - \frac{\partial f_-(\xi_-)}{\partial \xi_-}\right)$$

The derivative terms in parentheses on each side cancel, leaving

$$-\frac{Z_0}{v_0} = -\frac{\sqrt{\dfrac{L}{C}}}{\dfrac{l}{\sqrt{LC}}} = -\frac{L}{l}$$

which is an identity. Therefore, Eq. (6.3) is satisfied by the solutions. Making appropriate substitutions into Eq. (6.4),

$$\frac{\partial i}{\partial x} = -\frac{C}{l}\frac{\partial v}{\partial t}$$

$$-\frac{1}{v_0}\left(\frac{\partial f_+(\xi_+)}{\partial \xi_+} + \frac{\partial f_-(\xi_-)}{\partial \xi_-}\right) = -\frac{C}{l}Z_0\left(\frac{\partial f_+(\xi_+)}{\partial \xi_+} + \frac{\partial f_-(\xi_-)}{\partial \xi_-}\right)$$

$$-\frac{1}{v_0} = -\frac{C}{l}Z_0$$

$$-\frac{1}{\dfrac{l}{\sqrt{LC}}} = -\frac{C}{l}\sqrt{\frac{L}{C}}$$

which is also an identity, therefore the solutions also satisfy Eq. (6.4).

In showing that the wave solutions satisfy the transmission-line equations, we have used only the fact that the variables x and t in the arguments of f_+ and f_- must occur in the forms given by Eqs. (6.13) and (6.14). In particular, we have assumed nothing about the actual functions f_+ and f_-. As we will see later, the value of these functions is determined by the circuit in which the transmission line is placed. But, from a mathematical point of view, the actual function is irrelevant so long as the functional dependence on x and t is of the proper form. For example, these functions are valid solutions to the transmission-line equations

$$i(x, t) = \left(t - \frac{x}{v_0}\right)^2 \cos\left(t - \frac{x}{v_0}\right) - \sqrt{t - \frac{l-x}{v_0}}$$

$$v(x, t) = Z_0\left[\left(t - \frac{x}{v_0}\right)^2 \cos\left(t - \frac{x}{v_0}\right) + \sqrt{t - \frac{l-x}{v_0}}\right]$$

since the variables x and t appear in the correct combinations.

6.4 Interpretation of the Wave Solutions

The solutions to the telegrapher's equations describe traveling waves. The function $f_+(\xi_+)$ represents a wave traveling in the positive x direction, and is therefore called a **right-traveling** wave. Similarly, the function $f_-(\xi_-)$ describes a **left-traveling** wave which moves in the negative x direction. We will demonstrate these properties by example.

Example 6.1

Consider the function $f_+(\xi_+)$ in Figure 6.5. Plot this as a function of x and t, and show that it represents a right-traveling wave. Assume $v_0 = 0.5$ cm/s.

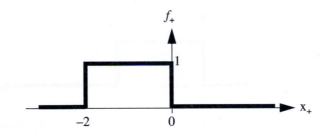

Figure 6.5 The $f_+(\xi_+)$ traveling-wave pulse for Example 6.1. The wave velocity in this example is $v_0 = 0.5$ cm/s.

The function $f_+(\xi_+)$ takes on the value 0 or 1, depending on the value of ξ_+. We can express the function algebraically by

$$f_+ = \begin{cases} 0, & \xi_+ < -2 \\ 1, & -2 < \xi_+ < 0 \\ 0, & 0 < \xi_+ \end{cases}$$

Since $\xi_+ = t - \dfrac{x}{v_0} = t - 2x$, the function f_+ can be expressed equivalently as

$$f_+(x,t) = \begin{cases} 0, & t-2x < -2 \\ 1, & -2 < t-2x < 0 \\ 0, & 0 < t-2x \end{cases} \tag{6.15}$$

Suppose we evaluate this function at $t = 0$:

$$f_+(x,0) = \begin{cases} 0, & -2x < -2 \\ 1, & -2 < -2x < 0 \\ 0, & 0 < -2x \end{cases}$$

$$= \begin{cases} 0, & x > 1 \\ 1, & 1 > x > 0 \\ 0, & 0 > x \end{cases}$$

This is plotted in Figure 6.6. At $t = 0$, the function is zero everywhere along the x-axis, except between $x = 0$ and $x = 1$, where it is 1.

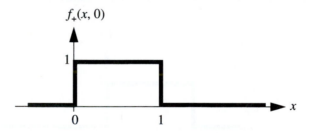

Figure 6.6 The $f_+(x, t)$ wave pulse at $t = 0$, plotted as a function of position.

To demonstrate the wave motion, we examine the x-dependence of the function a short time later. We evaluate f_+ after one second by substituting $t = 1$ in Eq. (6.15):

$$f_+(x,1) = \begin{cases} 0, & 1-2x < -2 \\ 1, & -2 < 1-2x < 0 \\ 0, & 0 < 1-2x \end{cases}$$

Evaluating the range constraints, this is equivalent to:

$$f_+(x,1) = \begin{cases} 0, & x > 1.5 \\ 1, & 1.5 > x > 0.5 \\ 0, & 0.5 > x \end{cases}$$

which is plotted in Figure 6.7.

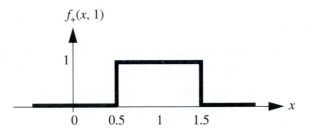

$f_+(x, 1)$

Figure 6.7 The $f_+(x, t)$ wave pulse at $t = 1$. Relative to its position at $t = 0$, it has shifted 0.5 cm to the right.

We see that the function has the same shape but has shifted along the x-axis. If we track corresponding parts of the pulse in time, we see that each has moved 0.5 cm in the 1-s interval; this is consistent with the given value of 0.5 cm/s for the wave velocity. Also, the pulse has shifted in the positive x-direction, hence f_+ is indeed a right-traveling wave.

We can plot the f_+ wave for any arbitrary time t using Eq. (6.15). In Figure 6.8 we show f_+ as a function of x and t at one-second intervals from $t = 0$ to $t = 4$ s. When plotted in this way, the interpretation of f_+ as a right-traveling wave, with a velocity of 0.5 cm/s, is quite apparent.

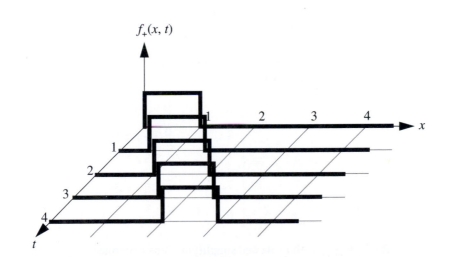

$f_+(x, t)$

Figure 6.8 Evolution of the f_+ traveling wave in time and space.

The next example illustrates the wave interpretation of f_+ with an asymmetric function.

Example 6.2

Consider the function $f_+(0, t)$ in Figure 6.9. Plot this as a function of x and show that it represents a right-traveling wave. Assume $v_0 = 0.25$ cm/s.

Note that the graph in Figure 6.9 illustrates the variation of f_+ in time, not position. We therefore need to determine how f_+ varies as x, from this information.

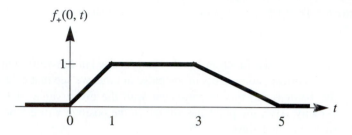

Figure 6.9 The f_+ traveling-wave pulse for Example 6.2, shown as a function of time at position $x = 0$.

We know that f_+ has the functional dependence on x and t as given in Eq. (6.13). Substituting in the known value of v_0:

$$f_+(\xi_+) = f_+\left(t - \frac{x}{v_0}\right) = f_+(t - 4x)$$

Since $\xi_+ = t$ when $x = 0$, the graph in Figure 6.9 also represents the functional dependence of f_+ on ξ_+. Therefore, we can express f_+ algebraically as:

$$f_+ = \begin{cases} 0, & \xi_+ < 0 \\ \xi_+, & 0 < \xi_+ < 1 \\ 1, & 1 < \xi_+ < 3 \\ \dfrac{5 - \xi_+}{2}, & 3 < \xi_+ < 5 \\ 0, & 5 < \xi_+ \end{cases}$$

Replacing ξ_+ with $t - 4x$ and simplifying, this becomes:

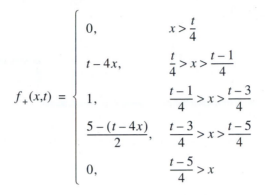

$$f_+(x,t) = \begin{cases} 0, & x > \dfrac{t}{4} \\[2ex] t - 4x, & \dfrac{t}{4} > x > \dfrac{t-1}{4} \\[2ex] 1, & \dfrac{t-1}{4} > x > \dfrac{t-3}{4} \\[2ex] \dfrac{5 - (t - 4x)}{2}, & \dfrac{t-3}{4} > x > \dfrac{t-5}{4} \\[2ex] 0, & \dfrac{t-5}{4} > x \end{cases}$$

This is plotted as a function of x in Figure 6.10, for both $t = 0$ and $t = 4$ s. Again we see the displacement along the positive x direction at increasing values of time.

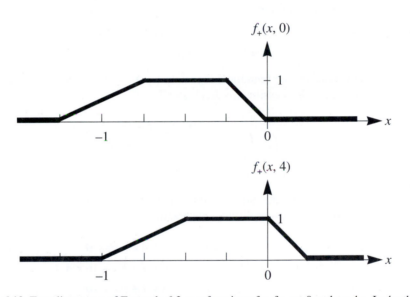

Figure 6.10 Traveling wave of Example 6.2 as a function of x, for $t = 0$ and $t = 4$ s. In 4 s the wave has moved 1 cm to the right, consistent with the given wave velocity $v_0 = 0.25$ cm/s and the interpretation of f_+ as a right-traveling wave.

We now consider a left-traveling wave.

Example 6.3

Consider the function $f_-(x, 0)$ in Figure 6.11. Show that it represents a left-traveling wave. Assume $1 = 4$ cm and $v_0 = 0.25$ cm/s.

Using the known functional form of $f_-(x, t)$, we can write the wave function as

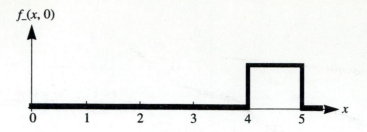

Figure 6.11 The f_- traveling-wave pulse for Example 6.3, shown as a function of position at time $t = 0$.

$$f_-\left(t - \frac{l-x}{v_0}\right) = f_-\left(t - \frac{4-x}{0.25}\right) = f_-(t + 4x - 16)$$

The graph tells us that $f_- = 1$ when $t = 0$ and x is between 4 and 5. Substituting $t = 0$ and the two limiting values of x into the argument of f_-:

$$t = 0, \qquad x = 4 \rightarrow t + 4x - 16 = 0$$
$$t = 0, \qquad x = 5 \rightarrow t + 4x - 16 = 4$$

This brackets the functional argument $t + 4x - 16$, where $f_- = 1$, between the values of 0 and 4. Expressing this algebraically,

$$f_-\left(t - \frac{l-x}{v_0}\right) = \begin{cases} 0, & t + 4x - 16 < 0 \\ 1, & 0 < t + 4x - 16 < 4 \\ 0, & 4 < t + 4x - 16 \end{cases}$$

To demonstrate the nature of this function, we determine its x-dependence at $t = 1$. Substituting and simplifying, we obtain

$$f_-(x, 1) = \begin{cases} 0, & 1 + 4x - 16 < 0 \\ 1, & 0 < 1 + 4x - 16 < 4 \\ 0, & 4 < 1 + 4x - 16 \end{cases}$$

$$f_-(x, 1) = \begin{cases} 0, & x < 3.75 \\ 1, & 3.75 < x < 4.75 \\ 0, & 4.75 < x \end{cases}$$

This is plotted in Figure 6.12. Again, the function has the same shape, but this time it has shifted along the x-axis in the negative direction. Also, by tracking corresponding parts of the pulse in time, we can confirm the interpretation of v_0 as a velocity, as the wave has moved –0.25 cm in the 1 s interval.

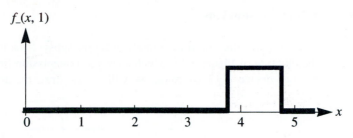

Figure 6.12 The f_- traveling-wave pulse for Example 6.3, shown as a function of position at time $t = 1$. The pulse has been displaced from its initial position by -0.25 cm.

The evolution of this left-traveling wave over the first four seconds is shown in Figure 6.13. Also shown in this figure is a dashed line, representing the position of the wavefront leading edge. The slope of this line (considering the time axis as the abscissa and the position axis the ordinate), dx/dt, is the wave velocity v_0.

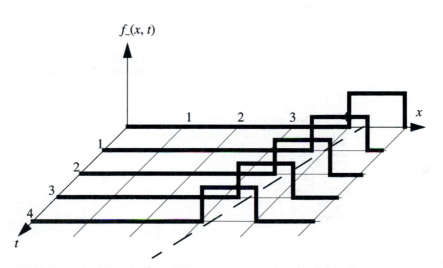

Figure 6.13 A plot showing the f_- traveling-wave as a function of time and position. The dashed line is a projection onto the t-x plane of the leading edge of the wavefront. The slope of this line corresponds to the wave velocity v_0.

These examples illustrate the traveling-wave nature of the functions f_+ and f_-. The difference in sign associated with the x-variable in the function arguments accounts for the opposite directions of propagation.

6.5 Reflections on Transmission Lines

Transmission lines are often used to connect different subsystems together. This is suggested in the diagram of Figure 6.14, which shows a transmission line between a driving circuit and a load device. By convention, we will always draw the driving circuit (known

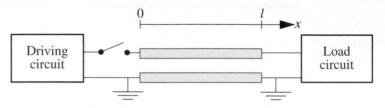

Figure 6.14 Transmission line connecting a driving circuit with a load. For example, the driving circuit could be a CMOS gate, and the load another gate or perhaps a peripheral device.

as the **source**) on the left, such that its output appears at $x = 0$. The load is drawn on the right, therefore the signal appearing at $x = l$ on the transmission line is presented directly to the load. The switch in this diagram, while actually part of the driving circuit, is shown explicitly to highlight the transient nature of the problem. When the switch is closed, the signal will originate at the source, travel down the transmission line, and arrive T_D seconds later at the load.

While this description makes intuitive sense, it does not explain the form of the solutions to the transmission-line equations, which contain both right- and left-traveling waves. The right-traveling wave clearly originates from the source at the time the switch is closed, but where does the left-traveling wave come from?

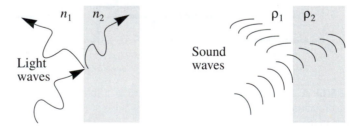

Figure 6.15 Reflections occur when a wave traveling in one medium encounters a discontinuity. Light waves traveling in a material with index of refraction n_1 will be partially reflected and partially transmitted at a boundary with a material of index n_2. Similarly, sound waves will reflect when they encounter a material of a different density.

The answer is that waves traveling along the transmission line can be **reflected** at either end. The left-traveling waves are reflections from the load end. Reflections of any kind of waves occur when there is a **discontinuity** in the medium. For example, light waves will reflect if the medium they are traveling in undergoes a change of refractive index (Figure 6.15). Sound waves reflect at a boundary where a material of different density is encountered. In the case of transmission lines, the discontinuity between the lossless line and another kind of device is the boundary which can cause reflections. Some of the wave energy will be reflected whenever there is an **impedance mismatch** between the transmission line and the devices connected to it.

To understand this, consider the circuit shown in Figure 6.16. Assuming that the switch has been open for a long time, the system is in a quiescent state. This means that the voltages and current along the transmission line are zero; it follows that $f_+ = f_- = 0$ prior to the switch closure. Just after the switch is closed, a right-traveling wave is launched at the left end of the line. The physical action of closing the switch at $t = 0$ is what causes the right-traveling wave to appear. Only a right-traveling wave can exist, because there has been no other physical action which would launch one. In particular, nothing has yet occurred on the load end which would launch a left-traveling wave.

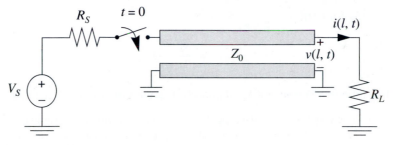

Figure 6.16 A lossless transmission line with characteristic impedance Z_0 connecting a series connected voltage source and source resistance, R_S, with a load resistor R_L. For $t < 0$, the system is quiet, therefore $f_+ = f_- = 0$. The relationships among R_S, Z_0, and R_L will determine the nature of the wave reflections at either end of the line.

The situation remains the same until $t = T_D$, when the right-traveling wave arrives at the right end of the line, $x = l$. The current and voltage at the end of the line are given by

$$i(l, t) = f_+\left(t - \frac{l}{v_0}\right) - f_-\left(t - \frac{l-l}{v_0}\right) = f_+(t - T_D) - f_-(t)$$

$$v(l, t) = Z_0\left[f_+\left(t - \frac{l}{v_0}\right) + f_-\left(t - \frac{l-l}{v_0}\right)\right] = Z_0[f_+(t - T_D) + f_-(t)]$$

However, the current $i(l, t)$ is also the current through the load resistor R_L, and $v(l, t)$ is the voltage across it. These two quantities are related by Ohm's law:

$$v(l, t) = R_L i(l, t)$$

Combining this with the above expressions,

$$Z_0(f_+(t - T_D) + f_-(t)) = R_L(f_+(t - T_D) - f_-(t)) \qquad (6.16)$$

Now, suppose we have chosen our load so that its resistance is exactly equal to the characteristic impedance of the transmission line, $R_L = Z_0$. In this case, Eq. (6.16) tells us that

$$Z_0(f_+(t - T_D) + f_-(t)) = Z_0(f_+(t - T_D) - f_-(t))$$

$$Z_0 f_-(t) = -Z_0 f_-(t)$$

which can only be satisfied if

$$f_-(t) = 0$$

The condition of $R_L = Z_0$ is known as a **matched load** and results in no reflection (i.e., no left-traveling wave arising from the boundary between the line and the load).

If, however, we have instead a **mismatched load**, $R_L \neq Z_0$, then $f_-(t) \neq 0$:

$$Z_0(f_+(t - T_D) + f_-(t)) = R_L(f_+(t - T_D) - f_-(t))$$

Solving for $f_-(t)$:

$$f_-(t) = \frac{R_L - Z_0}{R_L + Z_0} f_+(t - T_D) \tag{6.17}$$

Equation (6.17) reveals several important features of circuits containing transmission lines:

- The left-traveling wave can be expressed in terms of the right-traveling wave. This is consistent with our physical interpretation of the opposite traveling wave as a reflection.

- The left-traveling wave is a linear multiple of the right-traveling wave but is delayed by T_D, the length of time required for the wave to travel the length of the line.

- The magnitude of the reflection is a function of the relative difference between the load resistance and line impedance. The fraction reflected is given by

$$\Gamma_L = \frac{R_L - Z_0}{R_L + Z_0} \tag{6.18}$$

where Γ_L is called the **load reflection coefficient**. The remaining part of the wave energy is transmitted to the load. Note that Γ_L is bounded by ± 1. It can be positive or negative, depending on whether the load resistance is greater or less than the line impedance. If Γ_L is negative, the reflected wave is inverted (compared to the transmitted wave).

- Reflections will always occur if R_L is not equal to Z_0. For example, if the load is a capacitor, the ratio of voltage to current cannot even be expressed as a single number, let alone one equal to Z_0. Therefore, a capacitive load, such as the input of a logic gate, always results in reflections.

Reflections can also occur at the source. For example, the left-traveling wave, which had its leading edge at $x = l$ and $t = T_D$, will arrive at the left end of the line $x = 0$, at

$t = 2T_D$. Again, if there is an impedance mismatch between the source resistance and the line impedance, reflections will occur.

Reflections are undesirable in electronic circuitry because they increase the amount of time needed for the system to settle into steady state. The minimum time required for a signal to propagate from one stage to the next is given by T_D; if the impedances are matched, then this delay is the only effect of the line. With reflections, though, the traveling waves will continue to bounce back and forth, resulting in a change in the shape of the signal arriving at the load. Therefore, good design practice requires careful attention to impedance matching of various stages.

6.6 Transmission-Line Models

When solving circuits containing transmission lines, we generally do not need to know what is going on along the length of the line. We are primarily concerned with how the line interacts with other components in the circuit. Since the line is connected to these other components only at each end, it is sufficient to deal with only the currents and voltages at two points of the line: $x = 0$ and $x = l$. We also need to know how to treat the transmission line in the dc steady state. In this section we develop three models to simplify the solution of circuits containing lossless lines.

6.6.1 dc Model

The dc steady-state model of the lossless line is easily found from the n-lump approximation, Figure 6.17. After all transients have decayed away, each capacitor acts like an open circuit, while each inductor is a short circuit. Inserting these equivalent models, we arrive at the circuit shown in Figure 6.18.

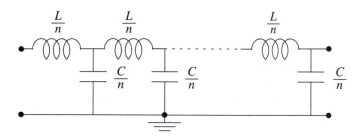

Figure 6.17 The n-lump model of the lossless line. The dc model is obtained by replacing each inductor and capacitor with its dc equivalent.

From this model, we see that in the dc steady state a lossless LC line acts like a short circuit along its length, and an open circuit between the line and the ground plane. This is clear when we consider the physical layout of the line, Figure 6.3(a), and the fact that we are neglecting the resistance of the conductor.

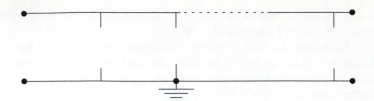

Figure 6.18 The dc steady-state model of a lossless transmission line.

6.6.2 Source End Model

We can derive an equivalent model for the source end ($x = 0$) of the transmission line from the solutions for $i(x, t)$ and $v(x, t)$ [Eqs. (6.5) and (6.6)], repeated here:

$$i(x, t) = f_+\left(t - \frac{x}{v_0}\right) - f_-\left(t - \frac{l - x}{v_0}\right) \tag{6.19}$$

$$v(x, t) = Z_0\left[f_+\left(t - \frac{x}{v_0}\right) + f_-\left(t - \frac{l - x}{v_0}\right)\right] \tag{6.20}$$

As a formal mathematical exercise, suppose we multiply Eq. (6.19) by Z_0 and subtract it from Eq. (6.20). The result is:

$$v(x, t) - Z_0 i(x, t) = 2Z_0 f_-\left(t - \frac{l - x}{v_0}\right)$$

Evaluating this expression for $x = 0$ gives

$$v(0, t) - Z_0 i(0, t) = 2Z_0 f_-(t - T_D) \tag{6.21}$$

Equation (6.21) expresses the relationship between the voltage and current at the source end of a transmission line.

Now consider the circuit of Figure 6.19. The two open terminals are assumed to be connected to something else, and the voltage between them is *defined* to be $v(0, t)$. Also, the current flowing into the resistor of value Z_0 through the upper terminal is *defined* to be $i(0, t)$, and the voltage source has been *defined* to have a value $2Z_0 f_-(t - T_D)$. Applying KVL around the loop:

$$v(0, t) - Z_0 i(0, t) = 2Z_0 f_-(t - T_D) \tag{6.22}$$

The two expressions, Eqs. (6.21) and (6.22), are precisely the same. Because they express the same relationship between the voltage $v(0, t)$ and current $i(0, t)$, we can use the circuit in Figure 6.19 as a model for the source end of the transmission line.

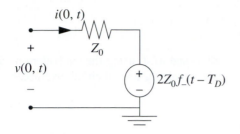

Figure **6.19** Equivalent-circuit model for the source end of the transmission line.

6.6.3 Load-End Model

To derive an equivalent model for the load end, we begin by multiplying Eq. (6.19) by Z_0 and adding it to Eq. (6.20). The result is:

$$v(x, t) + Z_0 i(x, t) = 2Z_0 f_+ \left(t - \frac{x}{v_0} \right)$$

Evaluating this expression for $x = l$ gives

$$v(l, t) + Z_0 i(l, t) = 2Z_0 f_+ (t - T_D) \tag{6.23}$$

Equation (6.23) also expresses the current-voltage relation for the circuit shown in Figure 6.20. Therefore this circuit is an equivalent model for the load end of the transmission line.

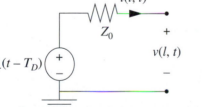
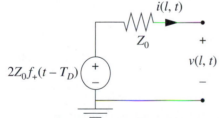

Figure **6.20** Equivalent-circuit model for the load end of the transmission line.

In both of these equivalent models, there is a voltage source whose value depends on the wave magnitude at $t - T_D$. We can derive expressions for the wave functions directly from the general solutions. Setting $x = 0$ in Eq. (6.6),

$$v(0, t) = Z_0 [f_+(t) + f_-(t - T_D)]$$

Solving for $f_+(t)$,

$$f_+(t) = \frac{1}{Z_0} v(0, t) - f_-(t - T_D) \tag{6.24}$$

Similarly, if we set $x = l$ in Eq. (6.6) and solve for $f_-(t)$ we obtain

$$f_-(t) = \frac{1}{Z_0}v(l, t) - f_+(t - T_D) \qquad\qquad (6.25)$$

It is now easy to compute $v(0, t)$, $i(0, t)$, $v(l, t)$ and $i(l, t)$ using the equivalent models and Eqs. (6.24) and (6.25). We will illustrate this by solving several circuit examples.

6.7 Solving Circuits with Transmission Lines

Solving circuits with transmission lines is a straightforward exercise. We will use the equivalent models of the source and load ends of the transmission lines to show how lossless lines affect switching events.

6.7.1 Transmission Line Matched at Both Ends

The simplest case is a matched load: the transmission line delivers a signal to a resistive load equal to the line impedance. Consider the circuit shown in Figure 6.21. In this circuit, the lossless line has a characteristic impedance of 50 Ω, and both the source and load resistances are also 50 Ω. Assume that the switch has been open for a long time before it is closed at $t = 0$. The objective is to determine the voltages $v(0, t)$ and $v(l, t)$.

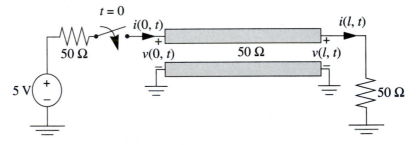

Figure 6.21 Transmission-line circuit with matched source and load impedances.

The first step is to determine the state of the circuit at the instant before the switch closes. Since the switch has been open for a long time before $t = 0$, the circuit has reached steady state. We can therefore insert the dc steady-state model of the transmission line and solve for the circuit variables. This is shown in Figure 6.22(a). From this diagram it is immediately apparent that the current through the transmission line is zero, and $v(0, t) = v(l, t) = 0$, for $t < 0$. Finally, using Eqs. (6.24) and (6.25), we can conclude that $f_+(0^-) = f_-(0^-) = 0$.

Next, we can use the dc steady-state model to find the behavior of the circuit as $t \to \infty$ [Figure 6.22(b)]. This is a straightforward voltage-divider circuit. By inspection, we see that $v(0, \infty) = v(l, \infty) = 2.5$ V.

At the instant the switch closes, a right-traveling f_+ wave begins propagating down the line. At the source end, $x = 0$, f_- must remain zero for at least the interval $t < 2T_D$, because a left-traveling wave can only be caused by a reflection at the load end, which will

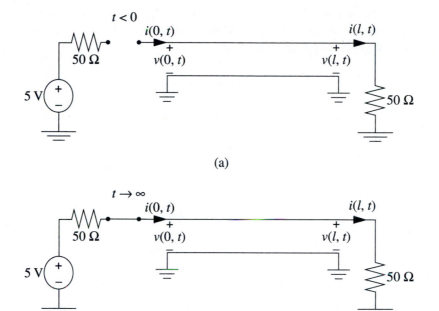

(a)

(b)

Figure 6.22 Transmission-line circuit of Figure 6.21 with dc steady-state model of the lossless line inserted. (a) Circuit for t < 0. (b) Circuit for t → ∞.

take twice the delay time to arrive back at the source. Therefore, for the interval $0 < t < 2T_D$, we can write

$$f_-(t - T_D) = 0$$

We redraw the circuit, starting from the source end, by inserting the equivalent circuit of the transmission line looking into the source end as given in Figure 6.19. This results in the circuit shown in Figure 6.23.

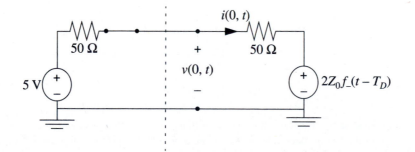

Figure 6.23 Transmission-line circuit of Figure 6.21, with equivalent model (looking into the source end) of the transmission line inserted. (The inserted model is the part to the right of the dotted line.)

We just reasoned out the fact that $f_-(t - T_D) = 0$, so the voltage source on the right side is equal to zero volts. The voltage at the source end of the transmission line is found by applying the voltage-divider rule:

$$v(0, t) = 5\frac{50}{50 + 50} = 2.5 \text{ V} \tag{6.26}$$

This value will be valid over at least the interval $0 < t < 2T_D$, after which we must consider the possibility of a left-traveling wave arriving back at the source end.

To solve for the voltage at the load end, we are going to need the value of the right-traveling wave $f_+(t - T_D)$ for use in the equivalent circuit. This can be found from Eqs. (6.24) and (6.26):

$$f_+(t - T_D) = \frac{1}{Z_0}v(0, t - T_D) - f_-(t - 2T_D)$$

$$= \frac{2.5}{50} - 0 = 0.05$$

This expression is valid for $T_D < t < 3T_D$.

We again redraw the circuit, this time by inserting the equivalent circuit of the transmission line looking into the load end, as given in Figure 6.20. The result is the circuit shown in Figure 6.24. Using the just-calculated value of $f_+(t - T_D) = 0.05$, we see that the voltage source in Figure 6.24 has a value of

$$2Z_0 f_+(t - T_D) = 2 \times 50 \times 0.05 = 5 \text{ V}$$

from which we calculate the load voltage:

$$v(l, t) = 5\frac{50}{50 + 50} = 2.5 \text{ V} \tag{6.27}$$

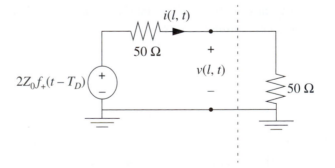

Figure 6.24 Transmission-line circuit of Figure 6.21, with equivalent model (looking into the load end) of the transmission line inserted. (The inserted model is the part to the left of the dotted line.)

This value is valid from $t = T_D$, since it takes one delay period for the right-traveling wave to reach the load end of the lossless line. [For $t < T_D$, $v(l, t) = 0$, as calculated from the initial state model.]

At this point we see that both ends of the transmission line have reached the dc steady-state value of 2.5 V. This is to be expected, since the load impedance is matched to the characteristic impedance of the line, preventing a reflection at the load. To verify this, we calculate the value of the left-traveling wave from Eq. (6.25) (for $t > T_D$):

$$f_-(t) = \frac{1}{Z_0}v(l, t) - f_+(t - T_D)$$

$$= \frac{2.5}{50} - 0.05 = 0$$

We see that the left-traveling wave has zero magnitude, which is the same as saying there is no reflection.

Finally, since no left-traveling wave is reflected from the load, nothing further changes in the circuit. The value of the right-traveling wave, f_+, is set up at the source end once the switch closes and does not vary again. So, the circuit does in fact reach dc steady state at $t = T_D$. Summarizing, we can express the transmission-line voltages as:

$$
\begin{aligned}
v(0, t) &= 0, & t < 0 \\
v(0, t) &= 2.5 \text{ V}, & t > 0 \\
v(l, t) &= 0, & t < T_D \\
v(l, t) &= 2.5 \text{ V}, & t > T_D
\end{aligned}
$$

The circuit response is plotted in Figure 6.25 (assuming a delay time of 1 μs).

6.7.2 Transmission Line with Matched Load Only

The next case is only slightly different from the last. Again, we will use a matched load, where the transmission line delivers a signal to a resistance equal to the line impedance. However, instead of an identical source resistance, we will consider a value of 10 Ω, as shown in Figure 6.26. All other values and conditions are the same as the previous example.

To determine the state of the circuit at the instant before the switch closes, we use the fact that the switch has been open for a long time before $t = 0$. This means that the circuit has reached steady state, and we can therefore insert the dc steady-state model of the transmission line [Figure 6.27(a)]. From this diagram, we see that the current through the transmission line is zero, and $v(0, t) = v(l, t) = 0$, for $t < 0$. Using Eqs. (6.24) and (6.25), we can solve for the wave functions, with the result that $f_+(0^-) = f_-(0^-) = 0$.

Next, we use the dc steady-state model to find the behavior of the circuit as $t \to \infty$ [Figure 6.27(b)]. Using the voltage divider rule, we see that

$$v(0, \infty) = v(l, \infty) = 5\frac{50}{10 + 50} = 4.17 \text{ V}$$

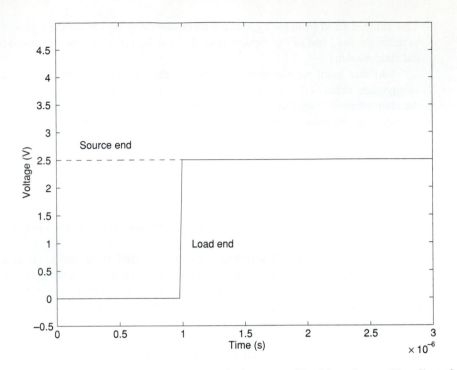

Figure 6.25 Transmission-line voltages for matched source and load impedances. The effect of the lossless transmission line is simply to delay the arrival of the signal at the load by a fixed period T_D.

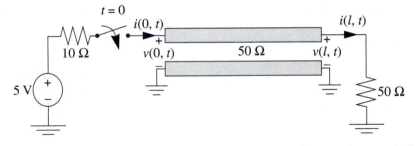

Figure 6.26 Transmission-line circuit with matched load impedance and unmatched source impedance.

A right-traveling f_+ wave begins propagating down the line as soon as the switch closes. At $x = 0$, $f_- = 0$ over at least the interval $t < 2T_D$, as before.

Inserting the equivalent-circuit model for the source end of the transmission line results in the circuit of Figure 6.28. The voltage source on the right is a short circuit (since the left-traveling wave $f_- = 0$), so we can write the voltage on the source end of the line as:

$$v(0, t) = 5\frac{50}{10 + 50} = 4.17 \text{ V} \tag{6.28}$$

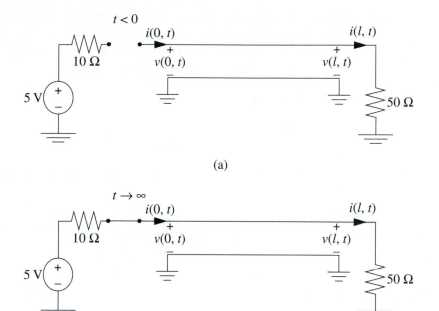

(a)

(b)

Figure 6.27 Transmission-line circuit of Figure 6.26 with dc steady-state model of the lossless line inserted. (a) Circuit for $t < 0$. (b) Circuit for $t \to \infty$.

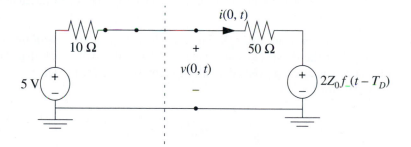

Figure 6.28 Transmission line circuit of Figure 6.26, with equivalent model (looking into the source end) of the transmission line inserted. (The inserted model is the part to the right of the dotted line.)

This is valid for values of t up to at least $2T_D$, after which we need consider the possibility of a reflected wave arriving back at the source end.

The value of the right-traveling wave $f_+(t - T_D)$ is found from Eqs. (6.24) and (6.28):

$$f_+(t - T_D) = \frac{1}{Z_0}v(0, t - T_D) - f_-(t - 2T_D)$$

$$= \frac{4.17}{50} - 0 = 0.0833$$

which is valid over the interval $T_D < t < 3T_D$. Redrawing the circuit again, using the load end model of the transmission line, gives us the circuit in Figure 6.29. Inserting the value of $f_+(t - T_D) = 0.0833$, we see that the voltage source in Figure 6.29 has a value of

$$2Z_0 f_+(t - T_D) = 2 \times 50 \times 0.083 = 8.33 \text{ V}$$

from which the load voltage is:

$$v(l, t) = 8.33 \frac{50}{50 + 50} = 4.17 \text{ V} \tag{6.29}$$

This value is reached once the right-traveling wave arrives at the load end at $t = T_D$.

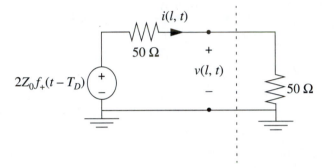

Figure 6.29 Transmission-line circuit of Figure 6.26, with equivalent model (looking into the load end) of the transmission line inserted. (The inserted model is the part to the left of the dotted line.)

As in the previous example, after one delay time both ends of the transmission line have reached the dc steady-state value of 4.17 V. We will check that the matched load impedance has again prevented a reflection, by solving for the left-traveling wave. Using Eq. (6.25) (for $t > T_D$):

$$f_-(t) = \frac{1}{Z_0} v(l, t) - f_+(t - T_D)$$

$$= \frac{4.17}{50} - 0.0833 = 0$$

Indeed, the left-traveling wave is zero, confirming the absence of reflection in this case as well. Summarizing, the transmission-line voltages are:

$$\begin{aligned} v(0, t) &= 0, & t &< 0 \\ v(0, t) &= 4.17 \text{ V}, & t &> 0 \\ v(l, t) &= 0, & t &< T_D \\ v(l, t) &= 4.17 \text{ V}, & t &> T_D \end{aligned}$$

The only effect of the different source resistance is a change in the division of voltage between R_S and R_L, which is apparent from the dc steady-state model. In principle, a left-traveling wave will reflect at the impedance mismatch between the transmission line and source impedance; this is not observed in this example, because no left-traveling wave is present (owing to the matching at the load end). Source-impedance matching becomes more important when the load impedance is not matched, as we will see next.

6.7.3 Transmission Line Unmatched at Both Ends

To observe reflections at both ends of the lossless line, both the source and load resistances R_S and R_L must be different from the characteristic impedance Z_0. This case is illustrated in Figure 6.30, where we have $R_S = 25\ \Omega$, $Z_0 = 50\ \Omega$, and $R_L = 100\ \Omega$. We will see that multiple reflections occur along this line, which has the effect of substantially distorting the signal received at the load.

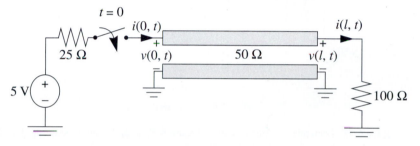

Figure 6.30 Transmission-line circuit with unmatched source and load resistances.

As before, we will assume the switch has been open for a long time before it is closed at $t = 0$. Inserting the dc steady-state model for a transmission line into the circuit results in the circuits shown in Figure 6.31. The top circuit, Figure 6.31(a), is valid for $t < 0$, before the switch has closed; the bottom circuit, Figure 6.31(b), corresponds to the dc steady state as $t \to \infty$. By inspection, we can solve these circuits and find that

$$v(0, 0) = v(l, 0) = 0$$
$$i(0, 0) = i(l, 0) = 0$$

$$v(0, \infty) = v(l, \infty) = 4.0\ \text{V}$$
$$i(0, \infty) = i(l, \infty) = 0.040\ \text{A}$$

We now insert the equivalent model for the source end of the transmission line. This results in the circuit of Figure 6.32. Just after the switch closes, the left-traveling wave f_- is zero (no chance to reflect yet), so initially the value of the right voltage source, $2Z_0 f_-(t - T_D)$, is zero. Solving for $v(0, t)$,

$$v(0, t) = 5\frac{50}{25 + 50} = 3.33\ \text{V} \tag{6.30}$$

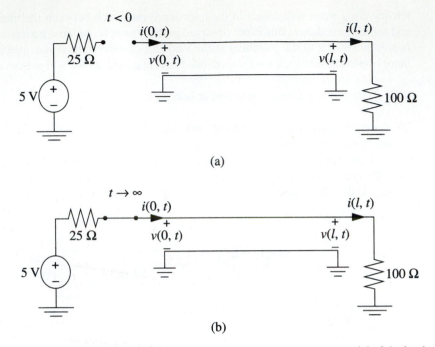

Figure 6.31 Transmission-line circuit of Figure 6.30 with dc steady-state model of the lossless line inserted. (a) Circuit for t < 0. (b) Circuit for t → ∞.

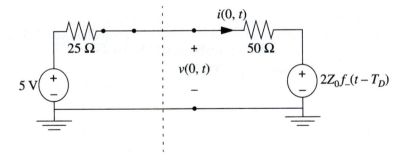

Figure 6.32 Transmission line circuit of Figure 6.30, with equivalent model (looking into the source end) of the transmission line inserted. (The inserted model is the part to the right of the dotted line.)

This value will remain valid over the interval $0 < t < 2T_D$, until a left-traveling wave arrives at the source end of the transmission line. Anticipating the next step, we use Eq. (6.24) to find the value of the right-traveling wave $f_+(t - T_D)$ for $T_D < t < 3T_D$:

$$f_+(t - T_D) = \frac{1}{Z_0}v(0, t - T_D) - f_-(t - 2T_D)$$

$$= \frac{3.33}{50} - 0 = 0.0666$$

We now examine the right end of the circuit. Inserting the load-end model, Figure 6.20, into the circuit of Figure 6.30 results in the circuit shown in Figure 6.33. The value of the voltage source is

$$2Z_0 f_+(t - T_D) = 2 \times 50 \times 0.0666$$

$$= 6.66$$

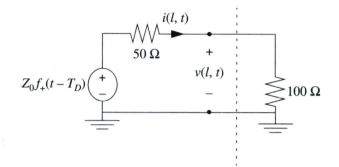

Figure 6.33 Transmission-line circuit of Figure 6.30, with equivalent model (looking into the load end) of the transmission line inserted. (The inserted model is the part to the left of the dotted line.)

From this, we calculate the load voltage as

$$v(l, t) = 6.66 \frac{100}{50 + 100} = 4.44 \text{ V}$$

valid for $T_D < t < 3T_D$.

At this point we observe that the voltage at the load end is not equal to the dc steady-state value of 4.0 V. We must continue to solve this circuit to see how subsequent wave reflections will bring the load voltage closer to the final value.

We first determine the magnitude of the left-traveling wave, f_-, using the load voltage just determined and Eq. (6.25):

$$f_-(t - T_D) = \frac{1}{Z_0} v(l, t - T_D) - f_+(t - 2T_D)$$

$$= \frac{4.44}{50} - 0.0666 = 0.0222$$

We now return to the circuit using the equivalent model for the source end of the transmission line, Figure 6.32, except that the value of the right-hand voltage source is

$$2Z_0 f_-(t - T_D) = 2 \times 50 \times 0.0222$$

$$= 2.22$$

Using this value, the voltage at the left end of the transmission line is:

$$v(0, t) = (5 - 2.22)\left(\frac{50}{25 + 50}\right) + 2.22 = 4.07 \text{ V}$$

This value is valid from the time the left-traveling wave first arrives at the source end, $t = 2T_D$, until $t = 4T_D$.

Again, we observe that the voltage on the line has not yet reached the dc steady-state value of 4.0 V. The arrival of the left-traveling f_- wave at the source end has changed the value of the voltage there. Because the source impedance (25 Ω) does not match the 50-Ω line impedance, there will be a reflection. We calculate the reflection by updating the value of the right-traveling f_+ wave using Eq. (6.24)

$$f_+(t - T_D) = \frac{1}{Z_0}v(0, t - T_D) - f_-(t - 2T_D)$$

$$= \frac{4.07}{50} - 0.0222 = 0.0593$$

which is valid for $3T_D < t < 5T_D$.

This process can be continued indefinitely, as the waves continue to reflect at both ends of the line. With time, the changes in voltage get smaller, resulting in a nearly constant transmission-line voltage approaching the dc steady state. Below, we carry out a few more iterations to illustrate this result.

At the right end of the line, the "new" f_+ wave arrives at $t = 3T_D$. The value of the voltage source in the load-end model is

$$2Z_0 f_+(t - T_D) = 2 \times 50 \times 0.0593$$

$$= 5.93$$

which results in a load voltage of

$$v(l, t) = 5.93\frac{100}{50 + 100} = 3.95 \text{ V}, \qquad 3T_D < t < 5T_D$$

Subsequent reflection at the load end launches a new left-traveling wave:

$$f_-(t - T_D) = \frac{1}{Z_0}v(l, t - T_D) - f_+(t - 2T_D)$$

$$= \frac{3.95}{50} - 0.0593 = 0.0198$$

When this wave arrives back at the source end, it changes the value of the voltage source in the source-end equivalent model to

$$2Z_0 f_-(t - T_D) = 2 \times 50 \times 0.0198$$

$$= 1.98$$

which in turn, results in a source-end voltage of

$$v(0, t) = (5 - 1.98)\left(\frac{50}{25 + 50}\right) + 1.98 = 3.99 \text{ V}, \qquad 4T_D < t < 6T_D$$

At this point the source end voltage is quite close to the dc steady-state value. The reflection at the source end changes the f_+ wave:

$$f_+(t - T_D) = \frac{1}{Z_0} v(0, t - T_D) - f_-(t - 2T_D)$$

$$= \frac{3.99}{50} - 0.0198 = 0.0601$$

which will arrive at the load end at $t = 5T_D$. The value of the voltage source in the load-end model is then

$$2Z_0 f_+(t - T_D) = 2 \times 50 \times 0.0601$$

$$= 6.01$$

Calculating the new voltage value at the load:

$$v(l, t) = 6.01 \frac{100}{50 + 100} = 4.01 \text{ V}, \qquad 5T_D < t < 7T_D$$

The time evolution of the voltages at source and load ends of the transmission line is plotted in Figure 6.34, again assuming a delay time of 1 μs. This plot illustrates the back-and-forth nature of the wave travel along the lossless line. It also shows how the unmatched source and load impedances have resulted in a significant distortion of the desired signal. In this example, the load voltage first overshoots the dc steady-state value, and on subsequent reflections oscillates about it. This behavior is reminiscent of the gradual damping of voltage in an underdamped *RLC* circuit. Both the signal distortion and the lengthened delay are undesired phenomenon which must be considered in designing electronic systems.

6.8 Capacitively Loaded Transmission Line

So far, we have solved circuits with transmission lines that are terminated (i.e., connected to the load end) with resistances. In many instances, though, the loads are not resistors. For example, a transmission line connecting two CMOS gates, Figure 6.35, sees a resistance at the source end but a capacitance at the load end. The presence of the gate input capacitance at the end of the transmission line will result in reflections due to the inherent impedance mismatch evidenced by this combination.

Using the simple models of CMOS inverters, neglecting the power and ground lead inductance, we arrive at the equivalent circuit shown in Figure 6.36. In analyzing the pull-up transition of this circuit, we will make the usual assumptions: the switch has been open for a long time preceding $t = 0$, and the initial capacitor voltage and current are zero.

Making use of the results of the last example, we will make another assumption: the circuit has been designed such that the output resistance of the driving inverter is equal to the characteristic impedance of the lossless line. In symbols,

$$R_s = Z_0 \tag{6.31}$$

This matched source impedance is good design practice, as it will prevent reflections of the f_- left-traveling wave at the source end of the transmission line.

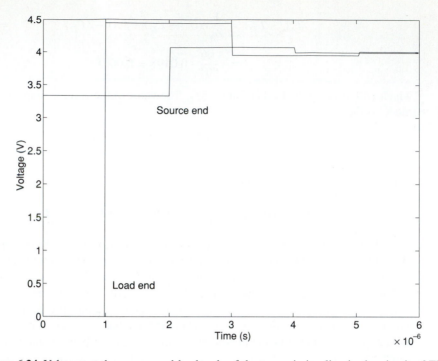

Figure 6.34 Voltages at the source and load ends of the transmission line in the circuit of Figure 6.30, where the impedances are mismatched at both ends. Multiple reflections occur on the line, resulting in significant signal distortion.

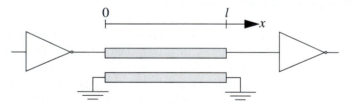

Figure 6.35 Transmission line connecting two CMOS inverter gates.

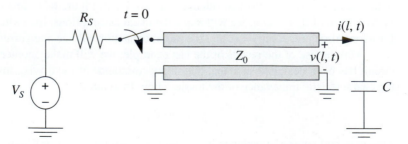

Figure 6.36 Equivalent circuit of two CMOS inverter gates connected by a transmission line. The lossless line is terminated by the input capacitance C of the driven inverter.

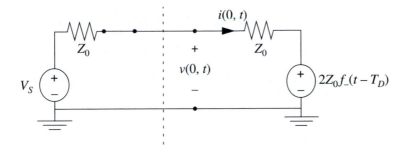

Figure 6.37 Transmission line circuit of Figure 6.36, with equivalent model (looking into the source end) of the transmission line inserted. (The inserted model is the part to the right of the dotted line.)

Inserting the source-end model of the transmission line, Figure 6.19, into the circuit, we have the equivalent model shown in Figure 6.37. Initially, the value of the source on the right side is zero (f_- must be zero, since nothing has happened at the load which would cause a left-traveling wave). Solving for $v(0, t)$:

$$v(0, t) = V_S\left(\frac{Z_0}{Z_0 + Z_0}\right) = \frac{1}{2}V_S \qquad 0 < t < 2T_D$$

We use this result, along with Eq. (6.24), to determine the right-traveling wave f_+:

$$f_+(t - T_D) = \frac{1}{Z_0}v(0, t - T_D) - f_-(t - 2T_D)$$

$$= \frac{1}{2Z_0}V_S$$

(6.32)

Turning to the load end, we insert the equivalent model of the other end of the transmission line, which results in the circuit shown in Figure 6.38.

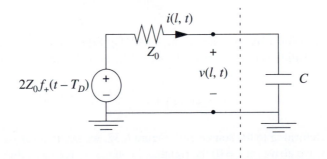

Figure 6.38 Transmission-line circuit of Figure 6.36, with equivalent model (looking into the load end) of the transmission line inserted. (The inserted model is the part to the left of the dotted line.)

The value of the voltage source in Figure 6.38 is

$$2Z_0 f_+(t - T_D) = 2Z_0 \frac{1}{2Z_0} V_S$$

$$= V_S$$

We recognize this circuit as a first-order RC circuit. Because the f_+ wave does not arrive at the load end until $t = T_D$, we express the initial condition as

$$v(l, T_D) = 0$$

The solution of this RC circuit is

$$v(l, t) = (v(l, T_D) - V_S)e^{-\frac{t - T_D}{Z_0 C}} + V_S$$

which, after inserting the initial condition, can be written

$$v(l, t) = V_S\left(1 - e^{-\frac{t - T_D}{Z_0 C}}\right), \qquad T_D < t < 3T_D \tag{6.33}$$

We see from Eq. (6.33) that the lossless line has simply delayed the pullup transition in capacitor voltage by the delay time T_D. Other than this rigid shift in the time axis, no other effect of the line is apparent at this point.

Continuing, we determine the value of the reflected, left-traveling f_- wave using Eqs. (6.25) and (6.33).

$$f_-(t) = \frac{1}{Z_0} v(l, t) - f_+(t - T_D)$$

$$= \frac{V_S}{Z_0}\left(1 - e^{-\frac{t - T_D}{Z_0 C}}\right) - \frac{V_S}{2Z_0}$$

$$= \frac{V_S}{Z_0}\left(\frac{1}{2} - e^{-\frac{t - T_D}{Z_0 C}}\right)$$

which is valid over the interval $T_D < t < 3T_D$. Equivalently, we can write an expression which is valid over the interval $2T_D < t < 4T_D$ by replacing the argument t with $t - T_D$:

$$f_-(t - T_D) = \frac{V_S}{Z_0}\left(\frac{1}{2} - e^{-\frac{t - 2T_D}{Z_0 C}}\right)$$

Returning to the source end, Figure 6.37, we see that for $t > 2T_D$ (the time at which the f_- wave arrives at $x = 0$) the righthand voltage source has value

$$2Z_0 f_-(t - T_D) = V_S\left(1 - 2e^{-\frac{t - 2T_D}{Z_0 C}}\right)$$

Solving for $v(0, t)$:

$$v(0, t) = (V_S - 2Z_0 f_-(t - T_D)) \left(\frac{Z_0}{Z_0 + Z_0} \right) + 2Z_0 f_-(t - T_D)$$

$$= \left(V_S - V_S \left(1 - 2e^{-\frac{t - 2T_D}{Z_0 C}} \right) \right) \left(\frac{1}{2} \right) + V_S \left(1 - 2e^{-\frac{t - 2T_D}{Z_0 C}} \right)$$

After some algebra, this is

$$v(0, t) = V_S \left(1 - e^{-\frac{t - 2T_D}{Z_0 C}} \right), \qquad 2T_D < t < 4T_D$$

This means that the voltage at the source end of the transmission line, which had been a constant value of $V_S/2$ during the first $2T_D$ seconds after the switching event, first jumps abruptly back to zero and then starts to rise towards the dc steady state value of V_S.

Since the driving-gate output resistance R_S is matched to the line impedance Z_0, we expect that the arrival of the f_- wave at the source end of the lossless line will not cause further reflections. To verify this, we use Eq. (6.24) to solve for the right-traveling wave f_+:

$$f_+(t) = \frac{1}{Z_0} v(0, t) - f_-(t - T_D)$$

$$= \frac{1}{Z_0} V_S \left(1 - e^{-\frac{t - 2T_D}{Z_0 C}} \right) - \frac{V_S}{Z_0} \left(\frac{1}{2} - e^{-\frac{t - 2T_D}{Z_0 C}} \right)$$

Evaluating,

$$f_+(t) = \frac{1}{2Z_0} V_S, \qquad 2T_D < t < 4T_D \qquad (6.34)$$

Comparing the values of f_+ given in Eqs. (6.32) and (6.34) (i.e., before and after $t = 2T_D$), we see that they are identical. There is no change in the magnitude of the right-traveling f_+ wave at $t = 2T_D$, confirming that the f_- wave is not reflected at the source end in this example.

Without any more reflections or other changes in the circuit, the expressions derived form a complete solution. We can summarize the voltages at each end of the line as follows:

$$v(0, t) = \begin{cases} 0, & t < 0 \\[2mm] \dfrac{V_S}{2}, & t < 2T_D \\[4mm] V_S \left(1 - e^{-\frac{t - 2T_D}{Z_0 C}} \right), & t > 2T_D \end{cases}$$

$$v(l, t) = \begin{cases} 0, & t < T_D \\ V_S\left(1 - e^{-\frac{t-T_D}{Z_0 C}}\right), & t > T_D \end{cases}$$

These responses are plotted in Figure 6.39.

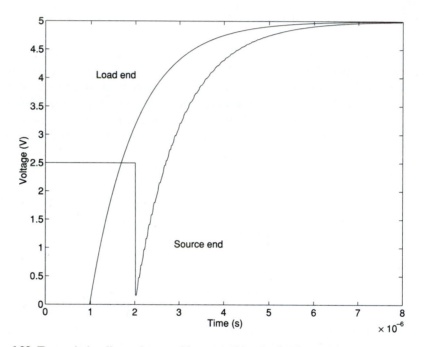

Figure 6.39 Transmission-line voltages with a capacitive load (Figure 6.36) and matched source impedance.

An important point of this example is that the matched source impedance prevents reflections at the source end. Reflections at a capacitive load are unavoidable, because the current-voltage relation for a capacitor is not resistive. Source matching at least prevents multiple reflections from further distortion of the signal.

Figure 6.40 illustrates typical signals found on a capacitively loaded transmission line where the source impedance is mismatched to the characteristic line impedance. The two graphs (a) and (b) correspond to the cases where $R_S < Z_0$ and $R_S > Z_0$. The signal distortions caused by reflections at both ends of the line are apparent. For this reason, circuits designed to drive off-chip transmission lines, where the delay time is significant, are generally built with an output impedance carefully matched to the transmission-line impedance.

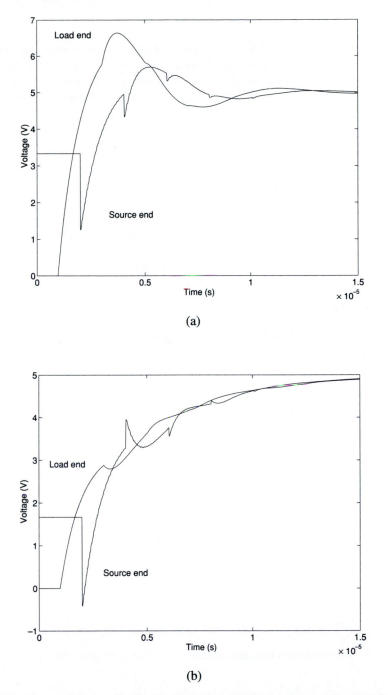

Figure 6.40 Voltage responses of capacitively loaded transmission lines with unmatched source impedances. (a) The source resistance is less than the line impedance. (b) The source resistance is greater than the line impedance.

6.9 Summary

Lossless transmission lines are widely used in electronic systems for clock and power distribution. Transmission of signals along these lines is affected by delays and distortions arising from the wave nature of the propagation. Problems can be minimized by careful design, which includes matching of source and load resistances to the characteristic line impedance.

Important concepts discussed in this chapter include:

- The relationship between voltage and current along a transmission line are governed by a set of coupled partial differential equations, known as the **telegrapher's equations**:

$$\frac{\partial v}{\partial x} = -\frac{L}{l}\frac{\partial i}{\partial t}$$

$$\frac{\partial i}{\partial x} = -\frac{C}{l}\frac{\partial v}{\partial t}$$

- The solutions to the telegrapher's equations are

$$i(x, t) = f_+\left(t - \frac{x}{v_0}\right) - f_-\left(t - \frac{l - x}{v_0}\right)$$

$$v(x, t) = Z_0\left[f_+\left(t - \frac{x}{v_0}\right) + f_-\left(t - \frac{l - x}{v_0}\right)\right]$$

where the functions

$$f_+\left(t - \frac{x}{v_0}\right) \text{ and } f_-\left(t - \frac{l - x}{v_0}\right)$$

describe right- and left-traveling waves.

- The **characteristic velocity** v_0 of the wave is given by

$$v_0 = \frac{1}{\sqrt{\left(\frac{L}{l}\right)\left(\frac{C}{l}\right)}} = \frac{l}{\sqrt{LC}}$$

where L and C are the total inductance and capacitance of a line of length l.

- The **characteristic impedance** of the line Z_0 is given by

$$Z_0 = \sqrt{\frac{L}{C}}$$

- The **delay time** of the loss line, T_D, is found from:

$$T_D = \frac{l}{v_0} = \sqrt{LC}$$

- Waves will **reflect** at a boundary. Reflections can be prevented by **matching** the circuit impedance to the line impedance.

- The source end of a transmission line can be modeled with this circuit:

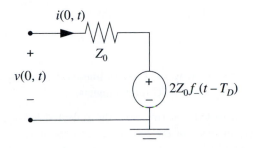

where $f_-(t - T_D)$ is found from

$$f_-(t - T_D) = \frac{1}{Z_0}v(l, t - T_D) - f_+(t - 2T_D)$$

- The load end of a transmission line can be modeled with this circuit:

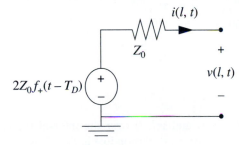

where $f_+(t - T_D)$ is found from

$$f_+(t - T_D) = \frac{1}{Z_0}v(0, t - T_D) - f_-(t - 2T_D)$$

- In the dc steady state, a transmission line acts like two resistanceless wires.

Problems

6.1 Typical representative dimensions for long interconnects on integrated circuits and printed circuit boards are shown below:

	length, l	width, w	height, h	gap, d
PC board	10 cm	500 μm	10 μm	100 μm
IC chip	1 cm	2 μm	1 μm	0.2 μm

Using these values, discuss why LC lossless line models are appropriate for PC board interconnects, but less appropriate for IC interconnects.

6.2 Equations (6.17) and (6.18) describe how the reflected left-traveling wave depends on the right-traveling wave in terms of the load reflection coefficient, Γ_L. Derive a similar relationship for the source end of the transmission line, and determine the formula for the corresponding **source reflection coefficient, Γ_S**.

6.3 Determine and plot $v(0, t)$ and $v(l, t)$ over the interval $0 \le t \le 12$ μs in the circuit below. Assume the delay time of the transmission line is $T_D = 2$ μs.

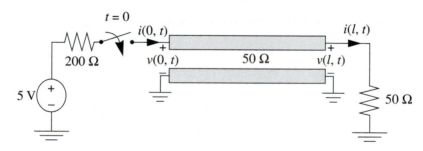

6.4 Determine and plot $v(0, t)$ and $v(l, t)$ over the interval $0 \le t \le 12$ μs in the circuit below. Assume the delay time of the transmission line is $T_D = 2$ μs.

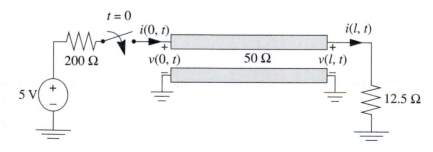

6.5 Determine and plot $v(0, t)$ and $v(l, t)$ over the interval $0 \le t \le 12$ µs in the circuit below. Assume the delay time of the transmission line is $T_D = 2$ µs.

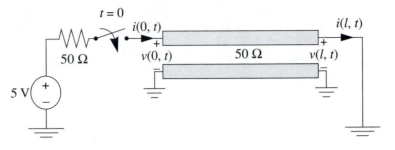

6.6 Determine and plot $v(0, t)$ and $v(l, t)$ over the interval $0 \le t \le 12$ µs in the circuit below. Assume the delay time of the transmission line is $T_D = 2$ µs.

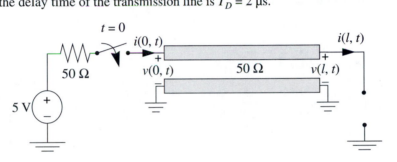

6.7 Determine and plot $v(0, t)$ and $v(l, t)$ over the interval $0 \le t \le 8$ µs in the circuit below. Assume the delay time of the transmission line is $T_D = 2$ µs.

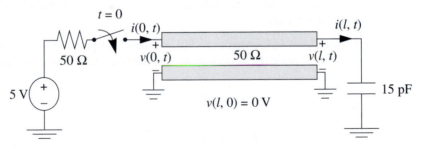

6.8 Determine and plot $v(0, t)$ and $v(l, t)$ over the interval $0 \le t \le 8$ µs in the circuit below. Assume the delay time of the transmission line is $T_D = 2$ µs.

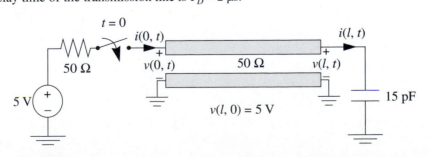

6.9 Determine and plot $v(0, t)$ and $v(l, t)$ over the interval $0 \le t \le 8$ μs in the circuit below. Assume the delay time of the transmission line is $T_D = 2$ μs.

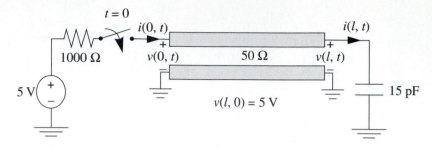

6.10 Determine and plot $v(0, t)$ and $v(l, t)$ over the interval $0 \le t \le 8$ μs in the circuit below. Assume the delay time of the transmission line is $T_D = 2$ μs, $R_S = 50$ Ω, and $v(l, 0) = 0$ V.

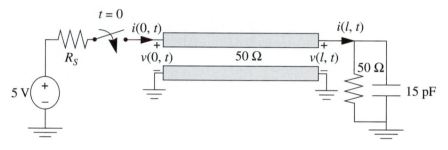

6.11 In Section 6.3.1 we showed that the characteristic velocity of a transmission line with a vacuum insulator is the speed of light. Suppose we have a transmission line whose width is equal to its distance above the ground plane, with vacuum as the insulator. Determine the characteristic impedance of such a line.

6.12 In the circuit below, the switch opens at $t = 0$. Determine and plot $v(0, t)$ and $v(l, t)$ over the interval $0 \le t \le 12$ μs. Assume the delay time of the transmission line is $T_D = 2$ μs.

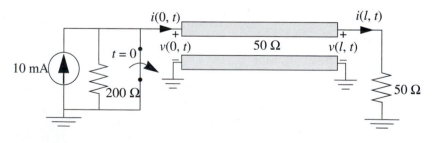

6.13 In the circuit below, the switch opens at $t = 0$. Determine and plot $v(0, t)$ and $v(l, t)$ over the interval $0 \le t \le 12$ µs. Assume the delay time of the transmission line is $T_D = 2$ µs.

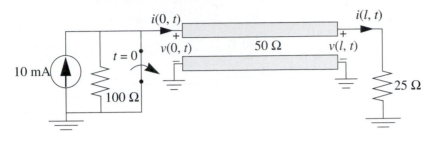

6.14 In the circuit below, the switch opens at $t = 0$. Determine and plot $v(0, t)$ and $v(l, t)$ over the interval $0 \le t \le 12$ µs. Assume the delay time of the transmission line is $T_D = 2$ µs.

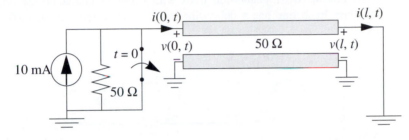

6.15 In the circuit below, the switch opens at $t = 0$. Determine and plot $v(0, t)$ and $v(l, t)$ over the interval $0 \le t \le 12$ µs. Assume the delay time of the transmission line is $T_D = 2$ µs.

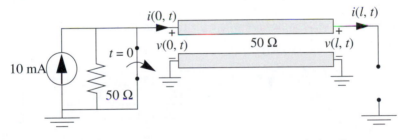

6.16 In the circuit below, the load resistance $R_L = 50$ Ω, the transmission-line wave velocity is 10^{10} cm/s; and the length of the line is $l = 50$ cm. Assume $v_C(0) = 1$ V and $v(x, 0) = 0$ V. Determine and plot $v(l, t)$.

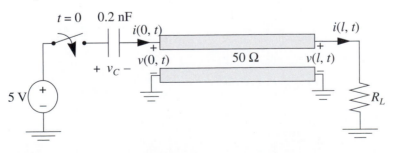

6.17 In the circuit below, the load resistance $R_L = 50 \ \Omega$, the transmission-line wave velocity is 10^{10} cm/s; and the length of the line is $l = 50$ cm. Assume $v(x, 0) = 0$ V. Determine and plot $v(l, t)$.

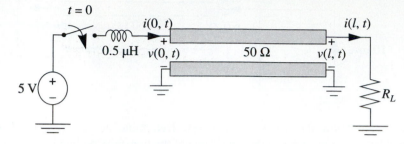

6.18 In the circuit below, two transmission lines with different characteristic impedances are chained together. Each line has a delay time of 1 μs. Determine $v_1(t)$, $v_2(t)$, and $v_3(t)$ for $0 \le t \le 8$ μs.

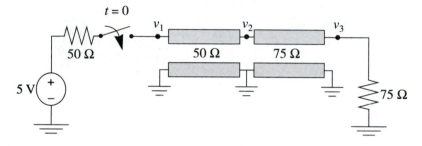

6.19 Repeat Exercise 6.18 for the case where the two transmission lines are reversed:

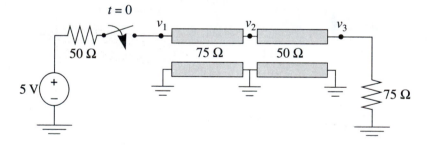

6.20 Determine the voltage response at the load for the three transmission-line circuits below. Describe the important differences between the three responses.

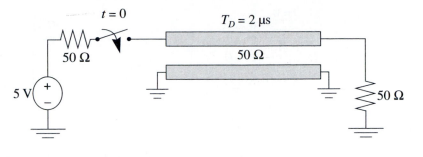

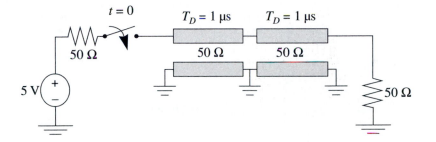

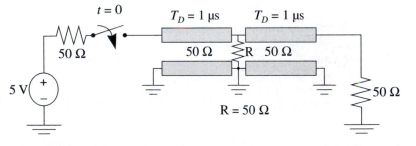

6.21 A standard characteristic impedance used for many systems is 50 Ω. Explain why it is not correct to simply use a 50-Ω resistor as a model for a 50-Ω transmission line.

6.22 In solving the problems of this chapter, the graphs indicate abrupt changes in voltage at either ends of the transmission lines, unlike the continuous curves obtained with RC and RLC circuits. Since the lossless lines have capacitance, these abrupt changes would appear at first glance to be unphysical. Reconcile these abrupt changes in voltage with the physical model of the transmission line and your understanding of circuits with capacitors.

6.23 An output gate (with equivalent series resistance 50 Ω) drives a 15-pF capacitive load from zero to 5 V through a printed-circuit-board interconnect. The interconnect has negligible resistance but equivalent lumped inductance and capacitance of 37.5 nH and 15 pF. If we treat the interconnect as a lossless LC line, we have the following model:

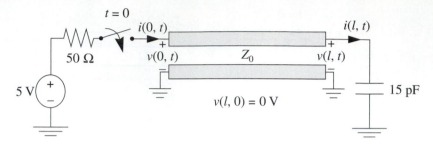

Alternatively, we could treat the interconnect as a single LC lump, resulting in this model:

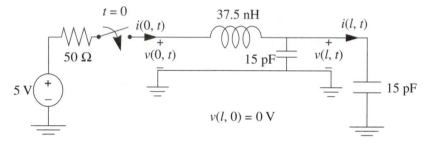

which we recognize as a series RLC circuit from Chapter 5. Solve for and plot the transient voltage across the load capacitor using both models. Does the lumped LC model accurately describe the effect of the transmission line?

6.24 An output gate (with equivalent series resistance 50 Ω) drives a 15-pF capacitive load from zero to 5 V through a matched, long printed-circuit-board interconnect. The delay time is 0.5 ns. Assuming the wave velocity is one-half the speed of light, determine the length and aspect ratio (width to gap) of the interconnect.

6.25 In Exercise 6.24, suppose the interconnect is made from copper, 5 μm thick and 20 μm wide. Determine the resistance of the interconnect.

6.26 Using the results of Exercises 6.24 and 6.25, calculate the transient response using: (a) an LC interconnect model, ignoring the series resistance; (b) a one-lump RC interconnect model, ignoring the series inductance; and (c) a series-RLC model, lumping the interconnect inductance and capacitance. Which model is most appropriate for this circuit?

7 Clock Skew and Signal Representation

7.1 Introduction

In studying the effects of interconnects on signal delay in digital systems we have essentially used the same method repeatedly: model a subset of the system using combinations of elementary circuit elements, derive a set of differential equations describing this combination, and solve the differential equations to obtain the signals as a function of time. This straightforward approach is known as **time-domain analysis**.

For many circuits, time-domain analysis is an easy and appropriate method for determining the signals. However, there is another approach, known as **frequency-domain analysis**, which provides alternative insights into the workings of the circuit. Frequency-domain analysis is especially suited to systems where the signals are periodic in time, such as the clock signals used to synchronize the progression of digital signals in a computer. It is also invaluable in understanding the workings of signal-processing systems, such as telecommunication circuits and stereos.

There are two main ideas in frequency-domain analysis. The first is that any periodic signal can be represented as a sum of sinusoids of different frequencies. The second is that the response of a circuit to a sinusoid can be determined in a particularly simple way. The result is that solving circuits using frequency-domain analysis is often, but not always, easier than the time-domain approach.

In this chapter we investigate the first of these two ideas, that signals can be represented as a sum of sinusoids. In the next chapter we will see how these signal representations are used to solve circuits. As a vehicle, we will use a square-wave clock signal commonly employed in computers for synchronization.

7.2 Clock Skew in Synchronous Logic Circuits

Thus far we have demonstrated only the delays and switching problems associated with **combinational logic** circuits. The steady-state output of a combinational circuit depends only on the values of its inputs. Such circuits, however, comprise just a small subset of digital logic. In a majority of modern integrated circuits most of the interesting and useful functions are realized using **sequential logic,** the outputs of which are functions of both the input and the previous state of the logic. Sequential logic circuits are, in general, composed of both combinational gates (NOT, NAND, NOR) and **flip-flops**. Flip-flops differ from combinational elements in that they "sample" and then hold their input signal on the rising or falling edges of a controlling clock signal. In this way the input data is stored for use in a subsequent operation. Sequential circuits whose inputs are sampled in accordance with a clock edge are said to be **synchronous**. A clocked flip-flop is an example of a synchronous logic circuit component.

As in combinational circuits, the higher the clock frequency, the faster the sequential logic switches from one state to the next. As we have seen, the switching speed is limited by how rapidly the signal voltages can transition between low and high threshold values. For synchronous logic circuits, a major factor limiting performance is the propagation delays of the clock signal.

As an example of the effects of clock propagation delay, consider the synchronous logic circuit shown in Figure 7.1. In this circuit, the clock C_0 is a symmetric square-wave signal distributed to the two flip-flops D_1 and D_2 through interconnects (symbolized by the shaded area). Because the interconnects have different lengths, the clock signals arrive at the flip-flops after different delays T_1 and T_2.

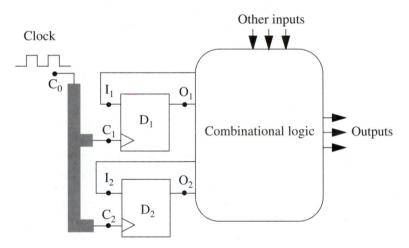

Figure 7.1 Example sequential logic circuit exhibiting clock skew. D_1 and D_2 are edge-triggered flip-flops driven by clock signals C_1 and C_2 and interconnected through a combinational logic circuit. The shaded areas represent interconnects which delay the arrival of the master clock signal C_0 to the two flip-flops by differing amounts.

A D-type flip-flop operates as follows: the output line (O_1 or O_2 in Figure 7.1) takes on the value of the input line (I_1 or I_2) at the time of the triggering event. In these devices the trigger event is defined by the rising edge of the clock signal (C_1 or C_2). In this circuit, the delays T_1 and T_2 are not equal, therefore the switching of the two flip-flops is not coincident. The difference between the arrival of the two clock signals, $T_2 - T_1$, is known as **clock skew**.

Consider the timing diagram for this circuit, shown in Figure 7.2. The top three curves illustrate the clock signals. The clock skew is seen as the difference between the arrival of the leading edge of the clock signal at the two flip-flops. The next line shows the input line I_1, which stays high through this example. After a delay of T_1, the D_1 flip-flop changes state so that the output line O_1 matches the logic high value of the input line I_1.

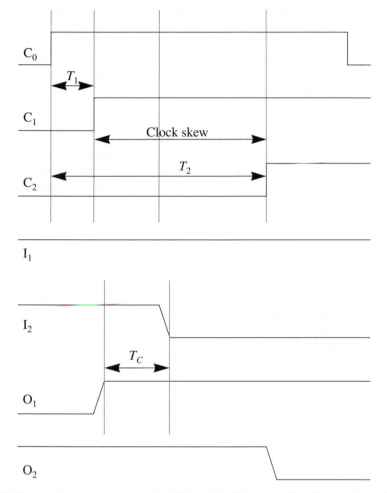

Figure 7.2 Timing diagram for the circuit of Figure 7.1. The output line O_1 goes high after timing delay T_1, which, after a delay of T_C through the combinational logic circuit, causes line I_2 to go low. Because of clock skew, the output line O_2 subsequently goes low, erroneously.

Now, suppose that the function of the combinational logic includes the operation $I_2 = \text{NOT}(O_1)$. This will cause the input line I_2 to go low, after a delay T_C through the combinational logic. Now, the value of the output line O_2 will depend on the length of the clock skew. In the diagram, the clock signal C_2 does not arrive until after the input I_2 changes, so the output line O_2 changes from high to low. If, however, the clock skew were less than T_C, the output line O_2 would not change state.

Note that a problem arises in this example, not because of the absolute clock delays, T_1 and T_2, between the master clock signal and the flip-flops, but because of the *relative* delay between the clock signals. Ideally, D_1 and D_2 should sample their respective inputs at precisely the same time. But, the signals propagate through the system at a finite speed, so this ideal situation can only be approximated. It is therefore necessary to estimate the clock skew so that erroneous switching events can be avoided.

The problem of predicting and minimizing clock skew is an important issue in IC design. In circuits with millions of components, it is imperative to know when the signals are going to arrive, but it is impractical to test every possible combination of logic states and signal overlap. Many methods of synthesizing clock systems have been developed; these techniques occupy a central place in the design of ICs.

Our task here can be stated as follows: given a periodic clock signal, distributed through interconnects of various lengths, what is the waveform of the signal when it arrives at different parts of the circuit? After a short discussion of periodicity, we will first approach this question using the familiar time-domain approach. We will see that even for a simple, one-lump RC interconnect approximation this approach is cumbersome and unwieldy. The alternate frequency-domain method will occupy this and the following chapter.

7.3 Periodic Signals

Careful circuit design ensures that all synchronous elements see the same rising and falling edges at approximately the same time. Designing a symmetric clock distribution layout for small integrated circuits is relatively easy. However, for large synchronous logic circuits, with millions of gates, the design of such a system is quite involved and time consuming. The complicated structure of large-scale clock distribution systems also makes them difficult to analyze. For example, a model of the clock distribution circuit of a fast microprocessor might need thousands or even millions of lumped-circuit elements for accurate results.

As we will see, the analysis of such a complicated model can be considerably eased by taking advantage of the fact that clock signals are **periodic**. A signal $v(t)$ is periodic if, for some positive constant T,

$$v(t) = v(t + T) \tag{7.1}$$

for all values of t. The smallest value of T which satisfies this condition is called the **period** of $v(t)$. The reciprocal of the period is called the **fundamental frequency** of the signal and is expressed in units of **hertz** (abbreviation: Hz), which are equivalent to s^{-1}. Sinusoidal signals which are integer multiples of the fundamental frequency are called **harmonics**. An example of a tremendously useful periodic signal is the symmetric square wave shown in Figure 7.3.

By definition, a periodic signal remains unchanged when shifted in time by one or more periods. This means that, strictly speaking, a periodic signal exists over the entire time domain $-\infty < t < \infty$. Another important observation is that, if we know the definition of a periodic signal over one period, we also know it over all time. We can construct a periodic function from a one-period segment by repeating reproductions of this segment ad infinitum on either side.

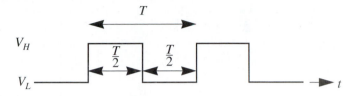

Figure 7.3 Symmetric periodic square-wave clock signal, switching between logic low and high values V_L and V_H.

7.4 Time-Domain Analysis of Clock Skew

7.4.1 Circuit Model

Consider the circuit shown in Figure 7.4. The clock generator, the output of which can be accurately modeled as a CMOS inverter gate, supplies a symmetric square wave signal with period T switching between V_L and V_H. This signal is coupled through an interconnect to the flip-flop. We wish to determine the waveform $v_C(t)$ which appears at the input of the flip-flop for various interconnect models (i.e., one-RC-lump, two-RC-lump, transmission-line, etc.).

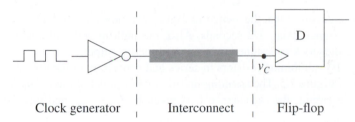

Figure 7.4 A flip-flop synchronized by a clock signal driven through an interconnect. The voltage signal arriving at the clock input of the flip-flop is $v_C(t)$.

The straightforward approach is to substitute appropriate circuit models for each of the three elements in Figure 7.4. We will carry out the analysis using the time-domain techniques developed in Chapter 2. Our objective here is to show that the straightforward approach is ugly and clumsy and is doomed for higher-order interconnect models.

To begin, we will use the inverter model (without inductance) for the clock, a single RC lump for the interconnect, and an input capacitance for the flip-flop. The resulting circuit is shown in Figure 7.5. In this circuit, the component symbols have their usual meanings:

R_{out} = pull-up and pull-down resistances

R_{int} = lumped interconnect resistance

C_{int} = lumped interconnect capacitance

C_{in} = flip-flop input capacitance

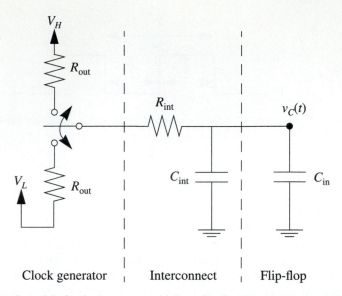

Figure 7.5 Circuit model of a clock generator driving a flip-flop through a single-RC-lump interconnect. The switch abruptly changes position every $T/2$ seconds.

It is understood that the switch in Figure 7.5 moves up and down in accordance with the
clock signal, every $T/2$ seconds. Also, the pull-up and pull-down resistances are equal,
since the clock is designed to have a symmetric output.

By combining resistors in series and capacitors in parallel, we can simplify the circuit of Figure 7.5. The resulting first-order RC model of the circuit is shown in Figure 7.6.
In this circuit, the capacitance C is given by $C_{in} + C_{int}$, and the resistance R is equal to
$R_{int} + R_{out}$.

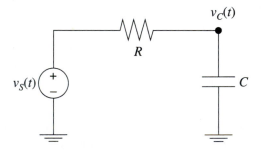

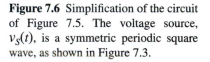

Figure 7.6 Simplification of the circuit
of Figure 7.5. The voltage source,
$v_S(t)$, is a symmetric periodic square
wave, as shown in Figure 7.3.

The voltage source $v_S(t)$ in Figure 7.6 is a periodic time-varying source given graph-
ically in Figure 7.3. We can express $v_S(t)$ formally as

$$v_S(t) = \begin{cases} V_H, & nT < t < \left(n + \dfrac{1}{2}\right)T \\ \\ V_L, & \left(n + \dfrac{1}{2}\right)T < t < (n+1)T \end{cases} \qquad (7.2)$$

where n is an integer. Although $v_S(t)$ is a time-varying source, it is a succession of constant values over different time intervals. Thus, we can apply the results of first-order RC circuit theory (Chapter 2) to this problem.

7.4.2 Qualitative Behavior

Suppose that the initial voltage across the capacitor is equal to logic low: $v_C(t) = V_L$. During the period $0 < t < T/2$, the source voltage is equal to V_H, and the capacitor voltage rises towards this value. At $t = T/2$, the source voltage switches to V_L; the capacitor voltage now decays exponentially toward this value. At the next switching event, $t = T$, the source voltage again goes high. However, the capacitor voltage at $t = T$ is slightly higher than V_L, so the extreme value reached at the next switching event ($t = 3T/2$) is above the previous maximum value (i.e., $v_C(3T/2) > v_C(T/2)$). This process continues until the capacitor voltage switches between two stable extreme values, which we will call v_l and v_h. This behavior is illustrated in Figure 7.7.

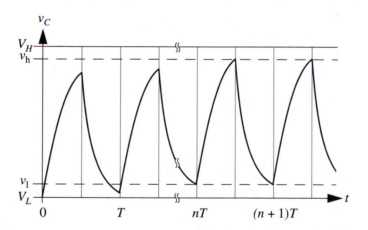

Figure 7.7 Qualitative response of the circuit of Figure 7.6, with $v_C(0) = V_L$. After a few cycles ("start-up transient"), the capacitor voltage oscillates between stable values of v_l and v_h.

Figure 7.7 illustrates an important general principle. Any linear circuit, when excited by a periodic signal, will have a response which eventually settles into a periodic form, with the same period. The shape of the "input" and "output" waveforms can differ (compare the square-wave source excitation with the exponential-sawtooth-shaped capacitor voltage in this example), but the frequency of the two waveforms is precisely the same. It takes a number of cycles before the circuit reaches this periodic state, called the **ac steady state**. This initial nonperiodic phase is called a **start-up transient** and is of limited interest compared to the periodic behavior.

7.4.3 Quantitative Analysis

We recognize the capacitor voltage in Figure 7.7 as being composed of exponential functions characteristic of first-order RC circuits. In general, the RC circuit response can be written as

$$v_C(t) = (v_C(t_0) - V_S)e^{-\frac{t-t_0}{t}} + V_S \tag{7.3}$$

where V_S is the (constant) source voltage, and $\tau = RC$ is the time constant.

We are particularly interested in the periodic portion of the response, after the start-up transient has decayed away. Our objective is to find analytic expressions for the extreme values v_l and v_h and for the curves connecting these values. From inspection of Figure 7.7 we can write

$$v_l = v_C(nT)$$

$$v_h = v_C\left(\left(n + \frac{1}{2}\right)T\right)$$

for sufficiently large values of n (i.e., after the start-up transient has decayed).

Consider a pull-up portion of the curve, $nT < t < (n + 1/2)T$. Over this interval the source voltage is equal to V_H. Substituting appropriate values into Eq. (7.3),

$$v_h = v_C\left(\left(n + \frac{1}{2}\right)T\right) = (v_C(nT) - V_H)e^{-\frac{\left(n + \frac{1}{2}\right)T - nT}{t}} + V_H$$

$$v_h = (v_l - V_H)e^{-\frac{T}{2t}} + V_H \tag{7.4}$$

Similarly, for a pull-down portion of the curve $(n + 1/2)T < t < (n + 1)T$, the source voltage is equal to V_L, and we can write

$$v_l = v_C((n + 1)T) = \left(v_C\left(\left(n + \frac{1}{2}\right)T\right) - V_L\right)e^{-\frac{(n + 1)T - \left(n + \frac{1}{2}\right)T}{\tau}} + V_L$$

$$v_l = (v_h - V_L)e^{-\frac{T}{2\tau}} + V_L \tag{7.5}$$

Equations (7.4) and (7.5) constitute two equations in two unknowns. They can be solved simultaneously by backward substitution to yield the following closed-form solutions for the capacitor voltage extrema:

$$v_h = \frac{\left(1 - e^{-\frac{T}{2\tau}}\right)\left(e^{-\frac{T}{2\tau}}V_L + V_H\right)}{\left(1 - e^{-\frac{T}{\tau}}\right)} \tag{7.6}$$

$$v_l = \frac{\left(1 - e^{-\frac{T}{2\tau}}\right)\left(e^{-\frac{T}{2\tau}}V_H + V_L\right)}{\left(1 - e^{-\frac{T}{\tau}}\right)} \tag{7.7}$$

These expressions can be used in the general RC solution to write equations describing the capacitor voltage as a function of time. For the pull-up portion, $nT < t < (n + 1/2)T$,

$$v_C(t) = \left(\frac{\left(1 - e^{-\frac{T}{2\tau}}\right)\left(e^{-\frac{T}{2\tau}}V_H + V_L\right)}{\left(1 - e^{-\frac{T}{\tau}}\right)} - V_H\right)e^{-\frac{t - nT}{\tau}} + V_H \tag{7.8}$$

and for the pull-down part, $(n + 1/2)T < t < (n + 1)T,$

$$v_C(t) = \left(\frac{\left(1 - e^{-\frac{T}{2\tau}}\right)\left(e^{-\frac{T}{2\tau}}V_L + V_H\right)}{\left(1 - e^{-\frac{T}{\tau}}\right)} - V_L\right)e^{-\frac{t - \left(n + \frac{1}{2}\right)T}{\tau}} + V_L \tag{7.9}$$

Example 7.1

Suppose a 5-V clock, operating at 100 MHz with an output resistance of 400 Ω, drives a flip-flop with an equivalent capacitance of 1 pF through an interconnect. The interconnect is modeled as a one-lump RC with $R_{int} = 100\ \Omega$ and $C_{int} = 1$ pF. The flip-flop triggers on a rising edge when the input reaches $V_{th} = 4.0$ V. Determine the clock skew between this and another flip-flop which is connected to the clock through an interconnect three times as long.

We model the two cases using the first-order interconnect approximation. For the short interconnect, the time constant is given by

$$\tau = (R_{out} + R_{int})(C_{in} + C_{int})$$

$$= (400 + 100)(1 + 1) \times 10^{-12}$$

$$= 10^{-9}\ s = 1\ ns$$

The clock period is equal to the reciprocal of the frequency, 10 ns. Inserting values into Eqs. (7.6) and (7.7):

$$v_{\mathrm{h}} = \frac{\left(1 - e^{-\frac{10}{2 \times 1}}\right)\left(e^{-\frac{10}{2 \times 1}}(0) + 5\right)}{\left(1 - e^{-\frac{10}{1}}\right)} = 4.967$$

$$v_{\mathrm{l}} = \frac{\left(1 - e^{-\frac{10}{2 \times 1}}\right)\left(e^{-\frac{10}{2 \times 1}}(5) + 0\right)}{\left(1 - e^{-\frac{10}{1}}\right)} = 0.033$$

The low-to-high transition time t_{lh} is defined by the condition $v_C(t_{\mathrm{lh}}) = V_{\mathrm{th}}$. Applying this condition to the pull-up expression Eq. (7.8),

$$V_{\mathrm{th}} = 4 = (0.033 - 5)e^{-\frac{t_{\mathrm{lh}}}{\tau}} + 5$$

$$t_{\mathrm{lh}} = 1.60 \text{ ns}$$

The resistance and capacitance of the long interconnect are each three times longer. The time constant for this case is

$$\tau = (R_{\mathrm{out}} + R_{\mathrm{int}})(C_{\mathrm{in}} + C_{\mathrm{int}})$$

$$= (400 + 300)(1 + 3) \times 10^{-12}$$

$$= 2.8 \times 10^{-9} \text{ s} = 2.8 \text{ ns}$$

Repeating the above calculations for this case, we find that

$$v_{\mathrm{h}} = 4.282$$

$$v_{\mathrm{l}} = 0.718$$

$$t_{\mathrm{lh}} = 4.07 \text{ ns}$$

The clock skew is given by the difference in transition times for the two flip-flops:

$$\mathrm{skew} = 4.07 - 1.60 = 2.43 \text{ ns}$$

which is nearly one-quarter of the entire clock period.

This example illustrates that while Eqs. (7.8) and (7.9) are correct, they are cumbersome for determining clock skew in that they provide little or no insight into the behavior of the circuit of Figure 7.6. To determine the effect of a different interconnect length on the system response requires involved calculations, which do not generalize to other interconnect lengths. Also, despite their unwieldy appearance, they represent only the simplest possible case, a first-order RC interconnect model. Previously, we have seen that accurate modeling of modern digital systems requires higher-order models, such as multiple-section RC ladder or series-RLC circuits. If we use these more accurate models to determine the clock skew, the result is a set of transcendental equations which cannot even be solved in closed form.

The point of this exercise is to demonstrate one of the shortcomings of time-domain analysis. In this example, we have not taken advantage of some subtle aspects of periodic functions, with the result that the time-domain solution of even a simple *RC* circuit with a periodic source appears inelegant and frightfully complicated. There must be a better way!

7.5 The Big Picture

A conceptual view of the circuit analysis process is illustrated in Figure 7.8. In the center is a box containing an arbitrary connection of components, representing the circuit. The function of the circuit is to take an input signal, the **excitation**, and transform it into an output signal, the **response**. Likewise, the objective of the circuit-solving process is to transform a representation of the excitation into an accurate representation of the output response.

Excitation: ⎍⎍⎍

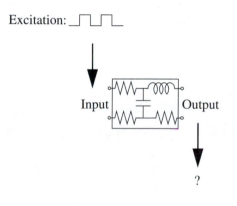

Input Output

?

Figure 7.8 Conceptual view of circuit analysis. The "excitation," a periodic signal, is applied to the input of a circuit. The objective is determine the resulting signal which appears at the circuit output.

The time-domain approach to solving the problem, Figure 7.9, uses the whole excitation signal (i.e., the square wave) as the input to a procedure. This procedure is symbolized by the machine in the center of Figure 7.9, in which turning the crank causes the differential equation corresponding to the circuit to be solved. Instead of an actual machine, we can use an algorithm which performs the same function, but automatic circuit-solving machines certainly do exist. Our algorithm for solving circuits in the time domain is to generate differential equations from the circuit and solve these differential equations subject to forcing functions and initial conditions imposed by the input signal.

As we have seen, even for relatively simple circuits, the crank can be difficult to turn when the input is not a constant dc signal. For more complicated circuits, and other, more sophisticated input signals, it is as if the bearings on the crank were rusted solid and the crank itself broken off.

Compare this to the frequency-domain approach, Figure 7.10, which consists of three steps. First, the input signal is decomposed into a series of sinusoidal signals at different frequencies, for reasons we will explain momentarily. It may seem incredible, but any electronic signal can be broken up in this way. Next, each of these individual sinusoidal signals is fed into "circuit-solver machines." The responses are then recombined (added together) to obtain the final output.

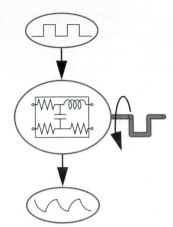

Figure 7.9 Time-domain circuit analysis. The process of "turning the crank" corresponds to finding the solution to the differential equation describing the circuit shown on the box. In time-domain analysis, the entire input signal is used as the input.

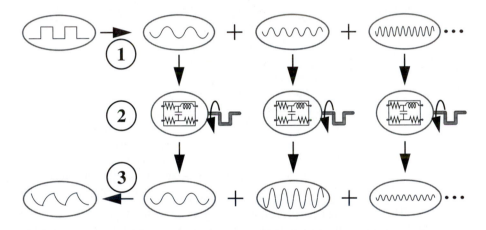

Figure 7.10 Frequency-domain circuit analysis. We first decompose the excitation signal (upper left) into a sum of sinusoidal signals (process ①). Next, each individual sinusoid is "cranked through" the circuit to obtain a partial output (process ②). Finally, the partial outputs (bottom row) are added together to produce the output (process ③). Although there appear to be more steps in this procedure than in time-domain analysis, solving problems in the frequency domain has many advantages.

Although the frequency-domain approach appears roundabout, it offers many advantages. Chief among these is the following: there exist very easy methods of solving circuits, but they only work for certain kinds of input signals. Sinusoidal functions are one of these kinds of signals. For sinusoidal inputs, it is as if the "circuit-solving machine" has lubricated bearings and an electric motor attached to the crank. These methods work so well that it is actually easier overall to make our arbitrary input signals look like sinusoids.

Another reason frequency-domain analysis is valuable is that by understanding how the circuit responds to simple signals at different frequencies, we can obtain insights into more complicated behaviors which are not apparent in the time domain. For example, describing how a radio selects out one station from many, or why a flute sounds different from a trumpet, requires the use of frequency-domain concepts.

In using a frequency-domain approach to solve for the time-domain behavior of clock circuits, we are, in essence, making a trade. The time-domain approach, for nonconstant excitation signals, is laborious and clumsy. The alternative frequency-domain method involves many more individual steps, but each of these steps is simpler. In many cases, especially for circuits of second order and higher, this is a worthwhile bargain.

The three steps in frequency-domain analysis, corresponding to processes ①, ②, and ③ in Figure 7.10, are called **Fourier analysis**, **ac steady-state circuit analysis**, and **superposition**. The first and last of these will occupy the remainder of this chapter. Techniques for analyzing the ac steady-state behavior of circuits will be covered in the next chapter.

7.6 Fourier Series Signal Representation

Consider a signal which can be represented by the periodic function $f(t)$. This function has period T (seconds) and **fundamental frequency** $f = 1/T$ (Hz). This function can be expressed as an infinite series of cosines, known as a **Fourier series**:

$$f(t) \cong A_0 + A_1 \cos(\omega t + \theta_1) + A_2 \cos(2\omega t + \theta_2) + A_3 \cos(3\omega t + \theta_3)$$

$$+ \cdots + A_n \cos(n\omega t + \theta_n) + \cdots$$

where $A_0, A_1, \theta_1, A_2, \theta_2, \ldots$ are constants, and ω, given by

$$\omega = \frac{2\pi}{T}$$

is the **fundamental angular frequency** in rad/s. We often refer to ω simply as the "frequency" when the context is clear. The **Hertzian frequency**, f, is related to the angular frequency through

$$\omega = 2\pi f$$

where f, which is not to be confused with the function $f(t)$, has the units of Hz. Although both Hz and rad/s are equivalent to s^{-1}, they are used in different ways and should not be interchanged.

(The symbol "$\cong$" is used in the above expression instead of an equals sign only because the series converges to the average value on either side of a discontinuity. For example, the clock signal function, Eq. (7.2), has a discontinuity at $t = 0$:

$$v_S(0^-) = V_L$$

$$v_S(0^+) = V_H$$

The Fourier series which we will develop for this function converges to

$$\frac{V_L + V_H}{2}$$

at this and similar discontinuities. With this understanding, we will dispense with the "$\cong$" symbol and use an equals sign from now on.)

The Fourier series representation of a function consists of an infinite number of terms. The first, given by A_0, is called the **dc component** because, as we will show, it is equal to the average value of the function over one period. Subsequent terms are sinusoidal functions with frequencies harmonically related to the fundamental frequency of the periodic function. We call the kth term, $\cos(k\omega + \theta_k)$, the "kth harmonic." The weight constants $A_1, A_2, \ldots$ are called **amplitudes** and have the same units as the signal function. θ_1, $\theta_2, \ldots$ are **phase** angles and are expressed in radians. Although the Fourier series has an infinite number of terms, in many cases a good approximation to the function is possible with only a few terms.

A more concise notation is

$$f(t) = A_0 + \sum_{n=1}^{\infty} A_n \cos(n\omega t + \theta_n) \tag{7.10}$$

For this series to be useful, we need to have expressions for the amplitude and phase constants. To develop these, we will find it convenient to put Eq. (7.10) into an alternate form. Using the trigonometric identity

$$\cos(\alpha + \beta) = \cos\alpha\cos\beta - \sin\alpha\sin\beta$$

we expand the nth harmonic as

$$A_n\cos(n\omega t + \theta_n) = A_n\cos n\omega t\cos\theta_n - A_n\sin n\omega t\sin\theta_n$$

$$= a_n\cos n\omega t + b_n\sin n\omega t$$

where the constants a_n and b_n are related to A_n and θ_n by

$$a_n = A_n\cos\theta_n$$

$$b_n = -A_n\sin\theta_n$$

$$A_n = \sqrt{a_n^2 + b_n^2}$$

$$\theta_n = \operatorname{atan}-\frac{b_n}{a_n}$$

Equation (7.10) can now be written as

$$f(t) = a_0 + \sum_{n=1}^{\infty} a_n\cos n\omega t + b_n\sin n\omega t \tag{7.11}$$

where $a_0 = A_0$.

Recall that the function $f(t)$ is periodic in T:

$$f(t) = f(t + T)$$

Suppose we evaluate the Fourier series at $t + T$:

$$f(t+T) = a_0 + \sum_{n=1}^{\infty} a_n \cos n\omega(t+T) + b_n \sin n\omega(t+T)$$

$$= a_0 + \sum_{n=1}^{\infty} a_n \cos n\omega\left(t + \frac{2\pi}{\omega}\right) + b_n \sin n\omega\left(t + \frac{2\pi}{\omega}\right)$$

$$= a_0 + \sum_{n=1}^{\infty} a_n \cos(n\omega t + 2\pi n) + b_n \sin(n\omega t + 2\pi n)$$

$$= a_0 + \sum_{n=1}^{\infty} a_n \cos n\omega t + b_n \sin n\omega t$$

$$= f(t)$$

This illustrates the result that the Fourier series for $f(t)$ is also periodic in T. This is a consequence of the fact that all of the sinusoidal terms are harmonics of the fundamental frequency, so they each are also periodic in T. Therefore, when performing calculations, we can use any convenient interval T units long that we wish.

7.6.1 Fourier Coefficients

We are now ready to determine explicit expressions for the constants in Eq. (7.11). The dc component is found by integrating both sides of Eq. (7.11) over an interval T. We will use the interval $0 < t < T$:

$$\int_0^T f(t)dt = \int_0^T a_0 dt + \int_0^T \left(\sum_{n=1}^{\infty} a_n \cos n\omega t + b_n \sin n\omega t \right) dt$$

Integrating term-by-term, this can be written in the equivalent form

$$\int_0^T f(t)dt = a_0 T + \sum_{n=1}^{\infty} a_n \int_0^T \cos n\omega t\, dt + \sum_{n=1}^{\infty} b_n \int_0^T \sin n\omega t\, dt$$

We observe that all of the integrals on the righthand side of this equation are integrals of sinusoids which are periodic in T. The integral of any sinusoid, when integrated over its period or an integer multiple of its period, is zero. Hence this equation reduces to

$$a_0 = \frac{1}{T} \int_0^T f(t)dt$$

a_0 is simply the average value of the function over one period. (Again, any interval T units long can be employed to determine a_0.)

To find the a_k coefficients, we first multiply both sides of Eq. (7.11) by $\cos k\omega t$, and then integrate from 0 to T:

$$\int_0^T f(t)\cos k\omega t\, dt = \int_0^T a_0 \cos k\omega t\, dt + \int_0^T \cos k\omega t \left(\sum_{n=1}^{\infty} a_n \cos n\omega t + b_n \sin n\omega t \right) dt$$

Interchanging the order of integration and summation:

$$\int_0^T f(t)\cos\ k\omega t\ dt = a_0 \int_0^T \cos\ k\omega t\ dt$$

$$+ \sum_{n=1}^{\infty} a_n \int_0^T \cos\ k\omega t\ \cos\ n\omega t\ dt \qquad (7.12)$$

$$+ \sum_{n=1}^{\infty} b_n \int_0^T \cos\ k\omega t\ \sin\ n\omega t\ dt$$

The first integral on the righthand side is equal to zero, since it is an integral of a sinusoid over an integer multiple of its period. We evaluate the integrals on the extreme righthand side using the trigonometric identity

$$\cos\alpha\sin\beta\ =\ \frac{1}{2}\sin(\alpha+\beta)+\frac{1}{2}\sin(\alpha-\beta)$$

Applying this identity to the nth integral,

$$b_n\int_0^T \cos\ k\omega t\ \sin\ n\omega t\ dt = \frac{1}{2}b_n\int_0^T \sin(k+n)\omega t\, dt + \frac{1}{2}b_n\int_0^T \sin(k-n)\omega t\, dt$$

Both of these integrals are zero, as they are integrals of sinusoids over a multiple of their periods. (The special case of $k = n$ in the second term results in a zero-valued integrand, so this case also yields zero.)

The middle term in Eq. (7.12) can be written as

$$\sum_{n=1}^{\infty} a_n\int_0^T \cos\ k\omega t\ \cos\ n\omega t\ dt = \sum_{\substack{n=1\\n\neq k}}^{\infty} a_n\int_0^T \cos\ k\omega t\ \cos\ n\omega t\ dt + a_k\int_0^T \cos\ k\omega t\ \cos\ k\omega t\ dt$$

To each of the cosine products we apply the trigonometric identity

$$\cos\alpha\cos\beta\ =\ \frac{1}{2}\cos(\alpha+\beta)+\frac{1}{2}\cos(\alpha-\beta)$$

which results in

$$\sum_{\substack{n=1}}^{\infty} a_n \int_0^T \cos k\omega t \, \cos n\omega t \, dt = \sum_{\substack{n=1 \\ n \neq k}}^{\infty} \frac{a_n}{2} \int_0^T \cos(k+n)\omega t \, dt$$

$$+ \sum_{\substack{n=1 \\ n \neq k}}^{\infty} \frac{a_n}{2} \int_0^T \cos(k-n)\omega t \, dt$$

$$+ \frac{a_k}{2} \int_0^T \cos 2k\omega t \, dt$$

$$+ \frac{a_k}{2} \int_0^T \cos 0 \, dt$$

The first three of these integrate to zero, for the usual reason. The last has an integrand of value unity, which integrates to $((a_k T)/2)$. This is the only nonzero term on the right hand side of Eq. (7.12), which evaluates to

$$\int_0^T f(t)\cos k\omega t \, dt = \frac{a_k T}{2}$$

Solving for a_k, the kth Fourier cosine coefficient is given by

$$a_k = \frac{2}{T} \int_0^T f(t)\cos k\omega t \, dt$$

The b_k sine coefficients are found similarly. We multiply both sides of Eq. (7.11) by $\sin k\omega t$ and integrate from 0 to T:

$$\int_0^T f(t)\cos k\omega t \, dt = \int_0^T a_0 \sin k\omega t \, dt + \int_0^T \sin k\omega t \left(\sum_{n=1}^{\infty} a_n \cos n\omega t + b_n \sin n\omega t \right) dt$$

As before, most of the integrals vanish, leaving only a single term. The final result is

$$b_k = \frac{2}{T} \int_0^T f(t)\sin k\omega t \, dt$$

To summarize, a periodic function $f(t)$ can be described by the Fourier series

$$f(t) = a_0 + \sum_{n=1}^{\infty} a_n \cos n\omega t + b_n \sin n\omega t \qquad (7.13)$$

where $\omega = \dfrac{2\pi}{T}$ and

$$a_0 = \frac{1}{T}\int_0^T f(t)\,dt \tag{7.14}$$

$$a_n = \frac{2}{T}\int_0^T f(t)\cos n\omega t\,dt \tag{7.15}$$

$$b_n = \frac{2}{T}\int_0^T f(t)\sin n\omega t\,dt \tag{7.16}$$

We illustrate the use of these formulae with several examples.

7.6.2 Square-Wave Clock Signal

The symmetric square-wave clock signal in Figure 7.11 has amplitude V and a frequency of 1 Hz. We wish to express this signal as a Fourier series.

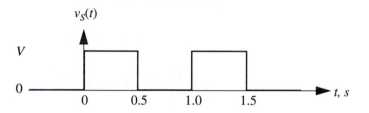

Figure 7.11 Symmetric square-wave clock signal with $f = 1$ Hz and amplitude V. The time origin has been chosen such that the signal (less the dc component) has odd symmetry.

We first find the dc component of the signal from Eq. (7.14):

$$a_0 = \frac{1}{T}\int_0^T v_S(t)\,dt$$

Inserting the value of the period $T = 1$, and the signal value,

$$a_0 = \frac{1}{1}\int_0^{0.5} V\,dt + \frac{1}{1}\int_{0.5}^{1} 0\,dt = \frac{V}{2}$$

The cosine coefficients of the Fourier series are found from Eq. (7.15) (with $\omega = 2\pi f = 2\pi$ rad/s):

$$a_n = \frac{2}{T}\int_0^T v_S(t)\cos n\omega t\, dt$$

$$= \frac{2}{1}\int_0^{0.5} V\cos 2\pi nt\, dt$$

$$= \frac{2V}{2\pi n}\sin 2\pi nt \Big|_0^{0.5}$$

$$= \frac{V}{\pi n}(\sin \pi n - \sin 0) = 0$$

All of the cosine coefficients are zero.

The sine coefficients are calculated from Eq. (7.16):

$$b_n = \frac{2}{T}\int_0^T v_S(t)\sin n\omega t\, dt$$

$$= \frac{2}{1}\int_0^{0.5} V\sin 2\pi nt\, dt$$

$$= -\frac{2V}{2\pi n}\cos 2\pi nt \Big|_0^{0.5}$$

$$= -\frac{V}{\pi n}(\cos \pi n - \cos 0)$$

$$= \frac{V}{\pi n}[1 - (-1)^n]$$

The term in brackets takes on either of two values. If n is an even integer, the term is zero; if n is odd, then the bracketed term is equal to 2. The sine coefficients are thus:

$$b_n = \begin{cases} 0, & n \text{ even} \\ \dfrac{2V}{\pi n}, & n \text{ odd} \end{cases} \tag{7.17}$$

Inserting the coefficient values into the Fourier series expression, the square wave can be expressed as:

$$v_S(t) = \frac{V}{2} + \sum_{n=1,3}^{\infty} \frac{2V}{\pi n}\sin 2\pi nt \tag{7.18}$$

An equivalent notation is

$$v_S(t) = \frac{V}{2} + \sum_{\substack{n=1 \\ n \text{ odd}}}^{\infty} \frac{2V}{\pi n} \sin 2\pi nt$$

At first exposure, it is disturbing to think that a series of sine functions, which are smooth, continuous curves, could possibly combine to form the angular, discontinuous square wave of Figure 7.11. Well, they do. To visualize this, we evaluate a few terms of Eq. (7.18) (with the amplitude V set to 1 volt):

$$v_S(t) = \frac{1}{2} + \frac{2}{\pi} \sin 2\pi t + \frac{2}{3\pi} \sin 6\pi t + \frac{2}{5\pi} \sin 10\pi t + \frac{2}{7\pi} \sin 14\pi t + \cdots$$

$$= 0.5 + 0.637 \sin 2\pi t + 0.212 \sin 6\pi t + 0.127 \sin 10\pi t$$

$$+ 0.091 \sin 14\pi t + \cdots$$

Note the decrease in value of the coefficients of the higher-order harmonics, owing to the n in the denominator of Eq. (7.17). Successive approximations to the square-wave function are shown in Figure 7.12. The graph in the upper left includes only the first two terms, the dc component and the fundamental-frequency component. It does not look much like a square wave. Adding the third harmonic term, as shown in the upper right, doesn't improve things very much. However, by the time the fifth harmonic is included (lower left), we see the approximation converging. Even with only the five terms through the seventh harmonic included (lower right) the outline of a square wave is definitely seen. In particular, the rapid rise and fall characteristic of a clock signal are faithfully reproduced in the approximation. As more terms are included, the series representation becomes more and more like the exact function.

The fact that all of the cosine coefficients are zero in this example is not an accident. Returning to Figure 7.12, we note that if the average value of the signal is subtracted out, the remaining function (i.e., the "ac" part) has odd symmetry. The sinusoidal terms in the Fourier series all have either even symmetry (the cosine terms) or odd symmetry (sine terms). The ac part of this signal function can be constructed only from Fourier components that have odd symmetry, else its own odd symmetry would be destroyed. It is also true that functions (less the dc component) with even symmetry have Fourier expansions containing only cosine functions. It is often helpful in deriving a Fourier series to choose a time origin such that the signal has even or odd symmetry, in order to simplify the calculation of the coefficients.

Since the sum of odd functions is itself odd, there is no way to construct a series containing only sines which does not have odd symmetry. Likewise, the sum of even functions is even, and a series containing only cosines must have even symmetry. Therefore if an ac signal is neither odd nor even, then its Fourier expansion must contain both cosine and sine terms.

7.6.3 Shifted Square Wave

In Figure 7.13 is a symmetric square-wave clock signal, similar to the one shown in Figure 7.11, but with even symmetry (less the dc component). This signal switches between V_L and V_H at a frequency of 100 MHz. The radian frequency is $\omega = 2\pi f = 2\pi 10^8$ rad/s, and the

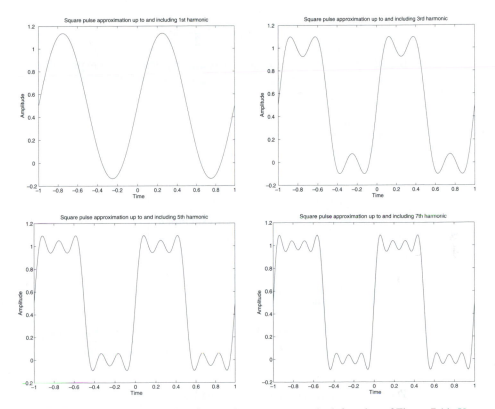

Figure 7.12 Fourier-series approximations to the square-wave clock function of Figure 7.11. *Upper left:* The dc and fundamental terms only. *Upper right:* After adding the third harmonic. *Lower left:* After adding the fifth harmonic. *Lower right:* Fourier series approximation with the dc, fundamental, and third, fifth, and seventh harmonics.

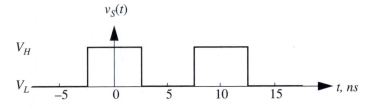

Figure 7.13 100-MHz symmetric square-wave clock signal, switching between V_L and V_H. The time origin has been chosen such that the ac part of the signal has even symmetry.

period is $1/f = 10$ ns. To express this signal as a Fourier series, we will find it convenient to perform our integrations over the interval -2.5 ns $< t < 7.5$ ns.

The dc component of the signal is found from Eq. (7.14). However, this is such a simple signal that it is not really necessary to perform the integration. We know that the dc component is the average value of the signal over one period; by inspection, the average value of this symmetric square wave is

$$a_0 = \frac{V_L + V_H}{2}$$

Because the ac part of the signal has even symmetry, we expect all the sine coefficients to be zero. We will verify this by calculating b_n from Eq. (7.16):

$$b_n = \frac{2}{T} \int_{-\frac{T}{4}}^{\frac{3T}{4}} v_S(t) \sin n\omega t \, dt$$

$$= \frac{2}{T} \left(\int_{-\frac{T}{4}}^{\frac{T}{4}} V_H \sin n\omega t \, dt + \int_{\frac{T}{4}}^{\frac{3T}{4}} V_L \sin n\omega t \, dt \right)$$

$$= -\frac{2}{T} \left(\frac{V_H}{n\omega} \cos n\omega t \Big|_{-\frac{\pi}{2\omega}}^{\frac{\pi}{2\omega}} + \frac{V_L}{n\omega} \cos n\omega t \Big|_{\frac{\pi}{2\omega}}^{\frac{3\pi}{2\omega}} \right)$$

$$= -\frac{2}{T} \left(\frac{V_H}{n\omega} \cos \frac{\pi n}{2} - \frac{V_H}{n\omega} \cos \left(-\frac{\pi n}{2} \right) + \frac{V_L}{n\omega} \cos \frac{3\pi n}{2} - \frac{V_L}{n\omega} \cos \frac{\pi n}{2} \right)$$

$$= 0$$

The cosine coefficients are found using Eq. (7.15):

$$a_n = \frac{2}{T} \int_{-\frac{T}{4}}^{\frac{3T}{4}} v_S(t) \cos n\omega t \, dt$$

$$= \frac{2}{T} \left(\int_{-\frac{T}{4}}^{\frac{T}{4}} V_H \cos n\omega t \, dt + \int_{\frac{T}{4}}^{\frac{3T}{4}} V_L \cos n\omega t \, dt \right)$$

$$= \frac{2}{T} \left(\int_{-\frac{T}{4}}^{\frac{T}{4}} V_H \cos n\frac{2\pi}{T} t \, dt + \int_{\frac{T}{4}}^{\frac{3T}{4}} V_L \cos n\frac{2\pi}{T} t \, dt \right)$$

In the second integral, we make the change of variables $t' = t - \frac{T}{2}$:

$$a_n = \frac{2}{T}\left(\int_{-\frac{T}{4}}^{\frac{T}{4}} V_H \cos n\frac{2\pi}{T} t\, dt + \int_{-\frac{T}{4}}^{\frac{T}{4}} V_L \cos n\frac{2\pi}{T}\left(t' + \frac{T}{2}\right) dt' \right)$$

$$= \frac{2}{T}\left(\int_{-\frac{T}{4}}^{\frac{T}{4}} V_H \cos n\frac{2\pi}{T} t\, dt + \int_{-\frac{T}{4}}^{\frac{T}{4}} V_L \cos \left(n\frac{2\pi}{T}t' + n\pi\right) dt' \right)$$

Examining the cosine function in the righthand integral, we recognize that it has two values, depending on whether n is even or odd:

$$\cos\left(n\frac{2\pi}{T}t' + n\pi\right) = \begin{cases} \cos n\frac{2\pi}{T}t', & n \text{ even} \\[2mm] -\cos n\frac{2\pi}{T}t', & n \text{ odd} \end{cases}$$

Therefore, for even values of n, the coefficients are

$$a_n = \frac{2}{T}\left(\int_{-\frac{T}{4}}^{\frac{T}{4}} V_H \cos n\frac{2\pi}{T} t\, dt + \int_{-\frac{T}{4}}^{\frac{T}{4}} V_L \cos n\frac{2\pi}{T} t'\, dt' \right)$$

$$= \frac{2}{T}(V_H + V_L) \int_{-\frac{T}{4}}^{\frac{T}{4}} \cos n\frac{2\pi}{T} t\, dt$$

$$= \frac{V_H + V_L}{\pi n} \sin n\frac{2\pi}{T} t \Big|_{-\frac{T}{4}}^{\frac{T}{4}}$$

$$= \frac{V_H + V_L}{\pi n}\left(\sin\frac{n\pi}{2} - \sin-\frac{n\pi}{2} \right)$$

$$= 0$$

For odd values of n,

$$a_n = \frac{2}{T}\left(\int_{-\frac{T}{4}}^{\frac{T}{4}} V_H \cos n\frac{2\pi}{T} t\, dt - \int_{-\frac{T}{4}}^{\frac{T}{4}} V_L \cos n\frac{2\pi}{T} t'\, dt' \right)$$

$$a_n = \frac{2}{T}(V_H - V_L)\int\limits_{-\frac{T}{4}}^{\frac{T}{4}} \cos n\frac{2\pi}{T}t\, dt$$

$$= \frac{V_H - V_L}{\pi n}\sin n\frac{2\pi}{T}t\,\Bigg|_{-\frac{T}{4}}^{\frac{T}{4}}$$

$$= \frac{V_H - V_L}{\pi n}\left(\sin\frac{n\pi}{2} - \sin-\frac{n\pi}{2}\right)$$

$$= \frac{V_H - V_L}{\pi n}\left(\sin\frac{n\pi}{2} - \sin-\frac{n\pi}{2}\right)$$

For successive odd values of n, $\sin\frac{n\pi}{2}$ alternately takes on values of $+1$ and -1. This is expressed algebraically as

$$\sin\frac{n\pi}{2} = (-1)^{\frac{n-1}{2}}, \qquad n \text{ odd}$$

Using this notation, the odd a_n coefficients are

$$a_n = \frac{V_H - V_L}{\pi n}\left((-1)^{\frac{n-1}{2}} - \left(-(-1)^{\frac{n-1}{2}}\right)\right)$$

$$= \frac{2}{\pi n}(V_H - V_L)(-1)^{\frac{n-1}{2}}$$

Putting these results together, we write the Fourier series for the function in Figure 7.13 as

$$v_S(t) = \frac{V_L + V_H}{2} + (V_H - V_L)\sum_{\substack{n=1 \\ n\ \text{odd}}}^{\infty}\frac{2}{\pi n}(-1)^{\frac{n-1}{2}}\cos(2\pi 10^8 nt) \qquad (7.19)$$

This expression can be compared to the Fourier representation of the same signal with the time axis coincident with a rising edge, as in Figure 7.11. Making appropriate modifications to Eq. (7.18):

$$v_S(t) = \frac{V_L + V_H}{2} + (V_H - V_L)\sum_{\substack{n=1 \\ n\ \text{odd}}}^{\infty}\frac{2}{\pi n}\sin(2\pi 10^8 nt) \qquad (7.20)$$

Two clock signals, with arbitrary angular frequency ω, are sketched in Figure 7.14 along with their respective Fourier expansions for later reference.

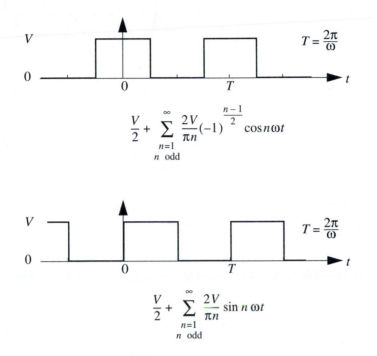

$$\frac{V}{2} + \sum_{\substack{n=1 \\ n \ \text{odd}}}^{\infty} \frac{2V}{\pi n}(-1)^{\frac{n-1}{2}} \cos n\omega t$$

$$\frac{V}{2} + \sum_{\substack{n=1 \\ n \ \text{odd}}}^{\infty} \frac{2V}{\pi n} \sin n\omega t$$

Figure 7.14 Comparison of two similar clock signals and their associated Fourier series representations.

7.6.4 Periodic Exponential Sawtooth

As a final example, we will derive the Fourier series expansion for the periodic exponential sawtooth signal illustrated in Figure 7.15. This is the signal we examined in Section 7.4 which appears at the input of a flip-flop. It arises from a symmetric square-wave clock signal driven through an *RC* interconnect.

(Before we begin, we recognize the fact that although this derivation is straightforward, it is long and error prone. We are deriving the Fourier series of this rather complicated signal so we can compare the series we obtain to the result of another, much easier method. Algebraphobics may wish to skip directly to the result on page 357.)

Symbolically,

$$v_C(t) = \begin{cases} (v_l - V_H)e^{-\frac{t}{\tau}} + V_H, & 0 < t < \dfrac{T}{2} \\[2em] (v_h - V_L)e^{-\frac{t-\frac{T}{2}}{\tau}} + V_L, & \dfrac{T}{2} < t < T \end{cases} \qquad (7.21)$$

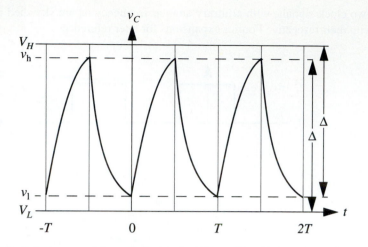

Figure 7.15 Periodic exponential sawtooth signal which oscillates between v_1 and v_h. This signal is encountered at the clock input of a flip-flop driven through an interconnect. Δ is defined in Eq. (7.24), below.

The values of the extrema are found by inserting t = $T/2$ and $t = T$ into these expressions:

$$v_h = (v_1 - V_H)e^{-\frac{T}{2\tau}} + V_H \qquad (7.22)$$

$$v_1 = (v_h - V_L)e^{-\frac{T}{2\tau}} + V_L \qquad (7.23)$$

Considering that this signal arises from the action of an *RC* circuit on a symmetric square wave, it is apparent that the magnitude of the two terms in parentheses in Eq. (7.21) are equal. We will define this magnitude as Δ:

$$\Delta = v_h - V_L = V_H - v_1 \qquad (7.24)$$

[The equality of these terms can also be shown explicitly using Eqs. (7.6) and (7.7); see Exercise 7.14.]

The dc component is found from Eq. (7.14):

$$a_0 = \frac{1}{T}\int_0^T v_C(t)dt$$

$$= \frac{1}{T}\int_0^{\frac{T}{2}}\left(-\Delta e^{-\frac{t}{\tau}} + V_H\right)dt + \frac{1}{T}\int_{\frac{T}{2}}^T\left(\Delta e^{-\frac{t-\frac{T}{2}}{\tau}} + V_L\right)dt$$

In the righthand integral, we make the change of variables $t' = t - \frac{T}{2}$:

$$a_0 = \frac{1}{T}\int_0^{\frac{T}{2}}\left(-\Delta e^{-\frac{t}{\tau}} + V_H\right)dt + \frac{1}{T}\int_0^{\frac{T}{2}}\left(\Delta e^{-\frac{t'}{\tau}} + V_L\right)dt'$$

$$= \frac{1}{T}\int_0^{\frac{T}{2}}(-\Delta + \Delta)e^{-\frac{t}{\tau}}dt + \frac{1}{T}\int_0^{\frac{T}{2}}(V_H + V_L)dt$$

$$= \frac{V_H + V_L}{2}$$

i.e., the average value of the two logic levels. (Given the symmetric nature of the sawtooth function, this result can be reasoned out directly.)

We next calculate the cosine coefficients from Eq. (7.15):

$$a_n = \frac{2}{T}\int_0^T v_C(t)\cos n\omega t \, dt$$

$$= \frac{2}{T}\int_0^{\frac{T}{2}}\left(-\Delta e^{-\frac{t}{\tau}} + V_H\right)\cos n\omega t \, dt + \frac{2}{T}\int_{\frac{T}{2}}^T\left(\Delta e^{-\frac{t-\frac{T}{2}}{\tau}} + V_L\right)\cos n\omega t \, dt$$

With a change of variables in the right integral and some minor rearrangement, this becomes

$$\frac{a_n T}{2} = \int_0^{\frac{T}{2}} V_H \cos n\omega t \, dt + \int_0^{\frac{T}{2}} V_L \cos n\omega\left(t + \frac{T}{2}\right)dt$$

$$+ \int_0^{\frac{T}{2}} -\Delta e^{-\frac{t}{\tau}}\cos n\omega t \, dt + \int_0^{\frac{T}{2}} \Delta e^{-\frac{t}{\tau}}\cos n\omega\left(t + \frac{T}{2}\right)dt$$

The first two integrals are zero, since the cosine functions have equal excursions above and below the t-axis over this integration interval. Making the substitution $T/2 = \pi/\omega$,

$$\frac{a_n T}{2} = \int_0^{\frac{T}{2}} -\Delta e^{-\frac{t}{\tau}}\cos n\omega t \, dt + \int_0^{\frac{T}{2}} \Delta e^{-\frac{t}{\tau}}\cos n\omega\left(t + \frac{\pi}{\omega}\right)dt$$

$$= \int_0^{\frac{T}{2}} -\Delta e^{-\frac{t}{\tau}}\cos n\omega t \, dt + \int_0^{\frac{T}{2}} \Delta e^{-\frac{t}{\tau}}\cos(n\omega t + n\pi)dt$$

(7.25)

The cosine function on the right has two cases, depending on the parity of n:

$$\cos(n\omega t + n\pi) = \begin{cases} \cos n\omega t, & n \text{ even} \\ -\cos n\omega t, & n \text{ odd} \end{cases}$$

If n is even, Eq. (7.25) becomes

$$\frac{a_n T}{2} = \int_0^{\frac{T}{2}} -\Delta e^{-\frac{t}{\tau}} \cos n\omega t \, dt + \int_0^{\frac{T}{2}} \Delta e^{-\frac{t}{\tau}} \cos n\omega t \, dt = 0$$

For odd values of n Eq. (7.25) is

$$\frac{a_n T}{2} = \int_0^{\frac{T}{2}} -\Delta e^{-\frac{t}{\tau}} \cos n\omega t \, dt - \int_0^{\frac{T}{2}} \Delta e^{-\frac{t}{\tau}} \cos n\omega t \, dt$$

$$= \int_0^{\frac{T}{2}} -2\Delta e^{-\frac{t}{\tau}} \cos n\omega t \, dt$$

Integrating,

$$\frac{a_n T}{2} = \int_0^{\frac{T}{2}} -2\Delta e^{-\frac{t}{\tau}} \cos n\omega t \, dt$$

$$= \frac{-2\Delta e^{-\frac{t}{t}}}{\frac{1}{\tau^2} + \omega^2 n^2} \left(-\frac{1}{\tau} \cos n\omega t + n\omega \sin n\omega t \right) \Bigg|_0^{\frac{T}{2}}$$

$$= \frac{-2\tau^2 \Delta e^{-\frac{t}{t}}}{1 + \omega^2 \tau^2 n^2} \left(-\frac{1}{\tau} \cos n\omega t + n\omega \sin n\omega t \right) \Bigg|_0^{\frac{T}{2}}$$

Evaluating at the upper and lower limits, and remembering that n is odd:

$$\frac{a_n T}{2} = \frac{-2\tau^2 \Delta}{1 + \omega^2 \tau^2 n^2} \left(e^{-\frac{T}{2\tau}} \left(-\frac{1}{\tau} \cos n\pi + n\omega \sin n\pi \right) - \left(-\frac{1}{\tau} + 0 \right) \right)$$

$$= \frac{-2\tau}{1 + \omega^2 \tau^2 n^2} \left(\Delta e^{-\frac{T}{2\tau}} + \Delta \right)$$

The term in parentheses can be evaluated by referring to Eqs. (7.23) and (7.24):

$$\Delta e^{-\frac{T}{2\tau}} + \Delta = (v_1 - V_L) + (V_H - v_1) = V_H - V_L$$

Inserting this into the previous expression, we finally obtain

$$a_n = -\frac{2}{n\pi}(V_H - V_L)\frac{\omega\tau n}{1 + \omega^2\tau^2 n^2} \tag{7.26}$$

To find the sine coefficients we insert the signal function into Eq. (7.16):

$$b_n = \frac{2}{T}\int_0^{\frac{T}{2}}\left(-\Delta e^{-\frac{t}{\tau}} + V_H\right)\sin n\omega t\, dt + \frac{2}{T}\int_{\frac{T}{2}}^{T}\left(\Delta e^{-\frac{t-\frac{T}{2}}{\tau}} + V_L\right)\sin n\omega t\, dt$$

Routine manipulation yields

$$\frac{b_n T}{2} = \int_0^{\frac{T}{2}}V_H\sin n\omega t\, dt + \int_0^{\frac{T}{2}}V_L\sin n\omega\left(t + \frac{T}{2}\right)dt$$

$$+ \int_0^{\frac{T}{2}}-\Delta e^{-\frac{t}{\tau}}\sin n\omega t\, dt + \int_0^{\frac{T}{2}}\Delta e^{-\frac{t}{\tau}}\sin n\omega\left(t + \frac{T}{2}\right)dt$$

$$\frac{b_n T}{2} = \int_0^{\frac{T}{2}}V_H\sin n\omega t\, dt + \int_0^{\frac{T}{2}}V_L\sin(n\omega t + n\pi)dt$$

$$+ \int_0^{\frac{T}{2}}-\Delta e^{-\frac{t}{\tau}}\sin n\omega t\, dt + \int_0^{\frac{T}{2}}\Delta e^{-\frac{t}{\tau}}\sin(n\omega t + n\pi)dt$$

$$\frac{b_n T}{2} = \int_0^{\frac{T}{2}}V_H\sin n\omega t\, dt + \int_0^{\frac{T}{2}}V_L(-1)^n\sin n\omega t\, dt$$

$$+ \int_0^{\frac{T}{2}}-\Delta e^{-\frac{t}{\tau}}\sin n\omega t\, dt + \int_0^{\frac{T}{2}}\Delta e^{-\frac{t}{\tau}}(-1)^n\sin n\omega t\, dt$$

$$\frac{b_n T}{2} = \int_0^{\frac{T}{2}}\left(V_H + (-1)^n V_L\right)\sin n\omega t \, dt + \int_0^{\frac{T}{2}}\left(-\Delta + (-1)^n\Delta\right)e^{-\frac{t}{\tau}}\sin n\omega t \, dt$$

Integrating:

$$\frac{b_n T}{2} = -\frac{1}{n\omega}(V_H + (-1)^n V_L)\cos n\omega t \Big|_0^{\frac{T}{2}}$$

$$+ \frac{(-\Delta + (-1)^n\Delta)e^{-\frac{t}{\tau}}}{\frac{1}{\tau^2} + \omega^2 n^2}\left(-\frac{1}{\tau}\sin n\omega t - n\omega\cos n\omega t\right)\Big|_0^{\frac{T}{2}}$$

$$\frac{b_n T}{2} = -\frac{1}{n\omega}(V_H + (-1)^n V_L)(\cos n\pi - 1)$$

$$+ \frac{(-1 + (-1)^n)\tau^2\Delta}{1 + \omega^2\tau^2 n^2}\left(e^{-\frac{T}{2\tau}}\left(-\frac{1}{\tau}\sin n\pi - n\omega\cos n\pi\right) - (-n\omega)\right)$$

$$\frac{b_n T}{2} = -\frac{1}{n\omega}(V_H + (-1)^n V_L)((-1)^n - 1)$$

$$+ \frac{(-1 + (-1)^n)\tau^2\Delta}{1 + \omega^2\tau^2 n^2}\left(-e^{-\frac{T}{2\tau}}n\omega(-1)^n + n\omega\right)$$

At this point we observe that the right side of this equation equals zero for even values of n. Therefore, all nonzero sine coefficients have odd n, and with this understanding we can write

$$\frac{b_n T}{2} = \frac{2}{n\omega}(V_H - V_L) + \frac{-2n\omega\tau^2}{1 + \omega^2\tau^2 n^2}\left(\Delta e^{-\frac{T}{2\tau}} + \Delta\right)$$

$$= \frac{2}{n\omega}(V_H - V_L) - \frac{2}{n\omega}\frac{\omega^2\tau^2 n^2}{1 + \omega^2\tau^2 n^2}(V_H - V_L)$$

$$b_n = \frac{\omega}{\pi}\frac{2}{n\omega}(V_H - V_L)\left(1 - \frac{\omega^2\tau^2 n^2}{1 + \omega^2\tau^2 n^2}\right)$$

$$b_n = \frac{2}{n\pi}(V_H - V_L)\frac{1}{1 + \omega^2\tau^2 n^2} \qquad (7.27)$$

Using these expressions for the coefficients, we can express the periodic sawtooth exponential signal with the Fourier series:

$$v_C(t) = \frac{V_H + V_L}{2}$$

$$+ (V_H - V_L) \sum_{\substack{n=1 \\ n \text{ odd}}}^{\infty} -\frac{2}{n\pi} \frac{\omega\tau n}{1 + \omega^2\tau^2 n^2} \cos n\omega t$$

$$+ (V_H - V_L) \sum_{\substack{n=1 \\ n \text{ odd}}}^{\infty} \frac{2}{n\pi} \frac{1}{1 + \omega^2\tau^2 n^2} \sin n\omega t \qquad (7.28)$$

If the two logic levels are $V_L = 0$ and $V_H = V$, then the series can be expressed in the slightly simpler form

$$v_C(t) = \frac{V}{2} + \sum_{\substack{n=1 \\ n \text{ odd}}}^{\infty} \left[-\frac{2V}{n\pi} \frac{\omega\tau n}{1 + \omega^2\tau^2 n^2} \cos n\omega t + \frac{2V}{n\pi} \frac{1}{1 + \omega^2\tau^2 n^2} \sin n\omega t \right] \qquad (7.29)$$

This series is plotted (for $V_L = 0$, $V_H = 5$, $\tau = 10^{-9}$ s, and $T = 10^{-8}$ s) in Figure 7.16, along with the exact curve. Only the fundamental and two harmonic terms were used in generating this graph, yet the agreement is excellent over most of the range.

Again, we stress that in electrical-circuit analysis, it is unusual to determine the analytic form of the Fourier series of a complicated function in this way! We have carried out this long derivation for the periodic exponential sawtooth signal only to make a point about the output of an *RC* circuit excited by a square wave. We need the expression in Eq. (7.29) derived directly from the sawtooth so we can compare it to the result of another, much simpler approach. We will see that ac steady-state analysis techniques allow us to find the output signal without these many pages of algebra.

7.7 Superposition

We have seen how to represent arbitrary periodic signals using Fourier series. In particular, we have derived series representations for the input (square wave) and output (exponential sawtooth) signals of a simple clock distribution circuit. In this section, we will show directly that the individual terms in these two representations are related through the differential equation describing the circuit. This will demonstrate the important concept of **superposition**, which is the summation of independent responses of a linear circuit to independent excitations.

7.7.1 Sinusoidal Steady-State Response of an *RC* Circuit

Recall from the "big picture" that the individual sinusoids which make up the composite input signal can be individually "processed" through the circuit. In the next chapter we

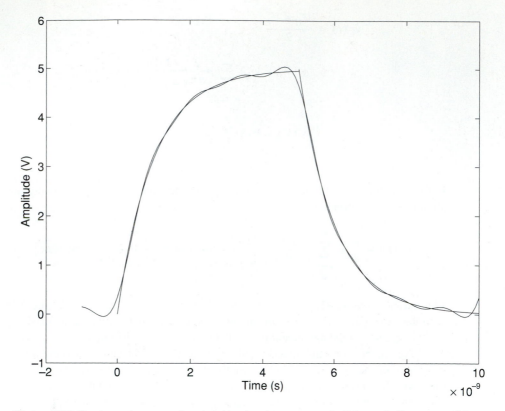

Figure 7.16 Fourier series approximation (dc plus three ac terms) of the periodic exponential saw-tooth signal. Note the excellent agreement to the exact function over most of the range.

will discuss an efficient means for doing this, but for now it is sufficient to use our standard time domain approach.

Suppose an *RC* circuit, Figure 7.17, is driven by a sinusoidal voltage source:

$$v_S(t) = V\cos\omega t$$

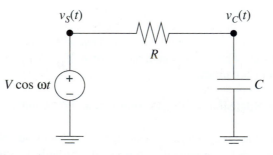

Figure 7.17 An *RC* circuit with a sinusoidal excitation. It is helpful to think of $v_S(t)$ as the input and $v_C(t)$ as the output.

What signal appears across the capacitor?

Applying KVL to this circuit, we can write

$$v_C(t) + RC\frac{d}{dt}v_C(t) = V\cos\omega t \qquad (7.30)$$

An educated guess for the capacitor voltage would be

$$v_C(t) = A\cos\omega t + B\sin\omega t$$

where A and B are constants to be determined. Substituting this into the differential equation, and using the time constant symbol τ for RC,

$$A\cos\omega t + B\sin\omega t + \tau\frac{d}{dt}[A\cos\omega t + B\sin\omega t] = V\cos\omega t$$

Differentiating,

$$A\cos\omega t + B\sin\omega t - A\omega\tau\sin\omega t + B\omega\tau\cos\omega t = V\cos\omega t$$

Equating coefficients of like terms:

$$A + B\omega\tau = V$$
$$B - A\omega\tau = 0$$

Solving for A and B,

$$A = \frac{1}{1 + \omega^2\tau^2}V$$

$$B = \frac{\omega\tau}{1 + \omega^2\tau^2}V$$

Therefore, the capacitor voltage in the circuit of Figure 7.17 is

$$v_C(t) = \frac{1}{1 + \omega^2\tau^2}V\cos\omega t + \frac{\omega\tau}{1 + \omega^2\tau^2}V\sin\omega t \qquad (7.31)$$

It is a simple matter to verify that Eq. (7.31) is indeed a solution to Eq. (7.30).

(Note that this problem has been set up with a strictly periodic source. If an initial condition were specified, a complete solution would include the transient component as well. This situation is considered in Exercise 7.25.)

We note in passing the following: with a sinusoidal excitation source, the process of solving the time-domain differential equation characterizing the circuit has become an exercise in algebra. This is a feature of frequency-domain analysis which we will explore more fully in the next chapter.

Example 7.2

Suppose the voltage source in Figure 7.17 is given by $v_S(t) = V\sin\omega t$. Determine the output signal $v_C(t)$.

Only the form of the source voltage has changed from the previous case; in particular, the same differential equation governs the circuit response:

$$v_C(t) + \tau \frac{d}{dt}(v_C(t)) = V \sin \omega t$$

If we assume the capacitor voltage to be

$$v_C(t) = A \cos \omega t + B \sin \omega t$$

and substitute this into the differential equation, we find

$$A \cos \omega t + B \sin \omega t + \tau \frac{d}{dt}[A \cos \omega t + B \sin \omega t] = V \sin \omega t$$

$$A \cos \omega t + B \sin \omega t - A \omega \tau \sin \omega t + B \omega \tau \cos \omega t = V \sin \omega t$$

Equating coefficients of like terms:

$$A + B \omega \tau = 0$$
$$B - A \omega \tau = V$$

Solving for A and B,

$$A = \frac{-\omega \tau}{1 + \omega^2 \tau^2} V$$

$$B = \frac{1}{1 + \omega^2 \tau^2} V$$

For a voltage input described by $v_S(t) = V \sin \omega t$, the capacitor voltage is thus

$$v_C(t) = \frac{-\omega \tau}{1 + \omega^2 \tau^2} V \cos \omega t + \frac{1}{1 + \omega^2 \tau^2} V \sin \omega t \qquad (7.32)$$

The action of the *RC* circuit in changing the sine input signal into the output signal given by Eq. (7.32) is depicted in Figure 7.18.

7.7.2 Multiple Sinusoidal Sources

According to the "big picture" theory, we should be able to take the individual sinusoids comprising a square-wave input signal (7.18), put each of them through the circuit-solving process (Figure 7.18) which yielded Eq. (7.32), and add up the results. What should come out is the periodic exponential sawtooth signal, Eq. (7.29). Let's see if this really works.

Our square-wave input signal is given in Figure 7.14:

$$v_S(t) = \frac{V}{2} + \sum_{\substack{n=1 \\ n \text{ odd}}}^{\infty} \frac{2V}{\pi n} \sin n \omega t \qquad (7.33)$$

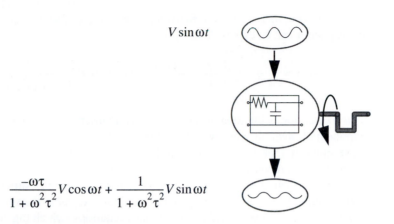

$$\frac{-\omega\tau}{1 + \omega^2\tau^2}V\cos\omega t + \frac{1}{1 + \omega^2\tau^2}V\sin\omega t$$

Figure 7.18 Pictorial representation of the effect of an *RC* circuit on a sinusoidal input voltage.

Expanding this to examine the individual terms:

$$v_S(t) = \frac{V}{2} + \frac{2V}{\pi}\sin\omega t + \frac{2V}{\pi 3}\sin 3\omega t$$

$$+ \frac{2V}{\pi 5}\sin 5\omega t + \cdots + \frac{2V}{\pi n}\sin n\omega t + \cdots \tag{7.34}$$

The first term in this expression is the dc component, $V/2$. If the voltage source in Figure 7.17 had this value, then the voltage appearing across the capacitor would also be $V/2$. Therefore, we can write the first term in our series for $v_C(t)$:

$$v_C(t) = \frac{V}{2} + \cdots$$

The next term in Eq. (7.34) is $(2V/\pi)\sin\omega t$. Examining Figure 7.18, we see that if this were the source voltage in the *RC* circuit, the output would be:

$$\frac{-\omega\tau}{1 + \omega^2\tau^2}\frac{2V}{\pi}\cos\omega t + \frac{1}{1 + \omega^2\tau^2}\frac{2V}{\pi}\sin\omega t$$

Therefore, the output appearing across the capacitor, from the dc and fundamental terms in the square-wave source, is

$$v_C(t) = \frac{V}{2} + \frac{-\omega\tau}{1 + \omega^2\tau^2}\frac{2V}{\pi}\cos\omega t + \frac{1}{1 + \omega^2\tau^2}\frac{2V}{\pi}\sin\omega t + \cdots$$

Likewise, the third harmonic term in the source, $(2V/\pi 3)\sin 3\omega t$, gives rise to a third harmonic term in the output. Adding this term to the dc and fundamental terms:

$$v_C(t) = \frac{V}{2} + \frac{-\omega\tau}{1 + \omega^2\tau^2}\frac{2V}{\pi}\cos\omega t + \frac{1}{1 + \omega^2\tau^2}\frac{2V}{\pi}\sin\omega t$$

$$+ \frac{-3\omega\tau}{1 + (3\omega)^2\tau^2}\frac{2V}{\pi}\cos 3\omega t + \frac{1}{1 + (3\omega)^2\tau^2}\frac{2V}{\pi}\sin 3\omega t + \cdots$$

Carrying this process out for every term in Eq. (7.34), we see that the voltage across the capacitor can be expressed as

$$v_C(t) = \frac{V}{2} + \sum_{\substack{n=1 \\ n \text{ odd}}}^{\infty} \left[-\frac{2V}{n\pi} \frac{\omega\tau n}{1 + \omega^2\tau^2 n^2} \cos n\omega t + \frac{2V}{n\pi} \frac{1}{1 + \omega^2\tau^2 n^2} \sin n\omega t \right]$$

This expression is **identical** to Eq. (7.29), the Fourier series representation of the periodic exponential sawtooth wave, which we know is the response of an *RC* circuit to the square-wave input. It appears that adding up the responses of the circuit to the individual sinusoids really does reconstruct the correct output signal.

The principle demonstrated by this example is known as **superposition**. We observe that the circuit response to the square-wave input is equal to the sum of the responses that would appear had the individual sinusoidal excitations (which together make up the input signal) been applied separately to the circuit. This is a very powerful and useful result which applies to all linear circuits.

Superposition is valuable because it enables us to break up complicated problems into simpler ones. For example, the periodic exponential sawtooth wave, which appears at the output of an inverter gate driven by a clock, might itself be applied to another part of the circuit. Deriving the output of even a simple *RC* circuit, with such a complicated input signal, is too much to ask of time-domain analysis techniques. Using a Fourier representation of the signal, though, reduces the problem to that of a multitude of sinusoidal inputs, which can be handled easily (and, as we will see in the next chapter, without using any differential equations at all). Superposition applies to all kinds of signals: sinusoids of the same or different frequencies; constant (dc) signals; signals of arbitrary shape.

Taking care, and with difficulty, we can also use superposition to find the transient response of a circuit. However, the transient response depends on the initial condition, which must be properly apportioned among the partial signals comprising the input in order to use superposition. Consequently, superposition is more useful in finding the dc and ac steady-state responses of a circuit.

7.7.3 Applying Superposition

Consider the three circuits in Figure 7.19. Each of these is an equivalent way of illustrating a square wave of period $T = 2\pi/\omega$ as the input to an *RC* circuit. Note that once the square wave is broken up into a series of sinusoids, we can use the Fourier representation in two ways. In the Figure 7.19(b) circuit, a single voltage source containing the sum of all terms drives the *RC* circuit. Alternatively, we can imagine an infinite number of voltage sources in series, as shown in Figure 7.19(c), where each source corresponds to one of the Fourier terms. It is this last representation which is most useful in applying superposition.

Writing the sum of source terms and the output terms as

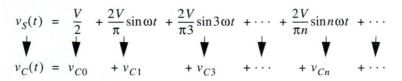

$$v_S(t) = \frac{V}{2} + \frac{2V}{\pi}\sin\omega t + \frac{2V}{\pi 3}\sin 3\omega t + \cdots + \frac{2V}{\pi n}\sin n\omega t + \cdots$$

$$v_C(t) = v_{C0} + v_{C1} + v_{C3} + \cdots + v_{Cn} + \cdots$$

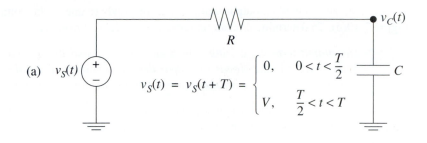

$$v_S(t) = v_S(t + T) = \begin{cases} 0, & 0 < t < \dfrac{T}{2} \\[2mm] V, & \dfrac{T}{2} < t < T \end{cases}$$

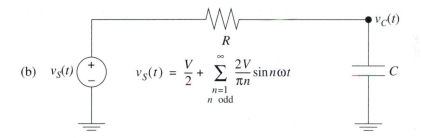

$$v_S(t) = \frac{V}{2} + \sum_{\substack{n=1 \\ n \text{ odd}}}^{\infty} \frac{2V}{\pi n} \sin n\omega t$$

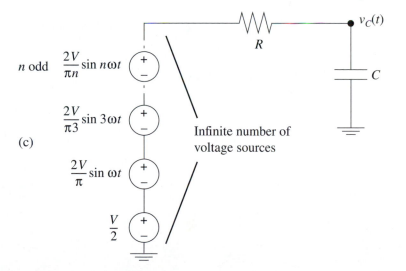

Figure 7.19 Completely equivalent depictions of a periodic square-wave source driving an *RC* circuit. (a) Source specified over two time intervals. (b) Source specified as a sum of sinusoids. (c) The circuit is driven by an infinite number of sources in series, which add up to a square wave.

we see that each individual output term, v_{Ck}, corresponds to one of the voltage sources in Figure 7.19(c). To determine this output term, we use the following rule:

> *The output term v_{Ck} arising from source v_{Sk} is found by replacing all other sources in the circuit with their **dead equivalents**. The individual responses v_{C0}, v_{C1}, ..., are then summed to find the total response.*

A "dead equivalent" of a source is just a zero-valued source. This means that independent voltage sources are replaced with short circuits; independent current sources are replaced with open circuits.

For example, to find the third harmonic term, we follow the procedure illustrated in Figure 7.20. Suppose we wish to find the output term $v_{C3}(t)$ due to excitation $(2V/\pi 3)\sin 3\omega t$. All other sources are replaced with short circuits and the circuit is solved for its periodic steady-state response. This procedure is repeated for each of the independent sources.

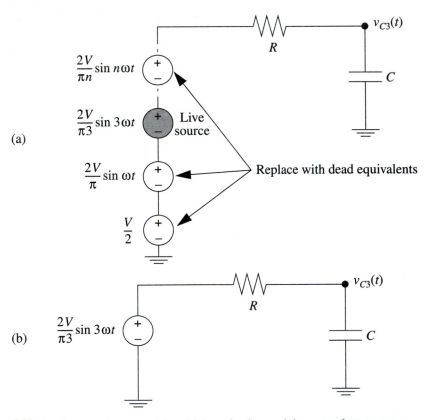

Figure 7.20 Application of superposition. (a) In a circuit containing several sources, one source is retained (for example, the shaded source), while the others are replaced with their dead equivalents. (b) The resulting circuit.

It is not necessary for the voltage-source terms to be in series to apply superposition. This is illustrated in Figure 7.21, which shows a circuit with two voltage sources, not in series, plus a current source. Each independent source contributes to the total response. The individual responses are found by replacing all other sources by their dead equivalents. Finally, the individual responses are superposed.

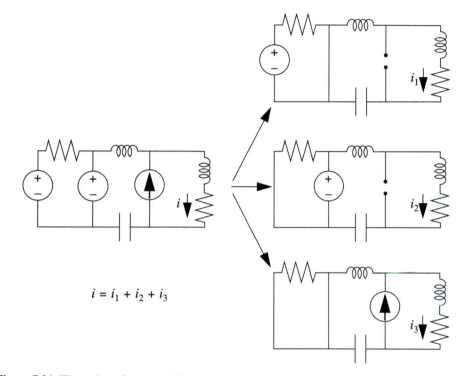

$$i = i_1 + i_2 + i_3$$

Figure 7.21 Illustration of superposition where voltage sources are not in series. All sources but one are replaced by their dead equivalents; the resulting responses from each such substitution are superposed.

Example 7.3

Use superposition to determine the voltage V in the dc circuit of Figure 7.22.

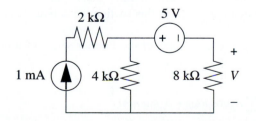

Figure 7.22 The dc circuit for the superposition problem of Example 7.3.

This circuit has two independent sources. First, we find the voltage V_1 arising from the current source, with the voltage source replaced by its dead equivalent:

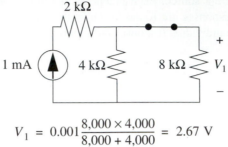

$$V_1 = 0.001 \frac{8,000 \times 4,000}{8,000 + 4,000} = 2.67 \text{ V}$$

Next, we find the voltage V_2 due to the voltage source, with the current source replaced by its dead equivalent:

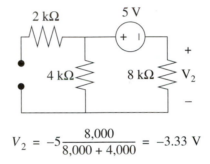

$$V_2 = -5 \frac{8,000}{8,000 + 4,000} = -3.33 \text{ V}$$

Finally, we superpose the two partial results:

$$V = V_1 + V_2 = -0.666 \text{ V}$$

7.8 Complex Form of the Fourier Series

The representation of signals using Fourier series is quite useful, but the form in which we have developed it could use some improvement. For example, we need to perform three separate integrations (for a_0, a_n, and b_n) to find the constants, and the series representation is anything but compact. Fortunately, there is an alternate representation, called the **complex form** of the Fourier series, which eliminates these difficulties. More importantly, it provides the foundation for another method of representing signals (**phasors**) which is extremely useful in ac steady-state circuit analysis. In this section we derive the complex form and apply it to the square wave.

We begin with the standard trigonometric form of the Fourier series of a periodic function, given by Eq. (7.11):

$$f(t) = a_0 + \sum_{n=1}^{\infty} a_n \cos n\omega t + b_n \sin n\omega t \qquad (7.35)$$

Using Euler's identity, the trigonometric functions in Eq. (7.35) can be written

$$\cos n\omega t = \frac{1}{2}(e^{jn\omega t} + e^{-jn\omega t})$$

$$\sin n\omega t = \frac{1}{2j}(e^{jn\omega t} - e^{-jn\omega t})$$

Substituting,

$$f(t) = a_0 + \sum_{n=1}^{\infty} \frac{a_n}{2}(e^{jn\omega t} + e^{-jn\omega t}) + \frac{b_n}{2j}(e^{jn\omega t} - e^{-jn\omega t})$$

$$= a_0 + \sum_{n=1}^{\infty} \frac{a_n - jb_n}{2}e^{jn\omega t} + \frac{a_n + jb_n}{2}e^{-jn\omega t}$$

Define the complex Fourier coefficient, f_n, as

$$f_n = \frac{a_n - jb_n}{2}$$

where n is a positive integer. The value of f_n, where n is a negative integer, can be determined by replacing n with $-n$ in the definitions of a_n and b_n:

$$a_{-n} = \frac{2}{T}\int_0^T f(t)\cos(-n\omega t)dt$$

$$= \frac{2}{T}\int_0^T f(t)\cos n\omega t\, dt$$

$$a_{-n} = a_n$$

$$b_{-n} = \frac{2}{T}\int_0^T f(t)\sin(-n\omega t)dt$$

$$= -\frac{2}{T}\int_0^T f(t)\sin n\omega t\, dt$$

$$b_{-n} = -b_n$$

Therefore,

$$f_{-n} = \frac{a_{-n} - jb_{-n}}{2}$$

$$= \frac{a_n + jb_n}{2}$$

The Fourier series can now be written

$$f(t) = a_0 + \sum_{n=1}^{\infty} f_n e^{jn\omega t} + f_{-n} e^{-jn\omega t}$$

We set f_0 equal to the dc value a_0; the series is thus

$$f(t) = f_0 + \sum_{n=1}^{\infty} f_n e^{jn\omega t} + f_{-n} e^{-jn\omega t}$$

which can be written in the compact form

$$f(t) = \sum_{n=-\infty}^{\infty} f_n e^{jn\omega t}$$

The coefficients f_n have an equally compact form. Starting from the definition of f_n

$$f_n = \frac{a_n - jb_n}{2}$$

$$= \frac{1}{2}\left(\frac{2}{T}\int_0^T f(t)\cos n\omega t\, dt - \frac{2j}{T}\int_0^T f(t)\sin n\omega t\, dt\right)$$

$$= \frac{1}{T}\int_0^T f(t)(\cos n\omega t - j\sin n\omega t)\, dt$$

Substituting the complex exponential forms of the cosine and sine functions:

$$f_n = \frac{1}{T}\int_0^T f(t)\left(\frac{1}{2}(e^{jn\omega t} + e^{-jn\omega t}) - \frac{j}{2j}(e^{jn\omega t} - e^{-jn\omega t})\right)dt$$

$$f_n = \frac{1}{T}\int_0^T f(t)e^{-jn\omega t}\, dt$$

There is just a single integral which defines all of the Fourier coefficients, instead of three as with the trigonometric form. However, unlike the trigonometric form, the coefficients will, in general, be complex numbers. (As before, the integration can be performed over any convenient interval, so long as it is one period long.)

In summary, the complex form of the Fourier series for a periodic function $f(t)$ is

$$f(t) = \sum_{n=-\infty}^{\infty} f_n e^{jn\omega t} \tag{7.36}$$

where the coefficients are

$$f_n = \frac{1}{T}\int_0^T f(t)e^{-jn\omega t}\,dt \qquad\qquad (7.37)$$

Example 7.4

Determine the complex Fourier series for the two periodic clock functions v_1 and v_2 in Figure 7.23.

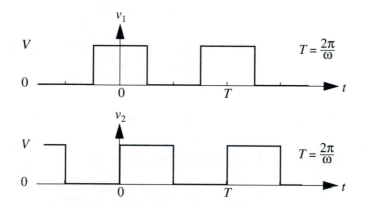

Figure 7.23 Square-wave clock signals for Example 7.4.

The first function is defined as

$$v_1(t) = \begin{cases} V, & -\dfrac{T}{4} < t < \dfrac{T}{4} \\[2ex] 0, & \dfrac{T}{4} < t < 3\dfrac{T}{4} \end{cases}$$

The complex Fourier coefficients are found from

$$f_n = \frac{1}{T}\int_{-\frac{T}{4}}^{3\frac{T}{4}} v_1(t)e^{-jn\omega t}\,dt$$

$$= \frac{1}{T}\int_{-\frac{T}{4}}^{\frac{T}{4}} V e^{-jn\omega t}\,dt$$

Carrying out the integration,

$$f_n = \frac{V}{T} \frac{1}{-jn\omega} e^{-jn\omega t} \Big|_{-\frac{T}{4}}^{\frac{T}{4}}$$

$$= \frac{V}{-2jn\pi} \left(e^{-\frac{jn\pi}{2}} - e^{\frac{jn\pi}{2}} \right)$$

$$= \frac{V}{n\pi} \sin \frac{n\pi}{2}$$

The value of the coefficient depends on the parity of n:

$$f_n = \begin{cases} 0, & n \text{ even} \\ \dfrac{V}{2}, & n = 0 \\ \dfrac{V}{n\pi}(-1)^{\frac{n-1}{2}}, & n \text{ odd} \end{cases}$$

The clock signal v_1 can therefore be expressed as

$$v_1(t) = \frac{V}{2} + \sum_{\substack{n=-\infty \\ n \text{ odd}}}^{\infty} \frac{V}{n\pi}(-1)^{\frac{n-1}{2}} e^{jn\omega t} \qquad (7.38)$$

The second clock signal is defined as

$$v_2(t) = \begin{cases} V, & 0 < t < \dfrac{T}{2} \\ 0, & \dfrac{T}{2} < t < T \end{cases}$$

The complex Fourier coefficients are found from

$$f_n = \frac{1}{T} \int_0^T v_2(t) e^{-jn\omega t} \, dt$$

$$= \frac{1}{T} \int_0^{\frac{T}{2}} V e^{-jn\omega t} \, dt$$

Integrating,

$$f_n = \frac{V}{T}\frac{1}{-jn\omega}e^{-jn\omega t}\Big|_0^{\frac{T}{2}}$$

$$= \frac{V}{-2jn\pi}(e^{-jn\pi} - 1)$$

$$= \frac{Vj}{2n\pi}(e^{-jn\pi} - 1)$$

Again, the value of the coefficient depends on the parity of n:

$$f_n = \begin{cases} 0, & n \text{ even} \\ \dfrac{V}{2}, & n = 0 \\ -\dfrac{Vj}{n\pi}, & n \text{ odd} \end{cases}$$

The complex Fourier series for the clock signal v_2 is

$$v_2(t) = \frac{V}{2} + \sum_{\substack{n=-\infty \\ n \text{ odd}}}^{\infty} -\frac{Vj}{n\pi}e^{jn\omega t} \qquad (7.39)$$

These expressions are equivalent to the trigonometric series representations of the two clock signals (see Exercise 7.5).

In both cases, we observe that the calculation of the complex Fourier coefficients is less onerous than for the trigonometric form of the series. We will make use of these expansions in the next chapter.

7.9 Summary

In this chapter we introduced the concept of **periodic signals**, defined by

$$v(t) = v(t + T)$$

The smallest value of T which satisfies this condition is called the **period** of $v(t)$. The reciprocal of the period is called the **fundamental frequency**. Sinusoidal signals which are integer multiples of the fundamental frequency are known as called **harmonics**.

Periodic signals can be written as sums of sinusoids, called **Fourier series**. The **trigonometric form** of the Fourier series is:

$$f(t) = a_0 + \sum_{n=1}^{\infty} a_n \cos n\omega t + b_n \sin n\omega t$$

where $\omega = 2\pi/T$ is the **fundamental radian frequency** of the signal. The **Fourier coefficients** are given by

$$a_0 = \frac{1}{T}\int_0^T f(t)dt$$

$$a_n = \frac{2}{T}\int_0^T f(t)\cos n\omega t\, dt$$

$$b_n = \frac{2}{T}\int_0^T f(t)\sin n\omega t\, dt$$

The constant a_0 is called the **dc component** and is equal to the average value of the signal over one period. In each case, the integration can be performed over any convenient interval one period long.

The Fourier series can also be written as

$$f(t) = A_0 + \sum_{n=1}^{\infty} A_n \cos(n\omega t + \theta_n)$$

where

$$A_n = \sqrt{a_n^2 + b_n^2}$$

$$\theta_n = \text{atan}-\frac{b_n}{a_n}$$

$$a_n = A_n \cos\theta_n$$

$$b_n = -A_n \sin\theta_n$$

The constants A_n and θ_n are called the **magnitude** and **phase** of the nth harmonic cosine function.

A third equivalent series representation is the **complex form** of the Fourier series:

$$f(t) = \sum_{n=-\infty}^{\infty} f_n e^{jn\omega t}$$

where the coefficients are given by

$$f_n = \frac{1}{T}\int_0^T f(t)e^{-jn\omega t}dt$$

Linear circuits can be solved using the principle of **superposition**. If a linear circuit has multiple excitation sources, superposition states that the response is equal to the sum of the responses of the circuit to each individual source. The individual responses can be found by replacing all but one excitation source with their **dead equivalents**. The dead equivalent of an independent voltage source is a short circuit; of an independent current source, an open circuit.

Problems

7.1 The clock signal in the circuit below operates between 0 V and 5 V and has a frequency of 50 MHz. The signal appearing as v_C at the input of the flip-flop is measured to oscillate between $v_l = 1$ V and $v_h = 4$ V. Assuming the circuit can be modeled with a first-order RC approximation, determine the value of the RC time constant.

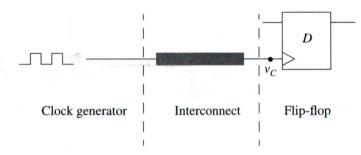

Clock generator | Interconnect | Flip-flop

7.2 A system similar to the one in Exercise 7.1 has an equivalent resistance of 100 Ω, and an equivalent capacitance of 80 pF. What is the maximum clock frequency which allows the input to the flip-flop to transition between 1 V and 4 V?

7.3 In the circuit below, $v_C(t)$ must rise above 4 V on each cycle. Determine the maximum value of the frequency f.

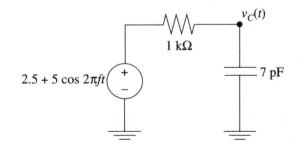

7.4 In the circuit below, $v_C(t)$ must rise above 4 V on each cycle. Determine the maximum value of the frequency f.

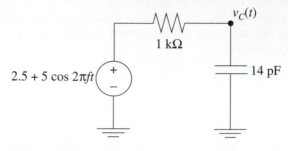

7.5 Show that the trigonometric series for the square-wave clock signals in Figure 7.14 are equivalent to the complex exponential representations, Eqs. (7.38) and (7.39).

7.6 Express the periodic function below as a trigonometric Fourier series.

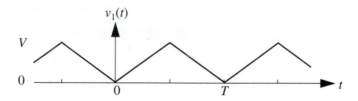

7.7 Express the periodic function below as a trigonometric Fourier series.

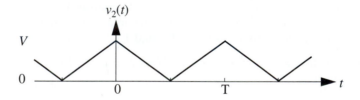

7.8 Express the periodic function below as a trigonometric Fourier series.

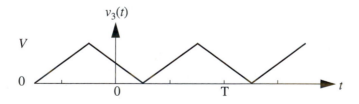

7.9 Express each of the signals in Exercises 7.6, 7.7, and 7.8 as a complex Fourier series.

7.10 Express the periodic function below as a trigonometric Fourier series.

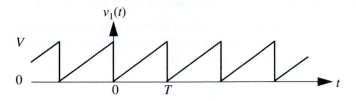

7.11 Express the periodic function below as a trigonometric Fourier series.

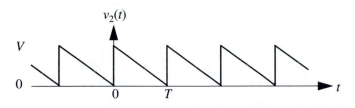

7.12 Express the periodic function below as a trigonometric Fourier series.

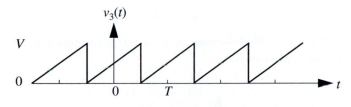

7.13 Express each of the signals in Exercises 7.10, 7.11, and 7.12 as a complex Fourier series.

7.14 Using Eqs. (7.6) and (7.7), show analytically that the periodic exponential sawtooth wave described by Eq. (7.21), is symmetric (i.e., $v_h - V_L = V_H - v_l$).

7.15 Each of the curved sections in the signal below is an exponentially decaying function with time constant τ. The extreme values are $\pm V$, and each segment decays toward zero. Express this signal as a trigonometric Fourier series.

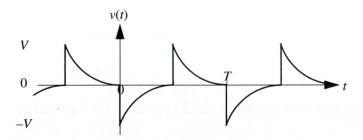

7.16 In Exercise 7.15, determine how many terms of the Fourier series are needed for the series to converge to within 1% of its true value at $t = 0.4T$.

7.17 Express the signal in Exercise 7.15 as a complex Fourier series.

7.18 The signal in Exercise 7.15 is the output of a simple circuit, whose input is a symmetric square-wave clock signal oscillating between 0 and *V.* Figure out what the circuit is.

7.19 Each of the curved sections in the signal below is an exponentially decaying function with time constant τ. The extreme values are $\pm V$, and each segment decays toward $\pm V/2$. Express this signal as a trigonometric Fourier series.

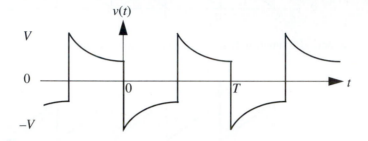

7.20 Express the signal in Exercise 7.19 as a complex Fourier series.

7.21 The periodic signal below, given by

$$v(t) = V|\sin \omega t|$$

is called a **full-wave rectified sinusoid**. Express $v(t)$ as a trigonometric Fourier series. Which harmonic has the highest amplitude?

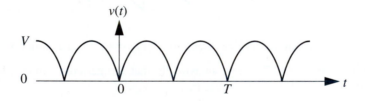

7.22 Express the signal in Exercise 7.21 as a complex Fourier series.

7.23 The periodic signal below, given by

$$v(t) = \begin{cases} V \sin \omega t, & 0 < t < \dfrac{T}{2} \\[2mm] 0, & \dfrac{T}{2} < t < T \end{cases}$$

is called a **half-wave rectified sinusoid**. Express $v(t)$ as a trigonometric Fourier series.

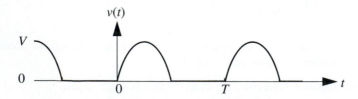

7.24 Express the signal in Exercise 7.23 as a complex Fourier series.

7.25 In the circuit below, the switch closes at $t = 0$; the capacitor voltage at the switch closure is given by $v_C(0) = v_{C0}$. Show that the capacitor voltage for $t > 0$ is given by

$$v_C(t) = \frac{V}{1 + \omega^2\tau^2}[\cos(\omega t + \varphi) + \omega\tau\sin(\omega t + \varphi)]$$

$$+ \left\{ v_{C0} - \frac{V}{1 + \omega^2\tau^2}[\cos\varphi + \omega\tau\sin\varphi] \right\} e^{-\frac{t}{\tau}}$$

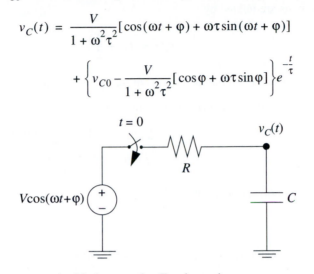

7.26 A signal can be represented with the complex Fourier series

$$v(t) = \sum_{n=-\infty}^{\infty} \frac{1}{(|2n| + 1)^2} e^{j2\pi nt}$$

(a) Determine the dc component of this signal.

(b) Express this signal as a trigonometric Fourier series.

7.27 Use a computer to plot the signal represented with the complex Fourier series

$$v(t) = \sum_{\substack{n=-\infty \\ n \text{ even}}}^{\infty} \frac{j\sqrt{n}}{(1 - n^2)} \frac{V}{\pi} e^{j\omega nt}$$

7.28 Use a computer to plot the signal represented with the complex Fourier series

$$v(t) = \sum_{n=-\infty}^{\infty} \frac{j^n|n - 1|}{|n|!} \frac{V}{\pi} e^{j\omega nt}$$

7.29 A function $f(t)$ has a complex Fourier expansion

$$f(t) = \sum_{n=-\infty}^{\infty} f_n e^{jn\omega t}$$

where the coefficients f_n are known. A related function, $g(t)$, has a complex Fourier expansion

$$g(t) = \sum_{n=-\infty}^{\infty} g_n e^{jn\omega t}$$

The two functions are related by

$$g(t) = f\left(t + \frac{T}{2}\right)$$

Express the g_n Fourier coefficients in terms of the f_n coefficients.

7.30 Use superposition to determine $v(t)$ in the circuit below.

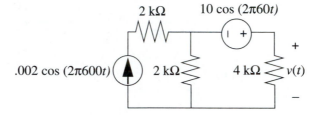

7.31 Determine the capacitor voltage $v_C(t)$ in the circuit below.

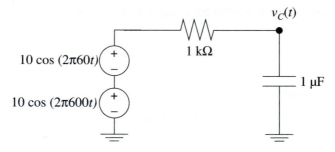

7.32 Determine the capacitor voltage $v_C(t)$ in the circuit below.

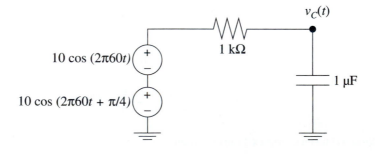

7.33 Determine the capacitor voltage $v_C(t)$ in the circuit below.

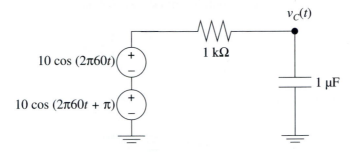

7.34 Determine the voltage $v(t)$ in the circuit below, where the source voltage $v_S(t)$ is given by the square wave signal

$$v_S(t) = \frac{V}{2} + \sum_{\substack{n=-\infty \\ n \text{ odd}}}^{\infty} -\frac{Vj}{n\pi} e^{jn\omega t}$$

with $\omega = 2$ krad/s. Plot the input and output signals.

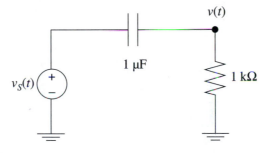

7.35 Determine the voltage $v(t)$ in the circuit below, where the source voltage $v_S(t)$ is given by the exponential sawtooth signal

$$v_S(t) = \frac{V}{2} + \sum_{\substack{n=-\infty \\ n \text{ odd}}}^{\infty} \left[-\frac{2V}{n\pi} \frac{\omega\tau n}{1+\omega^2\tau^2 n^2} \cos n\omega t + \frac{2V}{n\pi} \frac{1}{1+\omega^2\tau^2 n^2} \sin n\omega t \right]$$

with $\omega = 2$ krad/s and $\tau = 1.0$ ms. Plot the input and output signals.

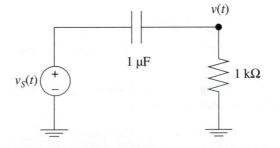

7.36 Show that the periodic exponential sawtooth signal, given by the trigonometric Fourier series in Eq. (7.29), has the complex Fourier series representation

$$v(t) = \frac{V}{2} + \sum_{\substack{n=-\infty \\ n \text{ odd}}}^{\infty} \frac{V}{jn\pi} \frac{1}{1 + jn\omega\tau} e^{jn\omega t}$$

7.37 Consider a periodic signal $f(t)$ with complex Fourier representation

$$f(t) = \sum_{n=-\infty}^{\infty} f_n e^{jn\omega t}$$

The complex Fourier coefficients f_n express the **spectral content** of the signal $f(t)$. One way to visualize this information is to plot $|f_n|$ as a function of $n\omega$. This kind of plot is known as an **amplitude spectrum**, and it tells us the relative weights of the different harmonics making up a periodic signal. For example, the sinusoid $V\cos(\omega t)$ has the spectrum

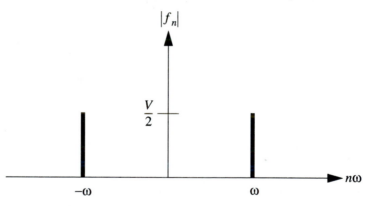

Plot the amplitude spectra of square-wave and periodic exponential sawtooth signals.

7.38 Suppose we have two sinusoidal signals at different frequencies:

$$v_C(t) = V_C \cos\omega_C t$$
$$v_m(t) = V_m \cos\omega_m t$$

From these signals we form a composite signal defined by

$$v_{AM}(t) = v_C(t)(1 + v_m(t))$$

(a) Suppose $V_m = 0.5V_C$, and $\omega_m = 0.2\omega_C$. Plot $v_{AM}(t)$. Can you explain the meaning of the subscript "AM"?

(b) Determine and plot the amplitude spectrum of $v_{AM}(t)$. Can you think of a practical application for mixing the two signals in this way?

7.39 Consider a periodic rectangular pulse train, with amplitude V and **duty cycle** α, as defined below:

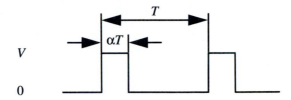

(a) Determine the complex exponential Fourier series for this function.

(b) Consider several such pulse trains with the same fundamental frequency but with different duty cycles. Plot the amplitude spectrum for this function, with $\alpha = 1/2, 1/3$, and $1/5$. From these spectra, can you conjecture anything about the relationship between the time-domain and frequency-domain representations of a signal?

7.40 The voltage across a capacitor in a certain circuit, $v_C(t)$, has a Fourier representation

$$v_C(t) = \sum_{n=-\infty}^{\infty} f_n e^{jn\omega t}$$

while the current through the capacitor, $i_C(t)$, is given by

$$i_C(t) = \sum_{n=-\infty}^{\infty} g_n e^{jn\omega t}$$

Determine the relationship between the Fourier coefficients f_n and g_n.

7.41 Consider the plots of the Fourier series for square-wave and exponential sawtooth signals given in Figures 7.12 and 7.16. Note that the three-term Fourier series for the sawtooth is a much closer approximation of the actual signal than the square-wave representation. Explain this result.

Clock Degradation and ac Steady-State Analysis

8

8.1 Introduction

In Chapter 7 we saw that periodic signals, such as the square-wave clock signals used to synchronize digital systems, can be represented as sums of sinusoidal signals. Furthermore, the principle of superposition allows us to treat the individual sinusoids comprising the clock signal independently of each other. These two techniques provide an alternate framework for solving circuits.

In this chapter we complete our introduction to this alternative method by discussing ac steady-state analysis. This is a method to predict the sinusoidal steady-state response of a linear circuit to a sinusoidal input. Because sinusoidal functions have some peculiar properties, we will be able to dispense entirely with the differential equations describing the circuits. Analysis of circuits in the frequency domain is thus reduced from a complicated problem in differential equations to an easier problem in algebra.

Because sinusoidal signals play a central role in frequency-domain analysis, we will first review some basic properties of sine and cosine functions. This will lead to the idea of phasor representation of signals, which is the key to ac steady state analysis. We will then apply our knowledge of the frequency domain to the problem of clock-signal degradation by interconnect distribution circuits.

8.2 Sinusoids and Phasors

8.2.1 Sinusoidal Signal Representation

Consider the sinusoidal voltage signal

$$v(t) = V\cos(\omega t + \varphi) \tag{8.1}$$

where V is the **amplitude**, $\omega = 2\pi/T$ is the **radian frequency**, and φ is the **phase angle**. (Owing to the periodicity of the cosine function, it is always possible to choose φ in the range $-\pi < \varphi < \pi$ such that the amplitude is nonnegative. We will assume that this is the case.) This signal has its first maximum at $t = -\varphi/\omega$, and subsequent maxima every $2\pi/\omega$ seconds. These attributes are illustrated in Figure 8.1.

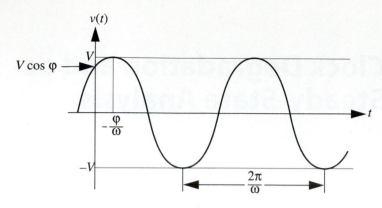

Figure 8.1 A sinusoidal signal described by $v(t) = V\cos(\omega t + \varphi)$.

Cosine and sine functions differ only by a relative phase difference of $\pi/2$ radians. Since a periodic signal stretches infinitely along the time axis, the distinction between the two is only a matter of where we choose to place the origin. We will find it convenient to express sinusoids in the form of Eq. (8.1) (i.e., cosine functions with a specified phase angle) rather than a mixture of sines and cosines.

It is important to keep track of relative differences in phase between sinusoids of the same frequency. For example, shown in Figure 8.2, are two sinusoids, both with frequency ω, but with different phases. The signal $v_1(t)$ is said to **lag** $v_2(t)$, since it is a time-delayed version of $v_2(t)$. (This terminology persists even if the two signals have different amplitudes.)

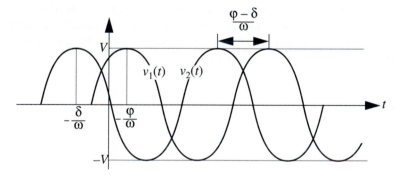

Figure 8.2 Two sinusoidal signals of the same frequency but different phase: $v_1(t) = V\cos(\omega t + \varphi)$ and $v_2(t) = V\cos(\omega t + \delta)$. $v_1(t)$ is said to lag $v_2(t)$.

Example 8.1

Determine the phase difference between capacitor voltage and current, if the voltage across a capacitor is given by

$$v_C(t) = V\cos(\omega t + \varphi)$$

Capacitor current is related to voltage by

$$i_C(t) = C\frac{d}{dt}v_C(t)$$

For a sinusoidal capacitor voltage, the current is therefore

$$i_C(t) = -\omega CV \sin(\omega t + \varphi)$$

$$= \omega CV \cos\left(\omega t + \varphi + \frac{\pi}{2}\right)$$

Therefore, the capacitor voltage lags the current by $\pi/2$ radians. This is a general result, which does not depend on the value of the capacitor or the circuit—only that the signals are sinusoidal. It is easy to show that a similar relationship applies to inductors (i.e., the inductor current lags the voltage).

8.2.2 Phasors

A key feature of linear time-invariant circuits is that they cannot change the frequency of a sinusoidal input signal. With this in mind, a signal like that of Eq. (8.1),

$$v(t) = V\cos(\omega t + \varphi)$$

can be characterized using only two quantities: the amplitude V and the phase φ. The fact that the functional form is a cosine with angular frequency ω is understood and thus need not be explicitly specified. We can notate this signal as

$$V \angle \varphi$$

which we call a **phasor**. A phasor is simply a shorthand way of representing a cosine function.

In working with phasors, it is helpful to recall the fundamental definitions of the trigonometric functions cosine and sine. Figure 8.3 illustrates how the cosine and sine functions can be generated by rotating a vector about the origin. The vector, the heavy arrow with magnitude V, is shown in its position at t = 0 at an inclination of φ radians. This vector rotates about the origin, completing one orbit every T seconds. The projection of the vector onto the x-axis generates the cosine function, and the projection onto the y-axis corresponds to the sine function. Also shown, in lighter ink, is the position of the vector one-quarter cycle later, when $\omega t = \pi/2$. The vector used in this construction can be thought of as a "rotating phasor."

An even more fruitful interpretation can be had if we consider the axes of the vector in Figure 8.3 to be the real and imaginary axes. The tip of the vector specifies a complex number $\mathbf{V}$, whose value at time t is given by

$$\mathbf{V} = V\cos(\omega t + \varphi) + jV\sin(\omega t + \varphi) \tag{8.2}$$

Applying the Euler identity, this can be written as

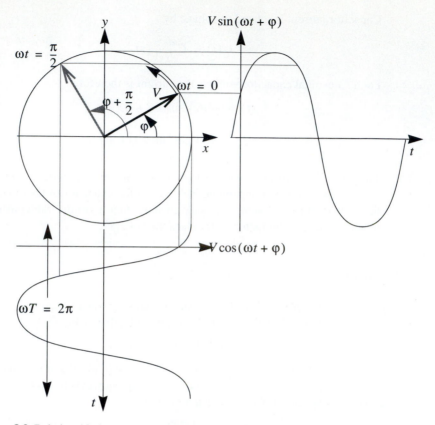

Figure 8.3 Relationship between a rotating phasor $V \angle \varphi$ and sinusoidal functions.

$$\mathbf{V} = V e^{j(\omega t + \varphi)}$$

or

$$\mathbf{V} = V e^{j\varphi} e^{j\omega t} \qquad (8.3)$$

Note that, if we agree that the point $\mathbf{V}$ in question rotates about the origin of the complex plane every $2\pi/\omega$ seconds, we can completely specify its location with just two numbers, V and φ. It is tempting to write

$$\mathbf{V} = V \angle \varphi \qquad (8.4)$$

which captures the fact that $\mathbf{V}$ has magnitude V (volts) and argument (phase) φ (in radians). Viewed this way, the phasor $V \angle \varphi$ can be interpreted as the position (at $t = 0$) of a vector in the complex plane rotating about the origin. Note that the factor $e^{j\omega t}$ present in Eq. (8.3) has been suppressed in Eq. (8.4). When working with phasor representations, this factor, which accounts for the sinusoidal variation at angular frequency ω, is *understood* to be present even when not explicitly written.

Given that we have plausible reasons for using the notation $V \angle \varphi$ for two different things, a signal $v(t)$ and a complex number $\mathbf{V}$, a question arises: what is the relationship between $v(t)$ and $\mathbf{V}$? Using the fact that a complex number can be written in the form

$$\mathbf{V} = \text{Re}(\mathbf{V}) + j\,\text{Im}(\mathbf{V})$$

it follows from Eq. (8.2) that

$$v(t) = V\cos(\omega t + \varphi) = \text{Re}(\mathbf{V}) \tag{8.5}$$

This correspondence between sinusoidal signals and complex numbers of the form $Ve^{j\varphi}e^{j\omega t}$ is the basis for a remarkable method of analyzing signals and circuits. Instead of representing sinusoidal signals with trigonometric functions, we pretend that they are complex exponentials with time dependence $e^{j\omega t}$. After performing mathematical manipulations which describe the effect of the circuit, we simply take the real part of the resulting expression and throw away the rest. Let's see how this works by revisiting Example 8.1.

Example 8.2

Determine the phase difference between capacitor voltage and current, if the voltage across a capacitor is given by

$$v_C(t) = V\cos(\omega t + \varphi)$$

We can describe the voltage across the capacitor using the phasor representation: $V \angle \varphi$. We will interpret this phasor as a complex number $\mathbf{V_C}$, which corresponds to the capacitor voltage v_C:

$$\mathbf{V_C} = V \angle \varphi = Ve^{j\varphi}e^{j\omega t}$$

The current is given by

$$i_C(t) = C\frac{d}{dt}v_C(t)$$

which becomes, in our frequency-domain representation,

$$\mathbf{I_C} = C\frac{d}{dt}\mathbf{V_C}$$

$$= C\frac{d}{dt}Ve^{j\varphi}e^{j\omega t}$$

where $\mathbf{I_C}$ is the phasor corresponding to the capacitor current. Differentiating,

$$\mathbf{I_C} = j\omega C Ve^{j\varphi}e^{j\omega t} \tag{8.6}$$

which can also be written as

$$\mathbf{I_C} = j\omega C \mathbf{V_C} \tag{8.7}$$

Making the substitution $j = e^{j\frac{\pi}{2}}$, Eq. (8.6) becomes

$$\mathbf{I_C} = \omega CV e^{j\left(\varphi + \frac{\pi}{2}\right)} e^{j\omega t} \tag{8.8}$$

Expressing this in magnitude-phase form,

$$\mathbf{I_C} = \omega CV \angle \left(\varphi + \frac{\pi}{2}\right) \tag{8.9}$$

We can compare Eq. (8.9) to the phasor representation of the capacitor voltage: $V \angle \varphi$. Even without going back to the time domain, we see that the phase of the current is advanced by $\pi/2$ compared to the voltage, in agreement with our previous result.

To obtain an explicit equation for the capacitor current, we simply take the real part of Eq. (8.8):

$$i_C(t) = \text{Re}(\mathbf{I_C})$$

$$= \text{Re}\left(\omega CV e^{j\left(\varphi + \frac{\pi}{2}\right)} e^{j\omega t} \right)$$

$$= \text{Re}\left(\omega CV \cos\left(\omega t + \varphi + \frac{\pi}{2}\right) + j\omega CV \sin\left(\omega t + \varphi + \frac{\pi}{2}\right) \right)$$

$$= \omega CV \cos\left(\omega t + \varphi + \frac{\pi}{2}\right)$$

which is precisely what we obtained in Example 8.1 by differentiating the time-domain expression for capacitor voltage.

Because phasors can be interpreted as complex numbers, all rules of complex arithmetic apply to phasors as well. This is handy when combining signals, as the next examples illustrate.

Example 8.3

Express the series connection of two voltage sources below as a single sinusoidal function.

The two sources, both operating at 1 MHz, can be represented with the phasors

$$\mathbf{V_1} = 3\angle-1.3$$
$$\mathbf{V_2} = 5\angle0.5$$

The value of the single equivalent source is given by the sum of the two voltages, which is represented as

$$\mathbf{V_1} + \mathbf{V_2} = (3\angle-1.3) + (5\angle0.5)$$

To add complex numbers, we must convert from polar to rectangular notation using the transformations

$$r\angle\theta = re^{j\theta} = x + jy$$
$$x = r\cos\theta, \qquad r = \sqrt{x^2 + y^2}$$
$$y = r\sin\theta, \qquad \theta = \text{atan}\,\frac{y}{x}$$

Using these formulas, the phasor sum is

$$\mathbf{V_1} + \mathbf{V_2} = (3\angle-1.3) + (5\angle0.5)$$
$$= 3\cos(-1.3) + j3\sin(-1.3) + 5\cos 0.5 + j5\sin(0.5)$$
$$= 5.19 - j0.49$$

Converting back to polar form,

$$\mathbf{V_1} + \mathbf{V_2} = \sqrt{5.19^2 + (-0.49)^2}\angle\,\text{atan}\,\frac{-0.49}{5.19}$$
$$= 5.21\angle-0.095$$

Finally, expressing this is as a time-domain function,

$$v(t) = 5.21\cos(2\pi 10^6 t - 0.095)$$

which, like the original expressions, is in volts.

Example 8.4

In Chapter 7 we found that an RC circuit driven by a source voltage $v_S = V\cos\omega t$ results in a capacitor voltage of

$$v_C(t) = \frac{V}{1 + \omega^2\tau^2}[\cos\omega t + \omega\tau\sin\omega t]$$

Express this voltage in the form

$$v_C(t) = A\cos(\omega t + \theta)$$

Rewriting the sine function as a shifted cosine,

$$v_C(t) = \frac{V}{1 + \omega^2\tau^2}\left[\cos\omega t + \omega\tau\cos\left(\omega t - \frac{\pi}{2}\right)\right]$$

Convert to phasor form:

$$\mathbf{V_C} = \frac{V}{1 + \omega^2\tau^2}\left[1\angle 0 + \omega\tau \angle -\frac{\pi}{2}\right]$$

Add the phasors in rectangular form:

$$\mathbf{V_C} = \frac{V}{1 + \omega^2\tau^2}[(1 + j0) + (0 - j\omega\tau)]$$

$$= \frac{V}{1 + \omega^2\tau^2}[1 - j\omega\tau]$$

Back to polar form:

$$\mathbf{V_C} = \frac{V}{1 + \omega^2\tau^2}\left[\sqrt{1^2 + (\omega\tau)^2}\angle\text{atan}\frac{-\omega\tau}{1}\right]$$

$$= \frac{V}{\sqrt{1^2 + \omega^2\tau^2}}\angle\text{atan}(-\omega\tau)$$

Expressing this as the time-domain signal,

$$v_C(t) = \frac{V}{\sqrt{1^2 + \omega^2\tau^2}}\cos(\omega t + \theta) \qquad (8.10)$$

where the phase angle θ is given by

$$\theta = \text{atan}(-\omega\tau)$$

8.3 Impedance and Admittance

8.3.1 Impedance of Basic Circuit Elements

In deriving the relationship between the sinusoidal current and voltage signals associated with a capacitor, we found that the phasors representing these signals are related through Eq. (8.7):

$$\mathbf{I_C} = j\omega C\mathbf{V_C}$$

which can be written in the equivalent form

$$\mathbf{V_C} = \frac{1}{j\omega C}\mathbf{I_C} \qquad (8.11)$$

If we were to write a similar relation for the phasors representing the current and voltage signals of a resistor (value R), we would find that

$$\mathbf{V_R} = R\mathbf{I_R} \qquad (8.12)$$

which follows from Ohm's law. Both of these expressions have the same general form: the voltage phasor is a constant times the current phasor. The constant of proportionality is given the symbol $\mathbf{Z}$, and is called the **impedance**. We define impedance as the ratio of voltage and current phasors:

$$\mathbf{Z} = \frac{\mathbf{V}}{\mathbf{I}} \qquad (8.13)$$

Because the magnitudes of the signal phasors $\mathbf{V}$ and $\mathbf{I}$ are expressed in the standard units of volts and amperes, we see that impedance has the same dimension as resistance, ohms. The impedance of a resistor is R; the impedance of a capacitor is

$$\mathbf{Z_C} = \frac{1}{j\omega C}$$

The impedance of an inductor can be found in the same manner Eq. (8.7) was derived for a capacitor.

Example 8.5

Determine the impedance of an inductor.

We first assume that the inductor current is given by the sinusoidal function

$$i_L(t) = I\cos(\omega t + \varphi)$$

In phasor form, this is $I\angle\varphi$, which we interpret as a complex number $\mathbf{I_L}$, corresponding to the inductor current i_L:

$$\mathbf{I_L} = I\angle\varphi = Ie^{j\varphi}e^{j\omega t}$$

The voltage is given by

$$v_L(t) = L\frac{d}{dt}i_L(t)$$

Using phasor notation,

$$\mathbf{V_L} = L\frac{d}{dt}\mathbf{I_L}$$

$$= L\frac{d}{dt}Ie^{j\varphi}e^{j\omega t}$$

where $\mathbf{V_L}$ is the phasor corresponding to the inductor voltage. Differentiating,

$$\mathbf{V_L} = j\omega LIe^{j\varphi}e^{j\omega t}$$

which can also be written as

$$\mathbf{V_L} = j\omega L \mathbf{I_L}$$

Impedance is defined as the ratio of the voltage and current phasors. Therefore, the impedance of an inductor is

$$\mathbf{Z_L} = j\omega L$$

The impedances of the three fundamental circuit elements are summarized in Table 8.1. Note that the impedance of a resistor is purely real, while the impedances of capacitors and inductors are purely imaginary. The significance of this is reflected in the fact that sinusoidal voltage and current signals associated with a resistor are always in phase, while the corresponding signals associated with a capacitor or inductor are always out of phase by exactly $\pi/2$ radians.

Table 8.1 Impedances of the three fundamental circuit elements. Note that impedance is a function of frequency for capacitors and inductors. The magnitude of impedance is measured in ohms, like resistance, but we must also take into account the phase angle of complex impedances.

Element	Impedance $\mathbf{Z} = \mathbf{V}/\mathbf{I}$
Resistance R (Ω)	R (Ω)
Capacitance C (F)	$\dfrac{1}{j\omega C}$ (Ω)
Inductance L (H)	$j\omega L$ (Ω)

The concept of impedance is valuable because it transforms the problem of determining the ac steady-state circuit response from a set of differential equations to a set of algebraic equations. Again, we will reconsider a previously solved problem to demonstrate the technique.

Example 8.6

Use phasor analysis to find the voltage across the capacitor in the RC circuit of Figure 8.4, where the source voltage is $v_S = V\cos\omega t$.

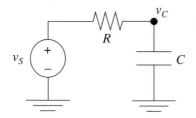

Figure 8.4 First-order RC circuit with sinusoidal excitation source v_S and time constant $\tau = RC$.

We begin by redrawing the circuit, replacing each element with its frequency-domain equivalent. That is, we use phasor notation for sources, and impedances for components. This results in the following circuit:

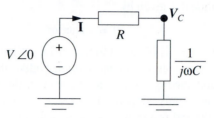

In this diagram the rectangular boxes represent impedances, which have current-voltage relationships governed by the phasor equation Eq. (8.13).

We solve this circuit as if the impedances were resistances. For example, the current phasor $\mathbf{I}$ is given by

$$\mathbf{I} = \frac{V\angle 0}{R + \dfrac{1}{j\omega C}}$$

which is the voltage divided by the equivalent impedance of the resistor and capacitor in series. Since impedances act like resistances, this equivalent is given by the sum of the two series impedances. The voltage (phasor) across the capacitor is given by the current multiplied by the impedance:

$$\mathbf{V_C} = \mathbf{I}\frac{1}{j\omega C}$$

$$= V\angle 0 \frac{\dfrac{1}{j\omega C}}{R + \dfrac{1}{j\omega C}}$$

$$= V\angle 0 \frac{1}{1 + j\omega RC}$$

$$= V\angle 0 \frac{1}{1 + j\omega\tau}$$

Converting the terms in the fraction to polar form:

$$\mathbf{V_C} = V\angle 0 \frac{1\angle 0}{\sqrt{1 + \omega^2\tau^2}\angle\text{atan}\dfrac{\omega\tau}{1}}$$

Evaluating,

$$\mathbf{V_C} = \frac{V}{\sqrt{1 + \omega^2\tau^2}}\angle -\text{atan}\,\omega\tau$$

Finally, converting back to the time-domain form, we have

$$v_C(t) = \frac{V}{\sqrt{1^2 + \omega^2 \tau^2}} \cos(\omega t + \theta)$$

where the phase angle θ is given by

$$\theta = -\mathrm{atan}\,\omega\tau$$

This result for $v_C(t)$ is identical to that of Eq. (8.10), which was originally obtained through a time-domain solution of the circuit. Unlike the time-domain method, which involved writing and solving a differential equation, the frequency-domain method with complex phasors uses only algebra. This feature is preserved for more difficult (i.e., higher-order) circuits and is a prime reason why frequency-domain analysis is often simpler despite having more steps.

8.3.2 Impedances in Series

In transforming our circuit into a frequency-domain representation, we make use of the fact that Kirchhoff's laws apply equally in the time and frequency domains. This is to be expected, since we are simply changing the way we represent the signals and components, which cannot change the underlying physical properties of the circuit.

Consider the two impedances in series, shown in Figure 8.5. The individual voltage drops across each impedance are given by Eq. (8.13):

$$\mathbf{V_1} = \mathbf{IZ_1}$$

$$\mathbf{V_2} = \mathbf{IZ_2}$$

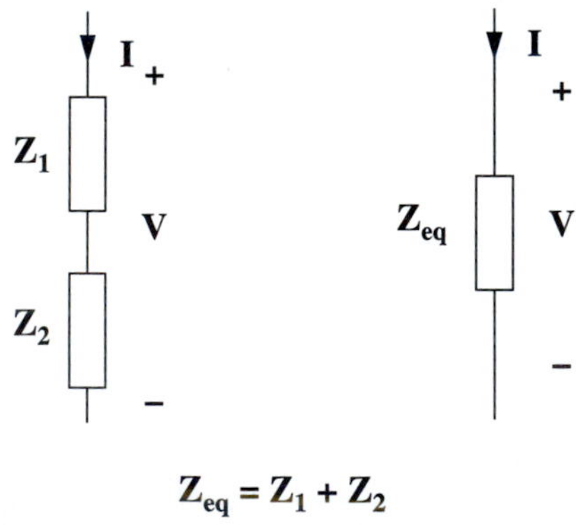

$$\mathbf{Z_{eq}} = \mathbf{Z_1} + \mathbf{Z_2}$$

Figure 8.5 Impedances in series. The equivalent impedance of elements in series is given by the sum of the individual impedances.

Applying KVL, the total voltage drop across the elements in series is

$$\mathbf{V} = \mathbf{V_1} + \mathbf{V_2}$$

$$= \mathbf{IZ_1} + \mathbf{IZ_2}$$

$$= \mathbf{IZ_{eq}}$$

where the equivalent impedance $\mathbf{Z_{eq}}$ is given by

$$\mathbf{Z_{eq}} = \mathbf{Z_1} + \mathbf{Z_2}$$

Naturally, this result extends to any number of impedances in series.

Impedance is a complex quantity, although it is NOT a phasor (i.e., the factor $e^{j\omega t}$, understood to be present in the phasors representing sinusoidal signals, is not part of the complex number representing impedance). As a complex number, we can draw it on the complex plane as an aid to interpreting the relationship between the individual components and the net magnitude and phase. For example, Figure 8.6 shows a resistor and inductor in series, with both time- and frequency-domain symbols, along with the complex representation of the equivalent impedance. We can express the impedance of the series combination in either rectangular form:

$$\mathbf{Z_{eq}} = R + j\omega L$$

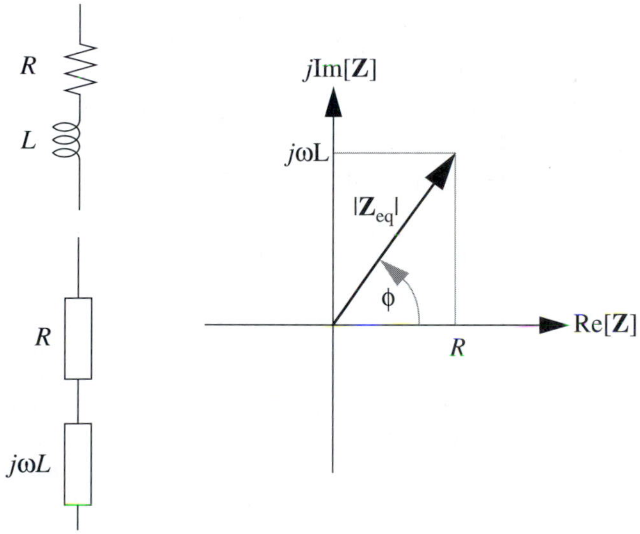

$$\mathbf{Z_{eq}} = R + j\omega L = |\mathbf{Z_{eq}}| \angle \phi$$

Figure 8.6 Impedance diagram for a resistor and inductor in series.

or polar form:

$$\mathbf{Z_{eq}} = |\mathbf{Z_{eq}}| \angle \phi$$

where the **magnitude** of the impedance, in ohms, is the positive root

$$|\mathbf{Z_{eq}}| = \sqrt{R^2 + \omega^2 L^2}$$

and the **phase** (also called the **argument**) is

$$\phi = \arg(\mathbf{Z}_{eq}) = \operatorname{atan}\frac{\omega L}{R}$$

which we will express in radians. Similar relationships apply to any series connection of impedances.

8.3.3 Impedances in Parallel: Admittance

Components in parallel can also be combined into an equivalent impedance. Consider the two impedances $\mathbf{Z}_1$ and $\mathbf{Z}_2$ connected in parallel, as shown in Figure 8.7. The individual currents through each element are given by

$$\mathbf{I}_1 = \frac{\mathbf{V}}{\mathbf{Z}_1}$$

$$\mathbf{I}_2 = \frac{\mathbf{V}}{\mathbf{Z}_2}$$

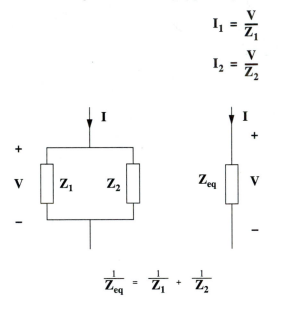

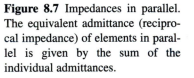

$$\frac{1}{\mathbf{Z}_{eq}} = \frac{1}{\mathbf{Z}_1} + \frac{1}{\mathbf{Z}_2}$$

Figure 8.7 Impedances in parallel. The equivalent admittance (reciprocal impedance) of elements in parallel is given by the sum of the individual admittances.

Using KCL, we can sum these individual currents into the total current $\mathbf{I}$ flowing through the parallel combination:

$$\mathbf{I} = \mathbf{I}_1 + \mathbf{I}_2$$

$$= \frac{\mathbf{V}}{\mathbf{Z}_1} + \frac{\mathbf{V}}{\mathbf{Z}_2}$$

$$= \mathbf{V}\left(\frac{1}{\mathbf{Z}_{eq}}\right)$$

where the equivalent reciprocal impedance $1/\mathbf{Z}_{eq}$ is given by

$$\frac{1}{\mathbf{Z}_{eq}} = \frac{1}{\mathbf{Z}_1} + \frac{1}{\mathbf{Z}_2}$$

Again, this result extends to any number of parallel impedances.

The reciprocal of impedance is given the symbol **Y** and is called the **admittance**:

$$\mathbf{Y} = \frac{1}{\mathbf{Z}} = \frac{\mathbf{I}}{\mathbf{V}}$$

Like impedance, admittance in general is a complex quantity which depends on frequency, thus it has both a magnitude, |Y|, and phase. The standard unit of admittance magnitude is the reciprocal ohm, known as a **siemens** (abbreviation: S), which is equal to an ampere/volt.

Following the definitions from Table 8.1, the admittances of the three basic circuit components are:

$$\mathbf{Y} = \frac{1}{R} \qquad\qquad \text{resistor}$$

$$\mathbf{Y} = j\omega C \qquad\qquad \text{capacitor}$$

$$\mathbf{Y} = \frac{1}{j\omega L} \qquad\qquad \text{inductor}$$

Using both impedance and admittance formulations, it is easy to combine complex combinations of elements into simpler equivalent forms. We need remember to add impedances of elements in series, and add admittances of elements in parallel. We illustrate this principle in the next example.

Example 8.7

Determine the voltage $v(t)$ in the circuit of Figure 8.8, if the current source is operating at 1 MHz with a magnitude of 5 mA ($i_S(t) = 0.005\cos(2\pi 10^6 t)$).

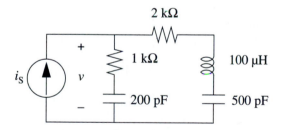

Figure 8.8 Third-order circuit for Example 8.7. The value of the current source is given by $i_S(t) = 0.005\cos(2\pi 10^6 t)$.

The first step is to redraw the circuit in the frequency domain, with a phasor representation of the current source, and impedances for each of the passive components:

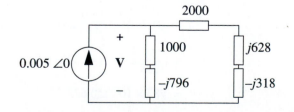

In determining these impedance values, we have used the formula in Table 8.1. For example, the impedance of the 200-pF capacitor is

$$\mathbf{Z_C} = \frac{1}{j\omega C} = -\frac{j}{2\pi f C}$$

$$= -\frac{j}{2\pi 10^6 200 \times 10^{-12}} = -j796$$

Next, we recognize two branches which consist of elements in series. We combine these elements by adding the individual impedances. The equivalent impedance of the two elements in series is

$$\mathbf{Z_1} = 1000 - j796$$

and for the three elements in series

$$\mathbf{Z_2} = 2000 + j628 - j318 = 2000 + j310$$

Redrawing the circuit with these equivalent impedances:

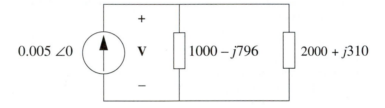

At this point we have two complex impedances in parallel. We want to combine these into a single equivalent impedance. To do so, we add the individual admittances. In order to find these we first express the impedances in polar form. Converting:

$$|\mathbf{Z_1}| = \sqrt{1000^2 + 796^2} = 1278$$

$$\arg(\mathbf{Z_1}) = \operatorname{atan} \frac{-796}{1000} = -0.672$$

$$\mathbf{Z_1} = 1278 \angle -0.672$$

Carrying out a similar procedure for the $2000 + j310\ \Omega$ impedance,

$$\mathbf{Z_2} = 2024 \angle 0.154$$

Taking the reciprocals, the admittances of the two series combinations are

$$\mathbf{Y_1} = \frac{1}{\mathbf{Z_1}} = \frac{1 \angle 0}{1278 \angle -0.672}$$

$$= \frac{1}{1278} \angle (0 - (-0.672))$$

$$= 783 \times 10^{-6} \angle 0.672$$

$$\mathbf{Y_2} = \frac{1}{\mathbf{Z_2}} = 494 \times 10^{-6} \angle -1.54$$

The two admittances in parallel add to a combined admittance of

$$\mathbf{Y_{eq}} = \mathbf{Y_1} + \mathbf{Y_2}$$

$$= (783 \times 10^{-6} \angle 0.672) + (494 \times 10^{-6} \angle -1.54)$$

$$= (612 \times 10^{-6} + j487 \times 10^{-6}) + (488 \times 10^{-6} + j75.7 \times 10^{-6})$$

$$= 1101 \times 10^{-6} + j412 \times 10^{-6}$$

$$= 1.18 \times 10^{-6} \angle 0.358$$

This results in a single equivalent impedance of

$$\mathbf{Z_{eq}} = \frac{1}{\mathbf{Y_{eq}}} = \frac{1}{1.18 \times 10^{-3} \angle 0.358}$$

$$= 851 \angle -0.358$$

Redrawing the circuit once more with a single equivalent impedance representing the network of resistors, capacitors, and inductors:

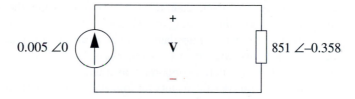

The voltage phasor is found by multiplying the current phasor and the equivalent impedance:

$$\mathbf{V} = \mathbf{IZ_{eq}}$$

$$= (0.005 \angle 0)(851 \angle -0.358)$$

$$= 4.26 \angle -0.358$$

Finally, we express the current source and voltage values as functions of time:

$$i_S(t) = 0.005 \cos 2\pi 10^6 t$$

$$v(t) = 4.26 \cos (2\pi 10^6 t - 0.358)$$

In this example we have found the response to a sinusoidal input of a circuit containing two capacitors and an inductor, a third-order circuit. This has been accomplished with scarcely more effort than a dc resistive circuit with the same number of elements would require.

8.3.4 Components of Impedance and Admittance

In the previous example we found the equivalent impedance of the circuit of Figure 8.8, containing three energy-storage elements, to be $851 \angle -0.358$ (when the excitation frequency is 1 MHz). This can be expressed in rectangular form as

$$\mathbf{Z_{eq}} = 797 - j298 \qquad (8.14)$$

In general, the complex impedance of an arbitrary connection of resistors, capacitors, and inductors can always be written as

$$\mathbf{Z_{eq}} = R(\omega) + jX(\omega) \qquad (8.15)$$

We call the real part of impedance the **resistive component**:

$$R(\omega) = \text{Re}[\mathbf{Z_{eq}}]$$

and the imaginary part of impedance the **reactive component**:

$$X(\omega) = \text{Im}[\mathbf{Z_{eq}}]$$

Both components have standard units of ohms. Note that the reactive component $X(\omega)$ is defined as a real number. Shorthand names for these two components are resistance and reactance, respectively. However, care must be taken in using the term "resistance," because the usual time-domain meaning of resistance is not a function of frequency, while the meaning here, the real part of impedance, can be a function of frequency.

In the example above, the resistive component is 797 Ω, and the reactive component is -298 Ω. The negative sign here is important, as it tells us the net reactive component is capacitive. To see this, consider the results for impedance of a single inductor and capacitor. For an inductor, we have

$$\mathbf{Z_L} = j\omega L$$

which is a purely imaginary quantity. Since both ω and L are positive quantities, the reactance of an inductor is positive:

$$X_L = \text{Im}[\mathbf{Z_L}]$$
$$= \text{Im}[j\omega L]$$
$$= \omega L$$
$$X_L > 0$$

For a capacitor, the impedance is

$$\mathbf{Z_C} = \frac{1}{j\omega C} = -\frac{j}{\omega C}$$

Again, both the frequency ω and the component value C are positive quantities, so it follows that the reactance of a capacitor is negative:

$$X_C = \text{Im}[\mathbf{Z_C}]$$

$$= \text{Im}\left[-\frac{j}{\omega C}\right]$$

$$= -\frac{1}{\omega C}$$

$$X_C < 0$$

This can also be seen from the complex impedance diagram. In Figure 8.9 we have plotted the impedance of the three basic circuit elements. Individually, these are always aligned along either the positive real axis (resistor), the positive imaginary axis (inductor), or the negative imaginary axis (capacitor).

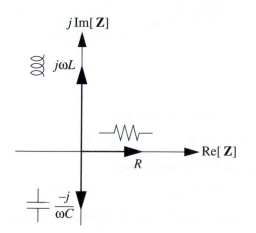

Figure 8.9 Impedance of basic circuit elements. The impedance of a resistor is purely real. Inductors have positive reactance, thus their purely imaginary impedance lies on the upper half of the imaginary **Z**-axis. Capacitors have negative reactance, so their impedance is on the lower half of the imaginary **Z**-axis.

Returning to our numerical example, we see that a reactance of −298 Ω at 1 MHz could also result from a capacitance of 534 pF:

$$X_C(\omega) = -\frac{1}{\omega C}$$

$$-298 = -\frac{1}{2\pi 10^6 C}$$

$$C = 534 \times 10^{-12}$$

Therefore, an equivalent circuit with same impedance would consist of a 534 pF capacitor in series with a 797 Ω resistor, Figure 8.10. This equivalent is only valid at 1 MHz, however, since the reactances of the capacitors and inductors are functions of frequency.

We also give names to the individual components of admittance. In general, an admittance can be written in the rectangular form

$$\mathbf{Y_{eq}} = G(\omega) + jB(\omega)$$

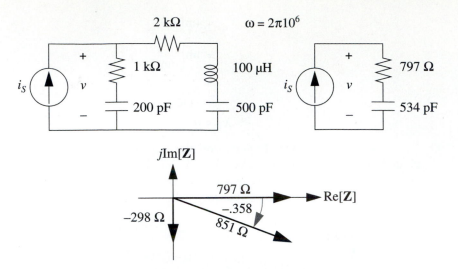

Figure 8.10 Equivalent impedances. Both circuits have the same equivalent impedance, 851 $\angle$–0.358, when the current source is operating at 1 MHz.

where the real part of admittance is known as the **conductance**:

$$G(\omega) \;=\; \text{Re}[\mathbf{Y_{eq}}]$$

and the imaginary part of admittance is called the **susceptance**:

$$B(\omega) \;=\; \text{Im}[\mathbf{Y_{eq}}]$$

Both conductance and susceptance have standard units of siemens (S), the same as admittance.

In dc circuits we are accustomed to using the term "conductance" to mean reciprocal resistance. This is true insofar as the impedance is purely real. However, in an ac situation where the admittance is complex, G and R are definitely NOT reciprocals of each other. (See Exercise 8.18.)

8.3.5 Asymptotic Behavior of Reactance

In Chapter 2 we learned that a capacitor acts as an open circuit in the dc steady state. Later we determined that an inductor acts as a short circuit in the dc steady state. Both of these behaviors can be understood on a more quantitative basis using our expressions for reactance.

Ignoring the $\pi/2$ difference in phase, the sinusoidal voltage and current associated with a capacitor are related through the susceptance by

$$|B_C| \;=\; \frac{|\mathbf{I}|}{|\mathbf{V}|} \;=\; \omega C$$

which we can solve for the current magnitude:

$$|\mathbf{I}| = \omega C |\mathbf{V}| \tag{8.16}$$

This says that for a given voltage magnitude, the current through the capacitor is directly proportional to the frequency ω. A dc signal can be considered to be a sinusoid with frequency zero:

$$v(t) = |\mathbf{V}| \cos \omega t \big|_{\omega = 0}$$

which is clearly a dc voltage of value $|\mathbf{V}|$. Hence, for a zero-frequency (dc) voltage, Eq. (8.16) tells us there is zero current through the capacitor, which is the equivalent of an open circuit.

But Eq. (8.16) actually tells us more. The reciprocal relation between capacitive reactance and frequency means there is a continuous reduction in the current through a capacitor as the frequency is reduced. Conversely, the capacitor "looks like" a short circuit for sufficiently high frequencies, since the current is directly proportional to frequency.

Similar reasoning holds for inductors. The magnitude of the inductive reactance is given by

$$|X_L| = \frac{|\mathbf{V}|}{|\mathbf{I}|} = \omega L$$

This expression quantifies our previous statement that an inductor acts as a short circuit for dc signals. If we take the limit as $\omega \to 0$, the voltage across the inductor also goes to zero, which is precisely what is meant by "short circuit." In the opposite limit, $\omega \to \infty$, the current approaches zero, so we can say that an inductor "looks like" an open circuit for sufficiently high frequencies.

8.4 Clock-Signal Analysis Using ac Steady-State Theory

In the last chapter we examined the effect on a square-wave clock signal by an interconnect. The circuit model, shown in Figure 8.11, used a first-order RC interconnect model. We can now extend this analysis to more complicated interconnect models, such as the two-section RC ladder, and the inverter model including package inductance.

8.4.1 First-Order RC Interconnect

]First, we recapitulate the results of the first-order model and express them in our phasor notation. The square-wave clock signal $v_S(t)$ operating between 0 and V is given by the Fourier-series expansion

$$v_S(t) = \frac{V}{2} + \sum_{\substack{n=1 \\ n \text{ odd}}}^{\infty} \frac{2V}{\pi n} \sin n\omega t \tag{8.17}$$

Expressing each of the sine terms as a cosine, this can be written as

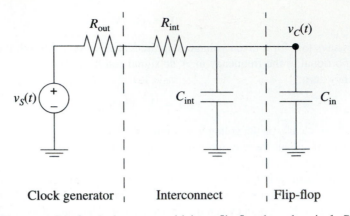

Clock generator   Interconnect   Flip-flop

Figure 8.11 Circuit model of a clock generator driving a flip-flop through a single-*RC*-lump interconnect.

$$v_S(t) = \frac{V}{2} + \sum_{\substack{n=1 \\ n \text{ odd}}}^{\infty} \frac{2V}{\pi n} \cos\left(n\omega t - \frac{\pi}{2}\right) \tag{8.18}$$

When this signal is passed through the *RC* circuit, the voltage appearing across the capacitor is

$$v_C(t) = \frac{V}{2} + \sum_{\substack{n=1 \\ n \text{ odd}}}^{\infty} \left[\frac{2V}{n\pi} \frac{1}{1 + n^2\omega^2 t^2}(\sin n\omega t - n\omega t \cos n\omega t)\right]$$

Applying the procedure we used in Example 8.4, this can be written (see Exercise 8.20) entirely with cosines as

$$v_C(t) = \frac{V}{2} + \sum_{\substack{n=1 \\ n \text{ odd}}}^{\infty} \left[\frac{2V}{n\pi} \frac{1}{\sqrt{1 + n^2\omega^2 t^2}} \cos(n\omega t + \theta_n)\right] \tag{8.19}$$

$$\theta_n = -\frac{\pi}{2} - \operatorname{atan} n\omega t$$

What do these expressions look like in phasor form? The *n*th source term, operating at angular frequency $n\omega$, has the time-domain representation

$$v_{Sn}(t) = \frac{2V}{\pi n} \cos\left(n\omega t - \frac{\pi}{2}\right)$$

and phasor representation

$$\mathbf{V}_{Sn} = \frac{2V}{\pi n} \angle -\frac{\pi}{2} \tag{8.20}$$

The corresponding nth term in the capacitor voltage is

$$v_{Cn}(t) = \frac{2V}{n\pi}\frac{1}{\sqrt{1+n^2\omega^2t^2}}\cos\left(n\omega t - \frac{\pi}{2} - \mathrm{atan}\, n\omega\tau\right)$$

Expressed as a phasor, this becomes

$$\mathbf{V_{Cn}} = \frac{2V}{n\pi}\frac{1}{\sqrt{1+n^2\omega^2t^2}}\angle\left(-\frac{\pi}{2} - \mathrm{atan}\, n\omega\tau\right) \tag{8.21}$$

The nth harmonic source term $\mathbf{V_{Sn}}$ in Eq. (8.20) is transformed into the nth harmonic output term $\mathbf{V_{Cn}}$ [Eq. (8.21)] through the action of the RC circuit. This circuit is depicted in frequency-domain form in Figure 8.12. Note that the impedance of the capacitor is $1/(jn\omega C)$; remember that the frequency of this harmonic is $n\omega$. If we apply the voltage-divider rule to this circuit, we find that

$$\mathbf{V_{Cn}} = \mathbf{V_{Sn}}\frac{\dfrac{1}{jn\omega C}}{R + \dfrac{1}{jn\omega C}}$$

$$\frac{\mathbf{V_{Cn}}}{\mathbf{V_{Sn}}} = \frac{1}{1 + jn\omega\tau} \tag{8.22}$$

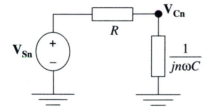

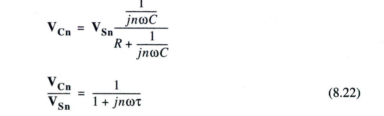

Figure 8.12 Frequency-domain circuit for the nth harmonic component of the input square wave. The angular frequency of this circuit is $n\omega$.

Expressed in polar form,

$$\frac{\mathbf{V_{Cn}}}{\mathbf{V_{Sn}}} = \frac{1}{\sqrt{1+n^2\omega^2t^2}}\angle-\mathrm{atan}\, n\omega\tau \tag{8.23}$$

Let's check that the phasor expressions in Eqs. (8.20) and (8.21) are consistent with this. (Recall that these terms were originally obtained from a time-domain solution of the RC circuit.) Taking the ratio of these two equations,

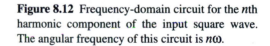

$$\frac{\mathbf{V_{Cn}}}{\mathbf{V_{Sn}}} = \frac{\dfrac{2V}{n\pi}\dfrac{1}{\sqrt{1+n^2\omega^2t^2}}\angle\left(-\dfrac{\pi}{2} - \mathrm{atan}\, n\omega\tau\right)}{\dfrac{2V}{\pi n}\angle-\dfrac{\pi}{2}}$$

$$\frac{\mathbf{V_{Cn}}}{\mathbf{V_{Sn}}} = \frac{1}{\sqrt{1 + n^2 \omega^2 t^2}} \angle -\mathrm{atan}\, n\omega\tau$$

which is the same as Eq. (8.23).

Now that we have confirmed that the phasor approach yields valid results for a first-order interconnect model, we will use a more accurate and difficult interconnect approximation.

8.4.2 Two-Section *RC* Ladder Interconnect

Using a more accurate two-*RC* section interconnect model results in the circuit of Figure 8.13. As before, the clock generator signal is a square wave with Fourier representation

$$v_S(t) = \frac{V}{2} + \sum_{\substack{n=1 \\ n\ \mathrm{odd}}}^{\infty} \frac{2V}{\pi n} \cos\left(n\omega t - \frac{\pi}{2}\right)$$

which can be expressed in phasor form as

$$\mathbf{V_S} = \mathbf{V_{S0}} + \mathbf{V_{S1}} + \mathbf{V_{S3}} + \cdots$$

$$\mathbf{V_{S0}} = \frac{V}{2}\angle 0$$

$$\mathbf{V_{Sn}} = \frac{2V}{\pi n}\angle -\frac{\pi}{2}$$

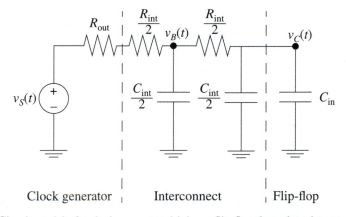

Figure 8.13 Circuit model of a clock generator driving a flip-flop through an interconnect modeled with two *RC* lumps.

After combining the resistors in series and capacitors in parallel, the simpler circuit shown in Figure 8.14 results. We have shown the circuit in the frequency domain, being driven by the *n*th source harmonic. The voltages across the two capacitors, $v_B(t)$ and $v_C(t)$, have Fourier expansions in phasor form

$$\mathbf{V_B} = \mathbf{V_{B0}} + \mathbf{V_{B1}} + \mathbf{V_{B3}} + \cdots$$

$$\mathbf{V_C} = \mathbf{V_{C0}} + \mathbf{V_{C1}} + \mathbf{V_{C3}} + \cdots$$

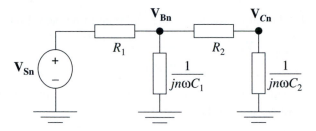

Figure 8.14 Frequency-domain equivalent circuit for the two-lump RC ladder interconnect approximation. The input source is nth harmonic component of the clock square wave. The angular frequency of the source in this circuit is $n\omega$.

We are interested in finding the phasor components of $\mathbf{V_C}$ so that we can reconstruct the time-domain signal $v_C(t)$.

Applying KCL at the $\mathbf{V_{Bn}}$ node,

$$\frac{\mathbf{V_{Sn}} - \mathbf{V_{Bn}}}{R_1} + \frac{\mathbf{V_{Cn}} - \mathbf{V_{Bn}}}{R_2} = jn\omega C_1 \mathbf{V_{Bn}}$$

To simplify the notation, we will temporarily drop the subscript "**n**" which indicates we are working with the nth harmonic. We must remember that the frequency of the harmonic under consideration is $n\omega$.

Rearranging this equation,

$$R_2(\mathbf{V_S} - \mathbf{V_B}) + R_1(\mathbf{V_C} - \mathbf{V_B}) = jn\omega R_1 R_2 C_1 \mathbf{V_B}$$

$$R_2\mathbf{V_S} + R_1\mathbf{V_C} = (R_1 + R_2 + jn\omega R_1 R_2 C_1)\mathbf{V_B} \qquad (8.24)$$

Applying the voltage-divider rule at the $\mathbf{V_C}$ node,

$$\mathbf{V_C} = \mathbf{V_B} \frac{\dfrac{1}{jn\omega C_2}}{R_2 + \dfrac{1}{jn\omega C_2}}$$

Transposing,

$$\mathbf{V_B} = \mathbf{V_C}(1 + jn\omega R_2 C_2) \qquad (8.25)$$

Substituting Eq. (8.25) in Eq. (8.24),

$$R_2 \mathbf{V_S} + R_1 \mathbf{V_C} = (R_1 + R_2 + jn\omega R_1 R_2 C_1)(1 + jn\omega R_2 C_2)\mathbf{V_C}$$

$$R_2 \mathbf{V_S} = \left(R_2 - n^2 \omega^2 R_1 R_2^2 C_1 C_2 + jn\omega \left(R_1 R_2 C_1 + R_2 C_2 (R_1 + R_2) \right) \right) \mathbf{V_C}$$

Solving for the output voltage phasor, and reinserting the subscript n,

$$\mathbf{V_{Cn}} = \mathbf{V_{Sn}} \frac{1}{\left(1 - n^2 \omega^2 R_1 R_2 C_1 C_2 \right) + jn\omega \left(R_1 C_1 + R_1 C_2 + R_2 C_2 \right)}$$

We will compare this result with the first-order interconnect approximation, using the numerical values previously assumed.

Example 8.8

A 5-V clock operating at 100 MHz with an output resistance of 400 Ω drives a flip-flop with an equivalent capacitance of 1 pF through an interconnect. The interconnect has a total lumped resistance of $R_{int} = 100\ \Omega$ and lumped capacitance $C_{int} = 1$ pF. Compare the clock-signal degradation using the one-lump and two-lump RC models.

We will start with the two-lump interconnect model. From the given parameters, the equivalent model values are

$$R_1 = R_{out} + \frac{R_{int}}{2} = 400 + 50 = 450$$

$$R_2 = \frac{R_{int}}{2} = 50$$

$$C_1 = \frac{C_{int}}{2} = 0.5 \times 10^{-12}$$

$$C_2 = C_{in} + \frac{C_{int}}{2} = (1 + 0.5) \times 10^{-12} = 1.5 \times 10^{-12}$$

$$\omega = 2\pi f = 2\pi 100 \times 10^6 = 6.28 \times 10^8$$

Inserting these values into Eq. (),

$$\mathbf{V_{Cn}} = \mathbf{V_{Sn}} \frac{1}{(1 - n^2 0.00666) + jn0.612} \tag{8.26}$$

Computing a few coefficients (with the clock amplitude $V = 5$ V):

$$n = 0: \qquad \mathbf{V_{S0}} = \frac{V}{2}\angle 0 = 2.5\angle 0$$

$$\mathbf{V_{C0}} = \mathbf{V_{S0}} \frac{1}{(1 - (0)^2 0.00666) + j(0)0.612}$$

$$= 2.5\angle 0$$

$$n = 1: \qquad \mathbf{V_{S1}} = \frac{2V}{\pi(1)} \angle -\frac{\pi}{2} = 3.18 \angle -\frac{\pi}{2}$$

$$\mathbf{V_{C1}} = \mathbf{V_{S1}} \frac{1}{(1-(1)^2 0.00666) + j(1)0.612}$$

$$= \left(3.18 \angle -\frac{\pi}{2}\right) \frac{1}{0.993 + j0.612}$$

$$= \frac{3.18 \angle -1.57}{1.17 \angle 0.552}$$

$$= 2.73 \angle -2.12$$

$$n = 3: \qquad \mathbf{V_{S3}} = \frac{2V}{\pi(3)} \angle -\frac{\pi}{2} = 1.06 \angle -\frac{\pi}{2}$$

$$\mathbf{V_{C3}} = \mathbf{V_{S3}} \frac{1}{(1-(3)^2 0.00666) + j(3)0.612}$$

$$= \left(1.06 \angle -\frac{\pi}{2}\right) \frac{1}{0.940 + j1.84}$$

$$= \frac{1.06 \angle -1.57}{2.06 \angle 1.10}$$

$$= 0.514 \angle -2.67$$

$$n = 5: \qquad \mathbf{V_{S5}} = \frac{2V}{\pi(5)} \angle -\frac{\pi}{2} = 0.636 \angle -\frac{\pi}{2}$$

$$\mathbf{V_{C5}} = \mathbf{V_{S5}} \frac{1}{(1-(5)^2 0.00666) + j(5)0.612}$$

$$= \left(0.636 \angle -\frac{\pi}{2}\right) \frac{1}{0.834 + j3.06}$$

$$= \frac{0.636 \angle -1.57}{3.17 \angle 1.31}$$

$$= 0.201 \angle -2.88$$

Putting these together, the clock signal as it appears on the far end of the loaded interconnect is

$$v_C(t) = 2.5 + 2.73 \cos(2\pi 100 \times 10^6 t - 2.12)$$

$$+ 0.514 \cos(2\pi 300 \times 10^6 t - 2.67)$$

$$+ 0.201 \cos(2\pi 500 \times 10^6 t - 2.88)$$

$$+ \cdots$$

Turning to the first-order interconnect model:

$$V_{Cn} = V_{Sn}\frac{1}{1 + jn\omega\tau}$$

where the time constant τ is the product of the total resistance (output plus interconnect) and the total capacitance (interconnect plus input). Inserting values,

$$V_{Cn} = V_{Sn}\frac{1}{1 + jn0.628} \tag{8.27}$$

The first two coefficients are:

$$n = 0: \qquad\qquad V_{S0} = \frac{V}{2}\angle 0 = 2.5\angle 0$$

$$V_{C0} = V_{S0}\frac{1}{1 + j(0)0.628}$$

$$= 2.5\angle 0$$

$$n = 1: \qquad\qquad V_{S1} = \frac{2V}{\pi(1)}\angle -\frac{\pi}{2} = 3.18\angle -\frac{\pi}{2}$$

$$V_{C1} = V_{S1}\frac{1}{1 + j(1)0.628}$$

$$= \frac{3.18\angle -1.57}{1.18\angle 0.561}$$

$$= 2.70\angle -2.13$$

The results of the two models are plotted in Figure 8.15. The driving square-wave signal, $v_S(t)$ is accurately approximated by the Fourier series, except near the discontinuities where the truncated series does not quite converge to the high and low logic values. The other curves, which are plots of $v_C(t)$ for the two interconnect models, clearly illustrate how the interconnect degrades the clock signal. For this length of interconnect, there is little difference between the first- and second-order RC interconnect models, as the two curves nearly coincide. (Note that both of these curves are quite smooth and do not suffer from the convergence problems we see in the Fourier series for the source signal. This is because as n increases, the higher-order coefficients for V_C decay in magnitude much more rapidly than the V_S coefficients. Hence, it takes many fewer terms to obtain an accurate representation of $v_C(t)$ than for $v_S(t)$.)

Let's extend this analysis to calculating clock skew between two flip-flops.

Example 8.9

Determine the clock skew between the two flip-flops in Figure 8.16, where the long interconnect is three times the length of the short interconnect, using the first- and second-order RC interconnect models. The clock, flip-flop, and short interconnect parameters are the same as in Example 8.8. Assume the flip-flops trigger on a rising edge when the input reaches $V_{th} = 4.0$ V.

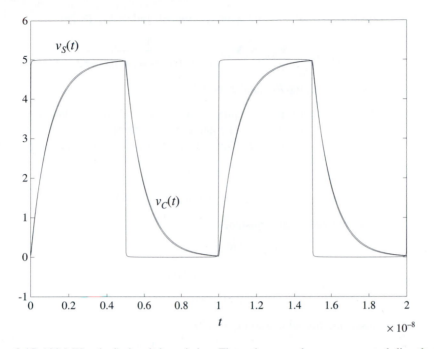

Figure 8.15 100-MHz clock-signal degradation. These three graphs were generated directly from the Fourier-series representations up to the 1001st harmonic. There is almost no difference between the first- and second-order interconnect models for this length of interconnect.

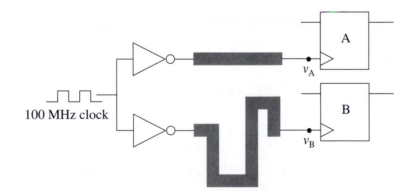

Figure 8.16 Clock skew example. The two flip-flops A and B are driven by a 100-MHz clock through separate interconnects, one three times the length of the other.

The signal appearing at flip-flop A is the same as the previous example. We found that the phasors describing v_A are given by Eq. (8.27),

$$\mathbf{V_{An}} = \mathbf{V_{Sn}}\frac{1}{1 + jn0.628}$$

for the first-order RC interconnect model, and by Eq. (8.26),

$$V_{An} = V_{Sn}\frac{1}{(1 - n^2 0.00666) + jn0.612}$$

for the two-lump RC ladder model.

To find the corresponding phasors for v_B, we need to calculate the appropriate values for the model parameters. The interconnect is three times as long, so the lumped resistance and capacitance are likewise three times that of the interconnect of flip-flop A:

$$R_{int,B} = 3R_{int,A} = 300$$

$$C_{int,B} = 3C_{int,A} = 3.0 \times 10^{-12}$$

The first-order time constant is thus

$$\tau_B = (R_{out} + R_{int,B})(C_{in} + C_{int,B})$$

$$= (400 + 300)(1 + 3) \times 10^{-12}$$

$$= 2.80 \times 10^{-9}$$

Inserting this value into Eq. (8.22):

$$V_{Bn} = V_{Sn}\frac{1}{1 + jn\omega\tau}$$

$$= V_{Sn}\frac{1}{1 + jn1.76}$$

For the two-lump ladder model, the parameters are

$$R_1 = R_{out} + \frac{R_{int,B}}{2} = 400 + 150 = 550$$

$$R_2 = \frac{R_{int,B}}{2} = 150$$

$$C_1 = \frac{C_{int,B}}{2} = 1.5 \times 10^{-12}$$

$$C_2 = C_{in} + \frac{C_{int,B}}{2} = (1 + 1.5) \times 10^{-12} = 2.5 \times 10^{-12}$$

Inserting these values into Eq. (),

$$V_{Bn} = V_{Sn}\frac{1}{1 - n^2\omega^2 R_1 R_2 C_1 C_2 + jn\omega(R_1 C_1 + R_1 C_2 + R_2 C_2)}$$

$$V_{Bn} = V_{Sn}\frac{1}{(1 - n^2 0.122) + jn1.62}$$

The signals described by these phasors are plotted in Figure 8.17. We see that the clock signal is even more degraded by the longer interconnect when it arrives at flip-flop B. However, the one-lump RC interconnect model somewhat overstates the extent of this degradation. The more accurate two-lump model predicts a larger swing in the voltage, which is important in protecting the circuit against the influence of noise. Also, the second-order model indicates the clock skew is 0.222 ns. The 0.248 ns predicted by the first-order model is too high by about 12%. If a computer manufacturer were to use the prediction of the first-order model, the clock distribution circuit design would be overly conservative, dictating that the system be run slower than the technology allows.

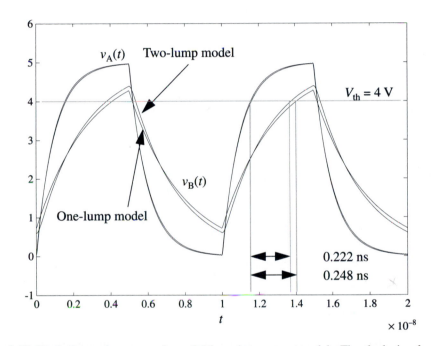

Figure 8.17 Clock skew using one- and two-RC-lump interconnect models. The clock signal arriving at flip-flop B is delayed relative to flip-flop B, but the amount of this skew is overstated by the first-order interconnect model.

8.4.3 Interconnect Model Including Package Inductance

In Chapter 5 we learned how to include the effects of package and power-supply inductance on gate switching speed. The same effects are present in clock distribution circuits, which lead to the series RLC circuit model shown in Figure 8.18. In this diagram, we have shown both the familiar time-domain circuit, driven by the square-wave clock generator $v_S(t)$, and the frequency-domain equivalent operating at the nth harmonic of the source frequency.

Applying the voltage-divider rule, we can write the voltage phasor $\mathbf{V_{Cn}}$ as

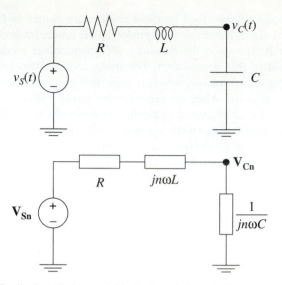

Figure 8.18 Clock distribution circuit model including package inductance. *Top:* Time domain circuit. *Bottom:* After transformation into the frequency domain.

$$\mathbf{V_{Cn}} = \mathbf{V_{Sn}} \frac{\dfrac{1}{jn\omega C}}{R + jn\omega L + \dfrac{1}{jn\omega C}}$$

$$= \mathbf{V_{Sn}} \frac{1}{(1 - n^2\omega^2 LC) + jn\omega RC}$$

We will discuss this result in detail in a later section of this chapter.

8.5 Transfer Functions

We have applied ac steady-state analysis to three circuit configurations: a first-order RC circuit, a two-section RC ladder circuit, and a series RLC circuit. In each case, we looked at the output $(v_C(t))$ obtained by driving the circuit from a sinusoidal source at angular frequency ω. At this point we will explore some common features of these problems and extend our analysis of second-order circuits.

8.5.1 Input-Output Relationship

For each of the three circuits we determined a relation between the input phasor $\mathbf{V_S}$ and the output phasor $\mathbf{V_C}$. These can be expressed as a ratio. For example, in the first-order circuit driven at frequency ω we found that

$$\frac{\mathbf{V_C}}{\mathbf{V_S}} = \frac{1}{1 + j\omega RC}, \qquad RC \text{ circuit} \qquad (8.28)$$

The ratio of an output phasor to an input phasor is called a **transfer function**. The transfer function for the two-section RC ladder circuit is

$$\frac{\mathbf{V_C}}{\mathbf{V_S}} = \frac{1}{1 - \omega^2 R_1 R_2 C_1 C_2 + j\omega(R_1 C_1 + R_1 C_2 + R_2 C_2)}, \qquad RC \text{ ladder circuit}$$

and for the series RLC circuit

$$\frac{\mathbf{V_C}}{\mathbf{V_S}} = \frac{1}{(1 - \omega^2 LC) + j\omega RC}, \qquad \text{series } RLC \text{ circuit} \qquad (8.29)$$

The transfer function describes how the circuit changes an input sinusoidal voltage (at frequency ω) into an output sinusoidal voltage. In the linear circuits we are concerned with, this output sinusoid is at the same frequency as the input. Therefore, the circuit, no matter how complicated, can do only two things to the input sinusoid: change its amplitude, or change its phase. In most cases we have a multitude of sinusoids (described by the Fourier representation), and the transfer function does different things to each of them.

To interpret the transfer function, consider the arbitrary circuit shown in Figure 8.19, with input signal $v_i(t)$ and output $v_o(t)$. We have shown both signals as voltages, but this need not be the case; transfer-function analysis is valid for current signals as well. Transforming this circuit into the frequency domain, the input and output signals are represented by the phasors $\mathbf{V_i}$ and $\mathbf{V_o}$, which by definition are related by the transfer function

$$\mathbf{H}(\omega) = \frac{\mathbf{V_o}}{\mathbf{V_i}} \qquad (8.30)$$

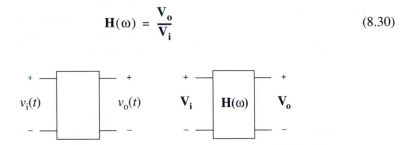

Figure 8.19 *Left:* Arbitrary circuit with input signal $v_i(t)$ and output signal $v_o(t)$ (sinusoids at frequency ω). *Right:* Frequency-domain depiction of this circuit. The input and output phasors are related by the transfer function $\mathbf{H}(\omega)$.

The transfer function is the ratio of two complex expressions, which in general are functions of frequency. Hence, the transfer function is also a complex expression, which can be expressed in magnitude-phase form:

$$\mathbf{H}(\omega) = |\mathbf{H}(\omega)| \angle \Theta(\omega) \qquad (8.31)$$

Using the transfer function, we can tell at a glance what the circuit does to an input sinusoid. For example, if the input voltage is

$$v_i(t) = A\cos(\omega t + \phi)$$

then the output signal is simply

$$v_o(t) = A|\mathbf{H}(\omega)|\cos(\omega t + \phi + \Theta(\omega))$$

Transfer functions contain a wealth of information about the circuit, but this information is not readily apparent when $\mathbf{H}(\omega)$ is expressed in complex form. To make the utility of the transfer function more transparent, we will separately graph the magnitude and phase components in a standard form, known as a **Bode plot**.

8.5.2 Bode Plots

Most transfer functions of interest vary by many orders of magnitude, and the range of frequencies we need to consider is likewise quite large. Thus, it is inconvenient to plot the components of the transfer function on a strictly linear scale. Instead, we will use a log-log plot for the magnitude, and a semilog plot for the phase.

The x-axis on both plots is log ω. (Common logarithms, with base 10, are always used.) Often, specifications for components and systems use the logarithm of the hertzian frequency f as the x-axis, which introduces a fixed offset into the curve. For historical reasons, the magnitude of the transfer function is expressed in units of **decibels** (abbreviation: dB) which are

$$|\mathbf{H}(\omega)|_{dB} = 20\log|\mathbf{H}(\omega)| \qquad (8.32)$$

Decibels are essentially a dimensionless unit, since they arise from taking the logarithm of a (dimensionless) ratio. Decibels originated in a time before calculators and have really outlived their usefulness, but since they are still encountered, a passing familiarity is warranted.

The Bode magnitude plot of the first-order RC transfer function, Eq. (8.28), is shown in Figure 8.20. The solid curve is the magnitude $|\mathbf{H}(\omega)|_{dB}$. The exact curve approaches two straight-line asymptotes at low and high frequencies. Except near the break point, $\omega \approx 1/\tau$, the magnitude is closely approximated by these asymptotes, which are useful aids in sketching this and other Bode plots. (A frequency where two straight-line asymptotes cross is called a **corner frequency**.) The magnitude (in dB) is given by

$$|\mathbf{H}(\omega)|_{dB} = 20\log\left(\left|\frac{1}{1 + j\omega\tau}\right|\right)$$

$$= 20\log\left(\frac{1}{\sqrt{1 + (\omega\tau)^2}}\right)$$

At low frequencies (i.e., where $\omega \ll 1/\tau$) the second term in the denominator is much less than the first: $(\omega\tau)^2 \ll 1$. Therefore, the magnitude is given approximately by

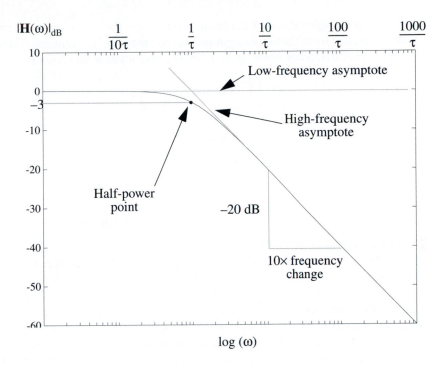

Figure 8.20 Bode plot of the magnitude of the first-order *RC* lowpass filter function.

$$|\mathbf{H}(\omega)|_{dB} \approx 20 \log\left(\frac{1}{\sqrt{1}}\right) = 0 \text{ dB}$$

Above the corner frequency, $\omega \gg 1/\tau$, the second term in the denominator dominates:

$$|\mathbf{H}(\omega)|_{dB} \approx 20 \log\left(\frac{1}{\sqrt{(\omega\tau)^2}}\right)$$

$$= 20 \log\left(\frac{1}{\omega\tau}\right)$$

$$= -20(\log(\omega) + \log(\tau))$$

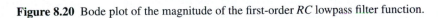

$$= \underbrace{-20}_{\text{slope}} (\underbrace{\log(\omega)}_{\substack{x\text{-axis} \\ \text{variable}}} - \underbrace{\log(1/\tau)}_{\substack{x\text{-axis} \\ \text{intercept}}})$$

When written in this form, we see that the magnitude in dB approaches a straight line of slope −20, which intercepts the log (ω) axis at log ($1/\tau$). Since a unit increase in $\log(\omega)$ corresponds to a 10× increase (one **decade**) in frequency, the slope of the high-frequency asymptote is −20 dB/decade.

In the neighborhood of the break point, where the two asymptotes cross, the exact curve deviates from the two straight-line approximations. The maximum deviation occurs at the corner frequency where $\omega = 1/\tau$. The magnitude at this point is

$$\left| \mathbf{H}\!\left(\frac{1}{\tau}\right) \right|_{dB} = 20\log\!\left(\left| \frac{1}{1+j} \right| \right)$$

$$= 20\log\!\left(\frac{1}{\sqrt{2}} \right)$$

$$= -20\log\!\left(2^{\frac{1}{2}} \right)$$

$$= -10\log(2) = -3.010 \text{ dB}$$

which is very close to –3 dB.

Over the range of frequency from $0 < \omega < 1/\tau$, the magnitude of the transfer function is within 3 dB of its maximum value. This means that a signal input in this frequency range will appear with approximately the same magnitude at the output, but higher frequencies will be severely attenuated. "Approximately" is understood to be "within $1/\sqrt{2}$" (approximately 0.707). We refer to this frequency range as the "3-dB bandwidth."

We see that the *RC* circuit, interpreted with its transfer function, allows signals of low frequency to pass through more or less unimpeded, while preventing the passage of high-frequency signals. We call circuits that discriminate signals on the basis of frequency **filters**. The *RC* circuit, where the output is taken across the capacitor, is an example of a **lowpass filter**, so named because it passes mainly the low-frequency components of the signal. "Bandwidth" refers to the range of frequencies selected by the filter.

Another name for the 3-dB bandwidth is the "half-power bandwidth." This term derives from the fact that the power available to a load device is proportional to the square of the voltage at the output of the system. The transfer-function magnitude has a maximum at some frequency ω_m; the power transmitted to the load device will likewise be a maximum at this frequency. At a frequency where the output voltage is reduced by the factor $1/\sqrt{2}$, the power has been reduced by the square, or 1/2. Hence a frequency at which the transfer-function magnitude is down by 3 dB (relative to its maximum value at ω_m) is called a **half-power point**.

The phase of the transfer function is given by

$$\Theta(\omega) = \arg(\mathbf{H}(\omega))$$

For the lowpass filter this is equal to

$$\Theta(\omega) = \arg\!\left(\frac{1}{1 + j\omega\tau} \right) = -\text{atan}\,\omega\tau$$

A Bode phase plot for the lowpass filter is shown in Figure 8.21. The phase asymptotically approaches zero at low frequencies and $-\pi/2$ at high frequencies. The behavior at low frequencies can be understood in terms of the dc behavior of a capacitor, an open circuit. We would expect no effect at all on the input signal as the frequency approaches zero, where the capacitor "looks like" an open circuit, in agreement with the Bode plot.

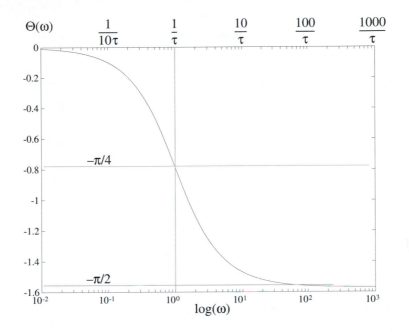

Figure 8.21 Bode plot of the lowpass filter phase $\Theta(\omega)$. At low frequencies the output phase unchanged from the input. At high frequencies the output voltage is $-\pi/2$ radians relative to the input. The phase shift at the corner frequency is $-\pi/4$.

8.6 *RLC* Circuits in the Frequency Domain

In Chapter 5 we saw that the time-domain behavior of circuits containing both inductors and capacitors could be fundamentally different than circuits where capacitors were the only energy-storage elements. In this section we revisit the topic of *RLC* circuits by investigating their behavior in the frequency domain.

8.6.1 Series *RLC* Circuit

Recall that the time-domain response of a series *RLC* circuit, Figure 8.22, consists of functions of the form e^{st}, where the natural frequencies s are found from the characteristic equation

$$s^2 + \frac{R}{L}s + \frac{1}{LC} = 0$$

The roots of this equation are

$$s = -\alpha \pm j\omega_d$$

where the damping factor α is

$$\alpha = \frac{R}{2L}$$

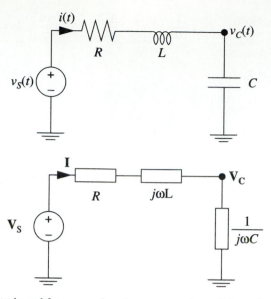

Figure 8.22 Time-domain and frequency-domain representations of the series *RLC* circuit.

and the damped and undamped natural frequencies are defined as

$$\omega_d = \sqrt{\omega_0^2 - \alpha^2}$$

$$\omega_0 = \frac{1}{\sqrt{LC}}$$

The time-domain transient response of the circuit is classified into one of four types, depending on the relative magnitude of α and ω_0:

$\alpha > \omega_0$:   Overdamped

$\alpha = \omega_0$:   Critically damped

$\alpha < \omega_0$:   Underdamped

$\alpha = 0$:   Undamped

The response of the series *RLC* circuit to a sinusoidal input is characterized by the transfer function in Eq. (8.29):

$$\mathbf{H}(\omega) = \frac{1}{(1 - \omega^2 LC) + j\omega RC}$$

Using the definitions above, this can be written as

$$\mathbf{H}(\omega) = \frac{1}{1 - \left(\dfrac{\omega}{\omega_0}\right)^2 + j2\dfrac{\alpha}{\omega_0}\dfrac{\omega}{\omega_0}}$$

Note that the ratio α/ω_0, which determines the damping in the transient response, appears in the denominator of this expression. We will use the symbol ζ for this ratio, known as the **dimensionless damping ratio**:

$$\zeta = \frac{\alpha}{\omega_0}$$

For the series *RLC*, this ratio is

$$\zeta = \frac{R}{2}\sqrt{\frac{C}{L}}.$$

With this definition, we can summarize the nature of the transient response as

$\zeta > 1$:   Overdamped

$\zeta = 1$:   Critically damped

$\zeta < 1$:   Underdamped

$\zeta = 0$:   Undamped

The transfer function is thus

$$\mathbf{H}(\omega) = \frac{1}{1 - \left(\dfrac{\omega}{\omega_0}\right)^2 + j2\zeta\left(\dfrac{\omega}{\omega_0}\right)}$$

Separating this into magnitude and phase components,

$$|\mathbf{H}(\omega)| = \frac{1}{\sqrt{\left(1 - \left(\dfrac{\omega}{\omega_0}\right)^2\right)^2 + 4\zeta^2\left(\dfrac{\omega}{\omega_0}\right)^2}} \tag{8.33}$$

$$\Theta(\omega) = -\text{atan}\frac{2\zeta\omega\omega_0}{\omega_0^2 - \omega^2} \tag{8.34}$$

The magnitude (in dB) is plotted in Figure 8.23 for various values of ζ. For values of $\zeta > 1/\sqrt{2}$, the magnitude is superficially similar to the *RC* Bode plot in that it passes low frequencies with unity gain (0 dB) and attenuates high frequencies. The rolloff at high frequencies, though, is steeper than for the *RC* lowpass filter. As $\omega_0 \to \infty$, Eq. (8.33) is dominated by the ω^4 term in the denominator:

$$|H(\omega)| \approx \frac{1}{\sqrt{\left(-\left(\dfrac{\omega}{\omega_0}\right)^2\right)^2}}, \qquad \omega \gg \omega_0$$

$$= \left(\frac{\omega_0}{\omega}\right)^2$$

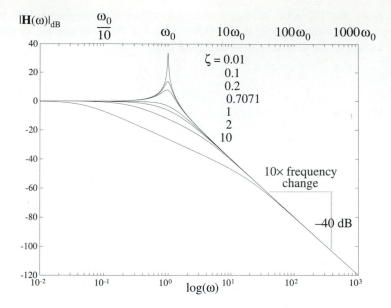

Figure 8.23 Transfer function for the series *RLC* circuit. Note the presence of a resonance peak, centered at approximately $\omega = \omega_0$ for $\zeta < 1/\sqrt{2}$.

Expressed in dB,

$$|\mathbf{H}(\omega)|_{dB} \approx 20 \log\left(\frac{\omega_0}{\omega}\right)^2$$

$$= 40 \log\left(\frac{\omega_0}{\omega}\right)$$

$$= -40(\log(\omega) - \log(\omega_0))$$

In this case, the high-frequency response decreases by 40 dB, a factor of 100, for every tenfold increase in signal frequency.

Much more interesting is the behavior of the transfer function near the corner frequency for $\zeta < 1/\sqrt{2}$. The magnitude of the transfer function has a peak, known as a **resonance peak**, which can be quite high for sufficiently small values of the damping ratio. The location of this peak is found by setting the derivative of the transfer-function magnitude to zero:

$$\frac{d}{d\omega}|\mathbf{H}(\omega)| = 0$$

$$\left(-\frac{1}{2}\right)\left(\frac{2\left(1 - \left(\frac{\omega}{\omega_0}\right)^2\right)\left(-\frac{2\omega}{\omega_0^2}\right) + \frac{8\zeta^2\omega}{\omega_0^2}}{\left(\left(1 - \left(\frac{\omega}{\omega_0}\right)^2\right)^2 + 4\zeta^2\left(\frac{\omega}{\omega_0}\right)^2\right)^{\frac{3}{2}}}\right) = 0$$

This constraint is satisfied when the numerator is equal to zero:

$$2\left(1 - \left(\frac{\omega}{\omega_0}\right)^2\right)\left(-\frac{2\omega}{\omega_0^2}\right) + \frac{8\zeta^2\omega}{\omega_0^2} = 0$$

$$\left(-\frac{2\omega}{\omega_0^2}\right)\left(1 - \left(\frac{\omega}{\omega_0}\right)^2 - 2\zeta^2\right) = 0$$

One root of this equation occurs at $\omega = 0$, where the transfer function approaches the low-frequency 0-dB asymptote. The other roots are at

$$\omega = \pm\omega_0\sqrt{1 - 2\zeta^2}$$

Real, nonnegative values of this root exist only for $\zeta < 1/\sqrt{2}$. This therefore defines the range of ζ where a resonant peak occurs; the magnitude reaches a maximum at frequency ω_m given by

$$\omega_m = \omega_0\sqrt{1 - 2\zeta^2} \tag{8.35}$$

The frequency of the peak magnitude approaches the undamped resonant frequency ω_0 as the damping in the circuit decreases (i.e., as ζ gets smaller).

The maximum magnitude at the resonant peak is found by inserting Eq. (8.35) into the magnitude expression, Eq. (8.33):

$$\left|\mathbf{H}(\omega_m)\right| = \frac{1}{\sqrt{\left(1 - \left(\frac{\omega_0\sqrt{1 - 2\zeta^2}}{\omega_0}\right)^2\right)^2 + 4\zeta^2\left(\frac{\omega_0\sqrt{1 - 2\zeta^2}}{\omega_0}\right)^2}}$$

$$\left|\mathbf{H}(\omega_m)\right| = \frac{1}{2\zeta\sqrt{1 - 2\zeta^2}} \tag{8.36}$$

Example 8.10

A series *RLC* circuit, with R = 200 Ω, L = 1 μH, and C = 10 pF, is driven by a variable-frequency sinusoid with an amplitude of 5 V. Determine the maximum voltage which appears across the capacitor, and the frequency at which this occurs.

The damping factor and undamped natural frequency are given by

$$\alpha = \frac{R}{2L} = 10 \times 10^6$$

$$\omega_0 = \frac{1}{\sqrt{LC}} = 100 \times 10^6$$

The dimensionless damping ratio is thus

$$\zeta = \frac{\alpha}{\omega_0} = 0.1$$

—well into the underdamped region, so a resonant peak exists. The peak will occur at the angular frequency

$$\omega_m = \omega_0\sqrt{1 - 2\zeta^2}$$

$$= 100 \times 10^6\sqrt{1 - 2(0.1)^2}$$

$$= 98.99 \times 10^6 \text{ rad/s}$$

which, when divided by 2π, corresponds to a hertzian frequency of 15.8 MHz. At this frequency, the transfer-function magnitude is

$$\left|\mathbf{H}(\omega_m)\right| = \frac{1}{2\zeta\sqrt{1 - 2\zeta^2}}$$

$$= \frac{1}{2(0.1)\sqrt{1 - 2(0.1)^2}}$$

$$= 5.05$$

Therefore, with a input signal magnitude of 5 V, the voltage across the capacitor reaches a peak of 25.3 V.

The fact that the voltage appearing across the capacitor is greater than the applied voltage is, at first, unnerving, as it seems to violate the "you can't get something for nothing" rule. This phenomenon is called **resonance** and is a common feature of underdamped circuits. Resonance arises from a coupling between the "natural" behavior of the circuit, which with no external input tends to oscillate at the damped natural frequency, and the input signal. When these two frequencies are close, the natural and impressed signals reinforce each other, causing the capacitor voltage amplitude to be larger than the applied signal. (In the next section we interpret resonance phenomena in terms of the energy stored in the capacitor and inductor, which provides additional insight into the process.)

Further clues to the behavior of the series *RLC* circuit are contained in the Bode phase plot, Figure 8.24. At very low frequencies the inductor "looks like" a short circuit, and the capacitor "looks like" an open circuit. We then expect the input voltage signal to appear unchanged across the capacitor. This is reflected in both the low-frequency magnitude (unity amplitude) and phase (zero radians) asymptotes.

The $-\pi/2$ phase shift at ω_0 is best understood by considering the current through the circuit. Referring to Figure 8.22, the current phasor is given by

$$\mathbf{I} = \frac{\mathbf{V_S}}{\mathbf{Z_{eq}}} = \frac{\mathbf{V_S}}{R + j\left(\omega L - \frac{1}{\omega C}\right)}$$

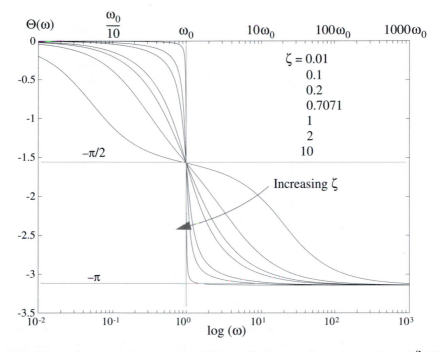

Figure 8.24 Phase of the second-order series *RLC* transfer function for various values of ζ. With less damping (lower ζ) the transition from $\Theta = 0$ to $\Theta = -\pi$ near $\omega = \omega_0$ becomes more abrupt.

$\mathbf{Z_{eq}}$ is the equivalent impedance of the resistor, inductor, and capacitor in series. The imaginary part of $\mathbf{Z_{eq}}$ is equal to zero when

$$\omega L = \frac{1}{\omega C}$$

$$\omega = \frac{1}{\sqrt{LC}} = \omega_0$$

In other words, the equivalent impedance of the series *RLC* circuit is purely resistive when the input signal frequency is the undamped natural frequency.[1]

The phase dependence of $\mathbf{Z_{eq}}$ above and below this frequency is depicted in Figure 8.25. Below ω_0, the capacitive reactance contributes more to the overall impedance than the inductive reactance. In this case, the phase angle of $\mathbf{Z_{eq}}$ is negative. Above ω_0, the net reactance is inductive, resulting in a positive phase. Exactly at ω_0, the inductive and capacitive reactances cancel, leaving only the resistor to contribute to $\mathbf{Z_{eq}}$.

Therefore, considering the voltage source to be the reference (i.e., the phase of $\mathbf{V_S}$ is zero) the phase of the equivalent impedance can be summarized as

---

[1]. Resonance is sometimes *defined* as the frequency where the equivalent impedance is purely real. Note, however, that the peak of the transfer-function magnitude occurs at ω_m, which is below ω_0. Often, though, the difference is slight, especially for little damping.

$$\mathbf{Z_{eq}} = \left|\mathbf{Z_{eq}}\right| \angle \phi = R + j\left(\omega L - \frac{1}{\omega C}\right)$$

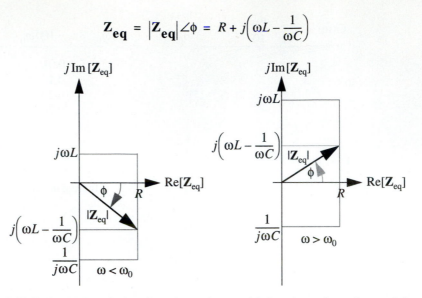

Figure 8.25 Series *RLC* equivalent impedance above and below the undamped natural frequency ω_0. *Left:* Below ω_0, the net reactance is negative (capacitive), resulting in a negative impedance phase angle. *Right:* Above ω_0, the net reactance is positive (inductive), so the impedance phase angle is positive. At $\omega = \omega_0$, the capacitive and inductive reactances cancel, leaving a purely resistive equivalent impedance.

$$\arg(\mathbf{Z_{eq}}) = \begin{cases} -\dfrac{\pi}{2}, & \omega \to 0 \\[2mm] 0, & \omega = \omega_0 \\[2mm] \dfrac{\pi}{2}, & \omega \to \infty \end{cases}$$

The current phasor is reciprocally related to the impedance, so its phase varies as

$$\arg(\mathbf{I}) = \begin{cases} \dfrac{\pi}{2}, & \omega \to 0 \\[2mm] 0, & \omega = \omega_0 \\[2mm] -\dfrac{\pi}{2}, & \omega \to \infty \end{cases}$$

Finally, the phase of the transfer function is the same as the capacitor voltage, which is related to the current phasor through

$$\mathbf{V_C} = \mathbf{V_S}\mathbf{H}(\omega) = \frac{1}{j\omega C}\mathbf{I} = \left(\frac{1}{\omega C}\angle -\frac{\pi}{2}\right)\mathbf{I}$$

$$\arg(\mathbf{V_C}) = \arg(\mathbf{H}(\omega)) = \arg(\mathbf{I}) - \frac{\pi}{2}$$

which means the transfer-function phase is

$$\arg(\mathbf{H}(\omega)) = \begin{cases} 0, & \omega \to 0 \\ -\dfrac{\pi}{2}, & \omega = \omega_0 \\ -\pi, & \omega \to \infty \end{cases}$$

in agreement with Figure 8.24.

At $\omega = \omega_0$, the circuit current and the source voltage are exactly in phase. This means that the product $v_S(t)i(t)$, representing the power delivered to the circuit by the voltage source, is positive throughout the entire cycle. The voltage source is continually supplying energy to the circuit, which can cause the peak amplitude to be greater than the source amplitude. At other frequencies, the product $v_S(t)i(t)$ is negative over some part of the cycle, meaning that some of the stored energy in the circuit is being returned to the source, so the amplitude of the oscillations is less.

8.6.2 Mechanical Analogy

A more familiar example of an underdamped system is a weight suspended on a spring. It can be shown that the displacement of the weight is described by a second-order differential equation of the same form as the *RLC* circuit. Consequently, an underdamped system with a spongy spring will have a resonant transfer function. In this case, the "input" is the up-and-down driving motion of the hand holding the spring, and the "output" is the motion of the weight.

You can demonstrate the three regions of the transfer functions using a weak spring or rubber band attached to a weight, as shown in Figure 8.26. First, if you extend the weight and release it, it will oscillate up and down at its damped natural frequency. Allow the oscillations to die out. Next, move the hand holding the spring up and down very slowly. The spring will hardly change its length at all, and as a result the weight moves up and down in sync with your hand. Here, the magnitude of the transfer function is unity, which is what we expect (from the transfer function) at low frequencies.

If you vibrate your hand very rapidly, faster than the speed of the natural oscillations, the weight will hardly move at all. The input oscillations have been attenuated, as predicted by the high-frequency behavior of the transfer function.

Finally, if you drive the system near the natural oscillation frequency, the weight will oscillate up and down considerably more than your hand. This is the resonant condition. Essentially, when the system is in resonance, the displacement caused by the hand motion adds to the natural, unforced oscillations. Because these two displacements are in phase, they add to a value greater than the input displacement.

In a digital electronic system, we do not want the signal lines to resonate at any harmonic of the clock frequency. Such resonances will greatly distort the signal and have the potential to cause false triggering of logic gates.

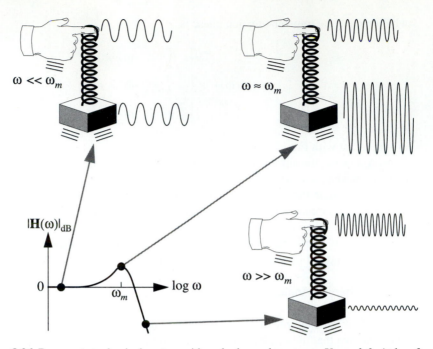

Figure 8.26 Resonant mechanical system with underdamped response. *Upper left:* At low frequencies, the weight moves up and down with approximately the same amplitude as the driving stimulus. *Upper right:* Near the resonant frequency, the weight moves up and down considerably more than the driving stimulus. *Lower right:* At high frequencies, the movement of the weight is attenuated.

8.6.3 Quality Factor and Bandwidth

In an *RLC* circuit, energy is stored in the magnetic field of the inductor, and in the electric field of the capacitor. We saw in Chapter 1 that the stored energies are given by

$$E_L = \frac{1}{2}Li_L^2$$

$$E_C = \frac{1}{2}Cv_C^2$$

Consider a series *RLC* circuit with a constant dc voltage source as the excitation and an infinitesimally small resistance. This is an approximation of the undamped *LC* circuit we solved in Chapter 5. We know that the capacitor voltage in this circuit oscillates at the undamped natural frequency:

$$v_C(t) = V\cos\omega_0 t$$

The inductor current is out of phase by $\pi/2$ with respect to the capacitor voltage:

$$i_L(t) = \omega_0 CV\cos\left(\omega_0 t + \frac{\pi}{2}\right)$$

Making the substitutions

$$\omega_0 = \frac{1}{\sqrt{LC}} \text{ and } \cos\left(\theta + \frac{\pi}{2}\right) = -\sin\theta,$$

the inductor current is

$$i_L(t) = -V\sqrt{\frac{C}{L}}\sin\omega_0 t$$

The total energy stored in the circuit is given by

$$E = E_L + E_C = \frac{1}{2}Li_L^2 + \frac{1}{2}Cv_C^2$$

Substituting the previous expressions for current and voltage,

$$\begin{aligned} E &= \frac{1}{2}L\left(-V\sqrt{\frac{C}{L}}\sin\omega_0 t\right)^2 + \frac{1}{2}C(V\cos\omega_0 t)^2 \\ &= \frac{1}{2}CV^2(\sin\omega_0 t)^2 + \frac{1}{2}CV(\cos\omega_0 t)^2 \\ &= \frac{1}{2}CV^2((\sin\omega_0 t)^2 + (\cos\omega_0 t)^2) \\ &= \frac{1}{2}CV^2 \end{aligned}$$

The total energy is constant because there is no resistance in the circuit to dissipate it. The energy is simply shuttled back and forth between the inductor and capacitor.

Now, suppose we introduce a finite resistance into the series circuit. The oscillations will gradually decay away as the energy in the circuit is dissipated as heat in the resistor. During one period, over which t advances by $2\pi/\omega$, the energy dissipated in the resistor is

$$\begin{aligned} E_R &= \int_0^{2\pi/\omega} Ri^2\,dt \\ &= R\int_0^{2\pi/\omega} \left(-V\sqrt{\frac{C}{L}}\sin\omega t\right)^2 dt \\ &= \frac{R}{L}CV^2 \int_0^{2\pi/\omega} \left(\frac{1}{2} - \frac{1}{2}\cos 2\omega t\right)dt \\ &= \frac{R\pi}{L\omega}CV^2 \end{aligned}$$

The amount of energy dissipated is directly proportional to the resistance.

Now, suppose the voltage source is sinusoidal. There will be energy transfer between the source and the circuit. When the source frequency matches the natural frequency of the LC circuit, the signals add in phase such that the stored energy increases on each cycle. The resistance limits this process by dissipating some of this energy. The height of the resonance peak is therefore a strong function of how much energy is dissipated on each cycle.

We quantify the energy dissipation with the parameter Q, called the **quality factor**. Q is defined as

$$Q = 2\pi\frac{\text{maximum stored energy}}{\text{energy dissipated per period}} \tag{8.37}$$

Q is generally evaluated at the undamped natural frequency. The name "quality factor" originates from the idea that a "higher-quality" circuit will lose less energy through resistive dissipation per cycle than a lower-quality circuit. Less energy loss per cycle results in a smaller value in the denominator of Eq. (8.37), which in turn results in a larger value of the quality factor. Despite this name, circuits with low Q-factors are quite useful and important and should not be construed as being shoddy or inferior.

For the series RLC circuit we have

$$Q = 2\pi\frac{E_L + E_C}{E_R}$$

$$= 2\pi\frac{\frac{1}{2}CV^2}{\frac{R}{L}\frac{\pi}{\omega_0}CV^2}$$

$$= \frac{\omega_0 L}{R}$$

The damping factor for this circuit is given by $\alpha = \dfrac{R}{2L}$, so we can write Q as

$$Q = \frac{\omega_0}{2\alpha} \tag{8.38}$$

Recall that we defined the dimensionless damping ratio as $\zeta = \alpha/\omega_0$ which means that a larger value of ζ corresponds to more damping. This will result in more energy dissipation in the resistor, lowering the quality factor. The relation between Q and ζ is

$$Q = \frac{1}{2\zeta} \tag{8.39}$$

If the damping coefficient is small, say $\zeta < 0.1$ (i.e., $Q > 5$), the peak in the transfer-function magnitude is approximately equal to Q:

$$|\mathbf{H}(\omega_m)| = \frac{1}{2\zeta\sqrt{1-2\zeta^2}} = \frac{Q}{\sqrt{1-2\zeta^2}} \approx Q$$

This provides a convenient method of determining Q in a sufficiently underdamped resonant system; it is simply the maximum value of the transfer-function amplitude.

The resonant series *RLC* circuit passes signals within a narrow band of frequencies near ω_m with a greater amplitude than other signals. A circuit with this characteristic is called a **bandpass filter** and is used extensively in communication systems. We characterize the width of the passband by the 3-dB bandwidth, which is the range of frequencies $\Delta\omega$ where the output magnitude is at least $1/\sqrt{2}$ times the peak value, as shown in Figure 8.27.

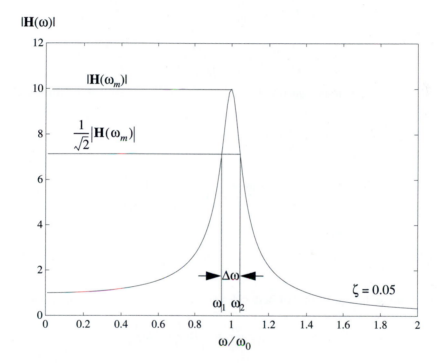

Figure 8.27 Transfer-function magnitude (plotted on linear axes) for an *RLC* bandpass filter with $\zeta = 0.05$ ($Q = 10$). $\Delta\omega$ is the 3-dB bandwidth.

The bandwidth $\Delta\omega$ is the difference between the upper and lower half-power frequencies, which are defined as

$$\left|\mathbf{H}(\omega_2)\right| \;=\; \left|\mathbf{H}(\omega_1)\right| \;=\; \frac{1}{\sqrt{2}}\left|\mathbf{H}(\omega_m)\right|$$

We will calculate the bandwidth for high-Q systems, where $Q \gg 1$, which means we can approximate the peak frequency ω_m by ω_0 [Eq. (8.35)]. At the half-power frequencies,

$$|\mathbf{H}(\omega)| = \frac{1}{\sqrt{2}}Q$$

$$\frac{1}{\sqrt{\left(1 - \left(\frac{\omega}{\omega_0}\right)^2\right)^2 + \frac{1}{Q^2}\left(\frac{\omega}{\omega_0}\right)^2}} = \frac{1}{\sqrt{2}}Q$$

Squaring both sides and rearranging,

$$Q^2 - 2Q^2\left(\frac{\omega}{\omega_0}\right)^2 + Q^2\left(\left(\frac{\omega}{\omega_0}\right)^2\right)^2 + \left(\frac{\omega}{\omega_0}\right)^2 = 2$$

$$\left(\left(\frac{\omega}{\omega_0}\right)^2\right)^2 + \frac{1 - 2Q^2}{Q^2}\left(\frac{\omega}{\omega_0}\right)^2 + \frac{Q^2 - 2}{Q^2} = 0$$

This is a quadratic equation in $(\omega/\omega_0)^2$. Applying the quadratic formula:

$$\left(\frac{\omega}{\omega_0}\right)^2 = -\left(\frac{1 - 2Q^2}{2Q^2}\right) \pm \sqrt{\left(\frac{1 - 2Q^2}{2Q^2}\right)^2 - \frac{Q^2 - 2}{Q^2}}$$

$$\left(\frac{\omega}{\omega_0}\right)^2 = 1 - \frac{1}{2Q^2} \pm \sqrt{\frac{1 + 4Q^2}{4Q^4}}$$

For $Q \gg 1$, this simplifies to

$$\left(\frac{\omega}{\omega_0}\right)^2 \approx 1 \pm \frac{1}{Q}$$

Taking the square root of both sides and applying the binomial approximation,

$$\frac{\omega}{\omega_0} = \sqrt{1 \pm \frac{1}{Q}} \approx 1 \pm \frac{1}{2Q}$$

The two half-power frequencies are

$$\omega_2 \approx \omega_0\left(1 + \frac{1}{2Q}\right)$$

$$\omega_1 \approx \omega_0\left(1 - \frac{1}{2Q}\right)$$

which results in a 3-dB bandwidth of

$$\Delta\omega = \omega_2 - \omega_1 \approx \frac{\omega_0}{Q}$$

The bandwidth is inversely proportional to the quality factor. It follows that the less damping in a circuit, the higher the quality factor, and the narrower the bandwidth. Q is therefore a measure of the sharpness of the bandpass filter; in the context of filter design, Q is also called the **selectivity**, since a higher Q corresponds to a sharper (i.e., more selective) passband.

8.6.4 Clock-Signal Degradation

We have seen that an *RLC* circuit can either attenuate or magnify the amplitude of a sinusoidal signal, depending on the quality factor and the source frequency. A square-wave clock signal in a digital system has components at many different frequencies. The filtering effect of an interconnect modeled by an *RLC* circuit can profoundly distort the shape of the clock signal, as illustrated in the next example.

Example 8.11

A series *RLC* circuit, with $R = 200\ \Omega$, and $C = 1$ pF, is driven by a 100-MHz square-wave with an amplitude of 5 V. Determine the voltage across the capacitor for inductance values of $L = 100\ \mu H$ and $280\ \mu H$.

The Bode magnitude plots of the transfer functions are illustrated in Figure 8.28. (Note that the frequency scale is in hertz, rather than rad/s.) The relevant parameters for the two cases are

L	$f_0 = \dfrac{1}{2\pi\sqrt{LC}}$	ζ	Q	$f_m = \dfrac{\omega_m}{2\pi}$	$\|\mathbf{H}(f_m)\|$
100 nH	503 MHz	0.316	1.58	450 MHz	1.77 (4.96 dB)
280 nH	301 MHz	0.189	2.65	290 MHz	2.75 (8.78 dB)

Both transfer functions have resonant peaks near one of the principal harmonics of the square-wave source signal. Figure 8.28 shows a series of vertical bars which represent the spectrum of the square wave, consisting of the odd harmonics of the fundamental. The horizontal position of the bars corresponds to the frequency, and the height is proportional to the Fourier coefficient. Recall that the Fourier series for the square wave is

$$v_S(t) = \frac{V}{2} + \sum_{\substack{n=1 \\ n \text{ odd}}}^{\infty} \frac{2V}{\pi n} \cos\left(n\omega t - \frac{\pi}{2}\right)$$

which has the phasor representation

$$\mathbf{V}_S = \mathbf{V}_{S0} + \mathbf{V}_{S1} + \mathbf{V}_{S3} + \cdots$$

$$\mathbf{V}_{S0} = \frac{V}{2}\angle 0$$

$$\mathbf{V}_{Sn} = \frac{2V}{\pi n}\angle -\frac{\pi}{2}$$

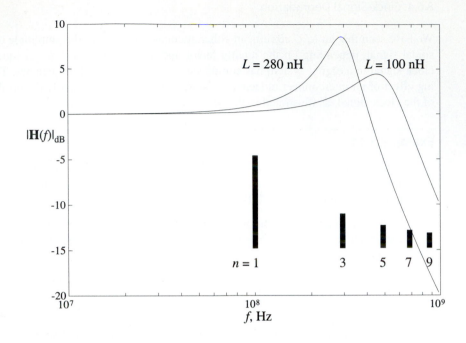

Figure 8.28 Series *RLC* transfer functions for Example 8.11. The heavy bars represent the weights of the harmonics making up the 100-MHz square-wave source signal.

Each of the n harmonics is inversely proportional to n; hence the length of the bars in Figure 8.28 decrease at higher frequencies.

Note that the two transfer functions each "select" one of the higher-order harmonics. With $L = 100$ nH, the resonance peak occurs at 450 MHz, close to the fifth harmonic (500 MHz) of the square wave. The amplitude of the transfer function at 500 MHz is

$$|\mathbf{H}(f)| \;=\; \frac{1}{\sqrt{\left(1-\left(\dfrac{f}{f_0}\right)^2\right)^2 + 4\zeta^2\left(\dfrac{f}{f_0}\right)^2}}$$

$$\left|\mathbf{H}(500\times10^6)\right| \;=\; \frac{1}{\sqrt{\left(1-\left(\dfrac{500}{503}\right)^2\right)^2 + 4(0.316)^2\left(\dfrac{500}{503}\right)^2}} \;=\; 1.57$$

This is higher than unity, which means the fifth harmonic appears in the output $v_C(t)$ more strongly than in the square-wave excitation. Similarly, with $L = 280$ nH, the transfer function has a resonance near the third harmonic of the square wave, 300 MHz. We therefore expect to see a strong component at this frequency in the output signal.

The capacitor voltages for the two cases are shown in Figure 8.29. Both responses are underdamped, as we expect from the subunity value of the damping ratio. Also, we see the appearance of either a strong third- or fifth-harmonic component, depending on the inductance value. This can be verified by calculating the period of these two harmonics:

$$T_3 = \frac{1}{3 \times 100 \times 10^6} = 3.33 \text{ ns}$$

$$T_5 = \frac{1}{5 \times 100 \times 10^6} = 2.00 \text{ ns}$$

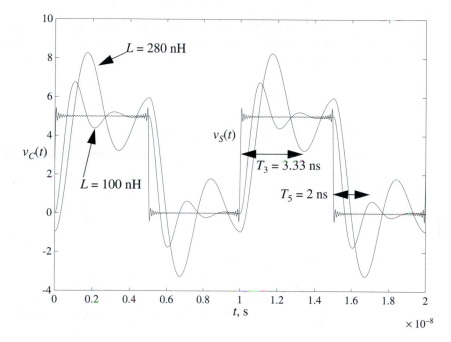

Figure 8.29 Square-wave source signal $v_S(t)$ and capacitor voltage $v_C(t)$ in Example 8.11 for two values of inductance. Each output signal has a pronounced harmonic corresponding to the peak in the transfer function.

The double-headed arrows in Figure 8.29 are scaled to the length of these two periods. We see that these are the periods of the damped sinusoid "ringing" in the underdamped responses.

8.7 ac Steady-State Analysis Using Complex Fourier Series

We have used the trigonometric form of the Fourier series to express our periodic signals. An alternate formulation is the complex exponential form, which is more compact and requires less effort to find the coefficients. In this section we illustrate the relationship between the complex exponential form of the Fourier series and the phasor representation of the signal.

Recall that a sinusoidal signal,

$$v(t) = V\cos(\omega t + \varphi)$$

has the phasor representation

$$\mathbf{V} = V\angle\varphi$$

where, by convention, we suppress the $e^{j\omega t}$ factor which describes the angular rotation of the phasor. We recover the time-domain signal from the phasor by reintroducing the $e^{j\omega t}$ and taking the real part:

$$v(t) = Re(\mathbf{V}e^{j\omega t})$$
$$= Re(Ve^{j\varphi}e^{j\omega t})$$
$$= V\cos(\omega t + \varphi)$$

The real part of a complex number can be found by taking the average of the number and its complex conjugate:

$$Re(\mathbf{V}e^{j\omega t}) = \frac{1}{2}\left(\mathbf{V}e^{j\omega t} + \overline{\mathbf{V}e^{j\omega t}} + \cdots + (\mathbf{V}e^{j\omega t})^*\right)$$

$$V\cos(\omega t + \varphi) = \underbrace{\frac{V}{2}e^{j\varphi}e^{j\omega t}}_{\text{CCW}} + \underbrace{\frac{V}{2}e^{-j\varphi}e^{-j\omega t}}_{\text{CW}} \tag{8.40}$$

where the notations "CCW" and "CW" stand for counterclockwise and clockwise (to be explained shortly).

When written in this form, the signal has a compelling graphical interpretation, illustrated in Figure 8.30. Instead of representing the cosine signal as the real part of a single, counterclockwise rotating phasor, (i.e., as shown in Figure 8.3), we can think of it as the vector sum of *two* complex conjugate phasors, shown as heavy arrows in the figure. The phasor $V/2 \angle\varphi$, shorthand for $(V/2)e^{j\varphi}e^{j\omega t}$, rotates counterclockwise (CCW) about the origin with angular frequency ω, while the CW phasor $V/2 \angle -\varphi$ rotates in the opposite direction. The vector sum of the two phasors is depicted by placing the tail of the CW phasor to the head of the CCW phasor (shown as the shaded, displaced version of the CW phasor in Figure 8.30), which lies along the real axis. Thus, despite the appearance of the imaginary unit j in Eq. (8.40), both sides this expression are purely real quantities.

The reason this interpretation of a signal as the sum of two complex-conjugate phasors is useful is that the complex exponential Fourier series gives us this representation automatically. To see this, recall that a signal $v(t)$ with period $T = 2\pi/\omega$ can be written as

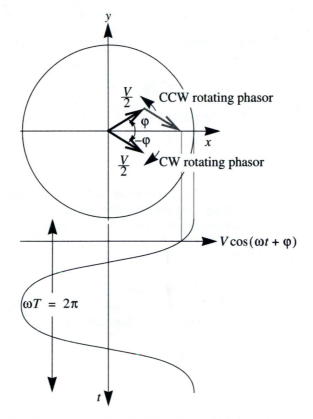

Figure 8.30 Relationship between a sinusoidal function and the two rotating phasors $V/2 \angle \varphi$ and $V/2 \angle{-}\varphi$.

$$v(t) = \sum_{n=-\infty}^{\infty} f_n e^{jn\omega t}$$

where

$$f_n = \frac{1}{T}\int_0^T f(t)e^{-jn\omega t}\,dt$$

The $\pm n$th terms in the Fourier series are

$$f_n e^{jn\omega t} \qquad \text{and} \qquad f_{-n} e^{-jn\omega t} \tag{8.41}$$

which combine to form the term $A_n \cos(n\omega t + \theta_n)$ in the trigonometric Fourier series.

Comparing the expressions in Eqs. (8.40) and (8.41), we see that the complex Fourier coefficients f_n and f_{-n} are the CCW and CW phasors of the nth harmonic of the signal. Put another way, a periodic signal with Fourier expansion

$$v(t) = A_0 + A_1\cos(\omega t + \theta_1) + \cdots + A_n\cos(n\omega t + \theta_n) + \cdots$$

can be expressed in the frequency domain as the sequence of phasors

$$\overbrace{A_0}^{\text{DC}} + \overbrace{A_1\angle\theta_1}^{\cos\omega t} + \cdots + \overbrace{A_n\angle\theta_n}^{\cos n\omega t} + \cdots$$

or, equivalently, as the sequence of phasors

$$\overbrace{f_0}^{\text{DC}} + \overbrace{f_1 + f_{-1}}^{\cos\omega t} + \cdots + \overbrace{f_n + f_{-n}}^{\cos n\omega t} + \cdots$$

where the complex numbers f_1, f_{-1}, and so on are the complex Fourier coefficients.

Performing ac steady-state analysis with complex Fourier series leads to an economy of notation and a minimum of steps, as this next example demonstrates.

Example 8.12

Determine the crosstalk induced into a quiet gate interconnect by an adjacent 100-MHz 5-V clock line, Figure 8.31. Assume the inverters have output resistance and input capacitance of 1 kΩ and 0.5 pF, the clock resistance is 500 Ω, and the coupling capacitance is 1 pF.

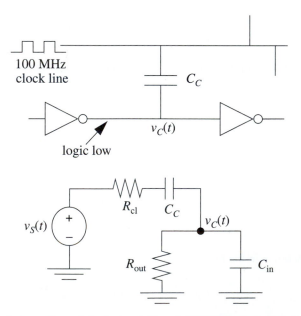

Figure 8.31 Unwanted coupling of clock signal. *Top:* A 100-MHz clock line signal is capacitively coupled into an interconnect between two logic gates. *Bottom:* Equivalent-circuit model.

The clock signal $v_S(t)$ is given by this complex Fourier series, derived in the last chapter:

$$v_S(t) = 2.5 + \sum_{\substack{n=-\infty \\ n \text{ odd}}}^{\infty} \frac{5}{jn\pi} e^{jn\omega t}$$

where $\omega = 2\pi 10^8$ rad/s is the fundamental radian frequency of the square wave. The transfer function, as a function of the harmonic frequency $n\omega$, is found by applying the voltage-divider rule:

$$\mathbf{H}(n\omega) = \frac{\mathbf{V_C}}{\mathbf{V_S}} = \frac{\left(\dfrac{R_{out}\dfrac{1}{jn\omega C_{in}}}{R_{out} + \dfrac{1}{jn\omega C_{in}}} \right)}{R_{cl} + \dfrac{1}{jn\omega C_C} + \left(\dfrac{R_{out}\dfrac{1}{jn\omega C_{in}}}{R_{out} + \dfrac{1}{jn\omega C_{in}}} \right)}$$

In this expression, the term in parentheses is the equivalent impedance of the parallel combination of R_{out} and C_{in}. After simplification, the transfer function can be written as

$$\mathbf{H}(n\omega) = \frac{jn\omega R_{out} C_{in}}{(1 - n^2\omega^2 R_{cl} R_{out} C_{in} C_C) + jn\omega(R_{cl}C_C + R_{out}C_C + R_{out}C_{in})}$$

Inserting the given component values,

$$\mathbf{H}(n\omega) = \frac{j0.314n}{(1 - 0.987n^2) + j1.26n}$$

The output function $v_C(t)$ is found by multiplying each of the phasors at frequency $\pm n\omega$ by the transfer function. Because the complex Fourier coefficients are identical to the phasors, we simply multiply the transfer function by each of the Fourier terms:

$$v_C(t) = 2.5 \times \mathbf{H}(0) + \sum_{\substack{n=-\infty \\ n \text{ odd}}}^{\infty} \left(\frac{5}{jn\pi} \times \mathbf{H}(n\omega) \right) e^{jn\omega t}$$

$$v_C(t) = \sum_{\substack{n=-\infty \\ n \text{ odd}}}^{\infty} \left(\frac{5}{jn\pi} \times \frac{j0.314n}{(1 - 0.987n^2) + j1.26n} \right) e^{jn\omega t}$$

Simplifying,

$$v_C(t) = \sum_{\substack{n=-\infty \\ n \text{ odd}}}^{\infty} \frac{0.5}{(1 - 0.987 n^2) + j1.26n} e^{jn\omega t}$$

This expression is in a convenient final form, as the terms can be readily summed using a computer. Note that this is a purely real expression; all of the complex terms appear as complex conjugates, so that in the summation all imaginary components cancel. Note the extent to which complex phasors and the complex Fourier series simplify the computation of the periodic steady-state response of a circuit. Similar computations using the trigonometric form are generally more tedious.

A graph of the crosstalk is shown in Figure 8.32. The coupling onto the interconnect, which is presumed to be quiet, is significant. Preventing this crosstalk onto signal lines is a major goal in designing clock distribution routes in ICs.

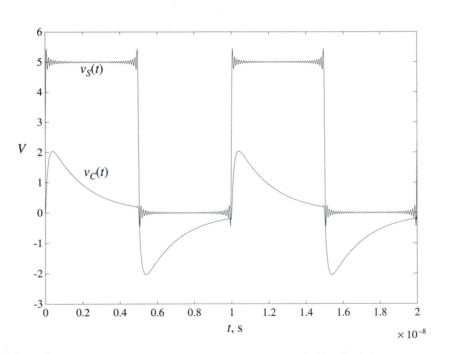

Figure 8.32 Crosstalk coupled onto a quiet interconnect from an adjacent clock line.

8.8 Summary

Periodic ac steady-state analysis, using phasors and frequency-domain representation of signals, simplifies the analysis of many circuits, such as those used to distribute clock signals within a digital system. A **phasor** is a way of representing a sinusoidal signal in the frequency domain:

$$V\cos(\omega t + \varphi) \Leftrightarrow V \angle \varphi$$

We interpret the phasor as a rotating vector in the complex plane:

$$\mathbf{V} = V \angle \varphi = V e^{j\varphi} e^{j\omega t}$$

Then the time-domain function is recovered by taking the real part of the complex phasor $\mathbf{V}$:

$$V\cos(\omega t + \varphi) = \mathrm{Re}(\mathbf{V})$$

The ratio of a voltage phasor to a current phasor associated with a circuit element is called the **impedance**:

$$\mathbf{Z} = \frac{\mathbf{V}}{\mathbf{I}}$$

which has the standard unit of ohms (Ω). The impedance of resistors, capacitors, and inductors are

$$\mathbf{Z_R} = R$$

$$\mathbf{Z_C} = \frac{1}{j\omega C}$$

$$\mathbf{Z_L} = j\omega L$$

Circuits are analyzed in the frequency domain by expressing sources and signals with phasors, components with their complex impedance, and applying Kirchhoff's laws. Impedances in series add, while reciprocal impedances, called **admittances**, add when in parallel.

A **transfer function** is the ratio of an output to an input phasor. For example, a circuit with a sinusoidal input signal $v_i(t) = A\cos(\omega t + \phi)$ (phasor $\mathbf{V_i}$) and output $v_o(t)$ (phasor $\mathbf{V_o}$) are related by the voltage transfer function

$$\mathbf{H}(\omega) = \frac{\mathbf{V}_o}{\mathbf{V}_i}$$

If the transfer function is expressed in magnitude-phase form:

$$\mathbf{H}(\omega) = |\mathbf{H}(\omega)| \angle \Theta(\omega)$$

then the output signal, in response to the input $v_i(t) = A\cos(\omega t + \phi)$, is

$$v_o(t) = A|\mathbf{H}(\omega)| \cos(\omega t + \phi + \Theta(\omega))$$

The magnitude of a transfer function is often expressed in **decibels** (dB), defined as

$$|\mathbf{H}(\omega)|_{dB} = 20 \log |\mathbf{H}(\omega)|$$

Plots of transfer-function magnitude and phase, as a function of log ω, are called **Bode plots**. Points on the Bode plot where straight-line asymptotes cross are called **corner frequencies**.

The series RLC circuit transfer function can be written as

$$\mathbf{H}(\omega) = \frac{1}{1 - \left(\dfrac{\omega}{\omega_0}\right)^2 + j2\zeta\left(\dfrac{\omega}{\omega_0}\right)}$$

where the **undamped natural frequency** ω_0 is

$$\omega_0 = \frac{1}{\sqrt{LC}}$$

ζ, the **dimensionless damping ratio**, is a measure of the damping in the circuit; it is defined as

$$\zeta = \frac{\alpha}{\omega_0}$$

where the **damping factor** α is

$$\alpha = \frac{R}{2L}$$

For sufficiently light damping, where $\zeta < 1/\sqrt{2}$, the circuit exhibits **resonance**: the transfer function has a peak, called a **resonance peak**, which exceeds unity. The frequency of this peak, ω_m, is given by

$$\omega_m = \omega_0\sqrt{1 - 2\zeta^2}$$

which approaches the undamped resonant frequency ω_0 as the damping in the circuit decreases.

The **quality factor** Q of a circuit containing energy-storage elements is defined as

$$Q = 2\pi\frac{\text{maximum stored energy}}{\text{energy dissipated per period}}$$

evaluated at the undamped natural frequency. Q is a measure of the losses in the resistive elements of a circuit. In a series RLC, Q is given by

$$Q = \frac{\omega_0}{2\alpha}$$

Q is also related to the circuit's **bandwidth**, the range of frequencies over which the input signal is passed with relatively little attenuation. The **half-power bandwidth** is defined as the frequency range $\Delta\omega$ where the transfer function is within 3 dB of its maximum; the series RLC circuit has a half-power bandwidth of

$$\Delta\omega = \frac{\omega_0}{Q}$$

Sinusoidal signals can also be represented as the sum of two counterrotating complex phasors. Viewed in this way, the component phasors are given by the coefficients of the complex Fourier series.

Problems

8.1 Determine the magnitude and phase of the following complex expressions:

$$3 + j4 \qquad\qquad 2.1 \times 10^6 + j1.2 \times 10^5$$

$$\frac{2.1 \times 10^6 + j1.2 \times 10^5}{3 \times 10^6 + j6 \times 10^6} \qquad\qquad \frac{1}{3 \times 10^6 + j6 \times 10^6}$$

$$\sin \omega t - j\cos \omega t \qquad\qquad \cosh x + j\sinh x$$

8.2 Determine the real and imaginary components of the following complex expressions:

$$5.1 \times 10^{-6} \angle 1.2 \qquad (5.6 \times 10^{-6} \angle 1.3) - j(4.0 \times 10^{-6} \angle 2.1)$$

$$\frac{2.1 \times 10^6 + j1.2 \times 10^5}{3 \times 10^6 + j6 \times 10^6} \qquad\qquad \frac{1.6 \angle 0.3}{3 \times 10^6 + j6 \times 10^6}$$

$$\frac{\sin \omega t - j\cos \omega t}{\cos \omega t + j\sin \omega t} \qquad\qquad \frac{r \angle \theta}{x + jy}$$

8.3 Express the signals below in phasor form.

$$v(t) = 3\cos\left(2\pi 10^6 t + \frac{\pi}{4}\right)$$

$$i(t) = 4\cos\left(2\pi 10^6 t + \frac{\pi}{6}\right) + 2\sin\left(2\pi 10^6 t - \frac{\pi}{4}\right)$$

$$v(t) = V\cos(\omega t + \varphi) - \frac{V}{2}\sin(\omega t - \varphi)$$

8.4 Given the two signals below:

$$v(t) = 3\cos\left(2\pi 10^6 t + \frac{\pi}{4}\right)$$

$$i(t) = 3\cos\left(2\pi 10^6 t + \frac{\pi}{4}\right) + 2\sin\left(2\pi 10^6 t - \frac{\pi}{6}\right)$$

Determine the amount by which the voltage lags the current.

8.5   Determine the equivalent voltage $v(t)$ in the circuit below.

$3\cos(2\pi10^6t-1.5)$ $\quad = \quad v(t)$

$5\cos(2\pi10^6t+0.3)$

8.6   Determine the equivalent voltage $v(t)$ in the circuit below.

$3\cos(2\pi10^6t-1.5)$

$5\cos(2\pi10^6t+0.3)$ $\quad = \quad v(t)$

$7\cos(2\pi10^6t-0.3)$

8.7   Consider two voltage sources in series:

$3\cos(2\pi f_1t-1.3)$

$\quad = \quad v(t)$

$5\cos(2\pi f_2t+0.5)$

The equivalent voltage of these two sources is given by $v(t)$. Determine the maximum value of $v(t)$ for (a) $f_1 = f_2$, and (b) $f_1 \neq f_2$.

8.8   Determine the current $i(t)$ in the circuit below. Show that the same result is obtained by using superposition in the time-domain and phasor methods. Which method is easier?

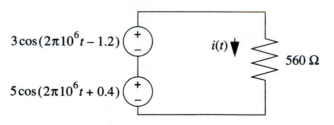

$3\cos(2\pi10^6t-1.2)$ $i(t)$ $560\ \Omega$

$5\cos(2\pi10^6t+0.4)$

8.9 Determine the current $i(t)$ in the circuit below.

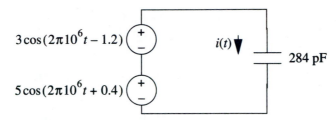

8.10 Determine the current $i(t)$ in the circuit below.

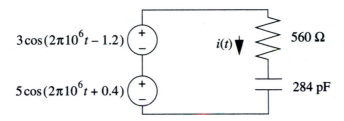

8.11 In Exercise 8.10, suppose the frequency of the 5-V source were increased to 2 MHz. Do you expect the magnitude of the current to increase or decrease?

8.12 Determine the current $i(t)$ in the circuit below.

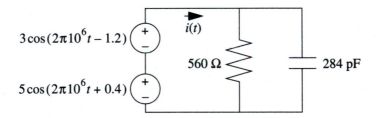

8.13 Determine the current $i(t)$ in the circuit below.

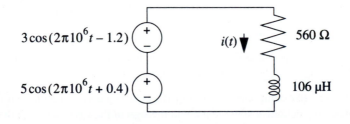

8.14 Find the equivalent impedance between the two terminals in the circuit below, at a frequency of 2 MHz.

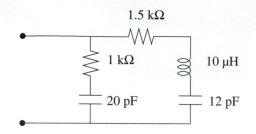

8.15 In the circuit of Exercise 8.14, determine the frequency at which the admittance is purely real.

8.16 In the circuit below, determine the frequency range over which the reactive component of the equivalent impedance is negative.

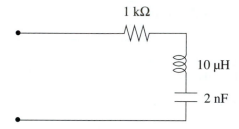

8.17 Determine the equivalent admittance between the two terminals in the circuit below as a function of frequency.

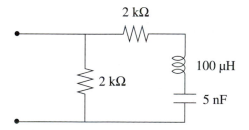

8.18 Given a complex impedance $R + jX$, determine the conductance G and susceptance B in terms of R and X, using the defining relation $\mathbf{Y} = 1/\mathbf{Z}$.

8.19 Suppose the reactive component of an impedance is negative at a particular frequency. Is it possible to make any definitive statement about the sign of the susceptance at that same frequency?

8.20 Show that a signal in the form of

$$v_C(t) = \frac{V}{2} + \sum_{\substack{n=1 \\ n \text{ odd}}}^{\infty} \left[\frac{2V}{n\pi} \frac{1}{1 + n^2 \omega^2 \tau^2} (\sin n\omega t - n\omega\tau \cos n\omega t) \right]$$

has the equivalent representation

$$v_C(t) = \frac{V}{2} + \sum_{\substack{n=1 \\ n \ \text{odd}}}^{\infty} \left[\frac{2V}{n\pi} \frac{1}{\sqrt{1 + n^2\omega^2\tau^2}} \cos(n\omega t + \theta_n) \right]$$

$$\theta_n = -\frac{\pi}{2} - \text{atan}\, n\omega\tau$$

Hint: It is helpful to determine the relation between atan(x) and atan($1/x$).

8.21 The circuit below illustrates a 5-V clock distribution circuit driving two flip-flops A and B. The long interconnect is five times the length of the short interconnect. The inverters and flip-flops have input capacitances and output resistances of 0.8 pF and 200 Ω; the short interconnect has a total lumped resistance of 150 Ω and lumped capacitance of 1.1 pF. The flip-flops trigger on a rising edge when the input reaches $V_{th} = 4.0$ V. Determine the clock skew using first- and second-order *RC* interconnect models for a clock frequency of 35 MHz.

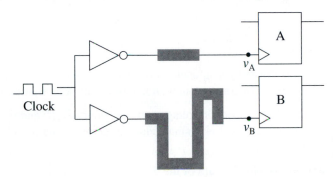

8.22 In Exercise 8.21, determine the approximate maximum clock frequency above which the circuit will not operate properly.

8.23 The circuit below illustrates a 5-V clock distribution circuit driving two flip-flops A and B. The long interconnect is seven times the length of the short interconnect. The inverters and flip-flops have input capacitances and output resistances of 0.8 pF and 200 Ω; the short interconnect has a total lumped resistance of 50 Ω and lumped capacitance of 1.1 pF. The series inductance associated with the power and ground package wires is 0.8 μH. The flip-flops trigger on a rising edge when the input reaches $V_{th} = 4.0$ V. Determine the clock skew using an *RLC* interconnect model for clock frequencies of 35 and 60 MHz.

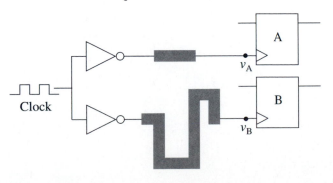

8.24 In Exercise 8.23, determine the consequence of neglecting to include the effect of the series inductance in calculating the clock skew at 35 MHz.

8.25 Consider a flip-flop driven by a square-wave clock through an interconnect. Determine the voltage-transfer function of the circuit, using a two-RC-lump model for the interconnect, and including the power and ground wire package inductance.

8.26 In Example 8.10 we considered the following series RLC circuit, with a resonant peak at $\omega_m = 98.99$ Mrad/s:

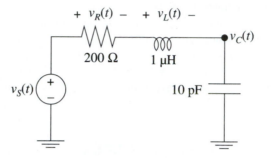

When this circuit is driven by a sinusoid at or near the resonant frequency, the amplitude of the voltage across the capacitor is more than five times the source amplitude.

(a) Solve for and plot the capacitor voltage and the voltages $v_R(t)$ and $v_L(t)$ across the resistor and inductor. Assume that $v_S(t)$ is a sinusoidal source operating at 100 Mrad/s with a peak amplitude of 5 V.

(b) On the same graph, plot the sum of the three component voltages $v_R(t)$, $v_L(t)$, and $v_C(t)$.

(c) Repeat these steps assuming the resistance is decreased to 100 Ω. Does it bother you that some of the voltages in this circuit are considerably larger than the source voltage?

8.27 Determine the transfer function $\mathbf{H}(\omega) = \mathbf{V}_o/\mathbf{V}_S$ of the circuit below.

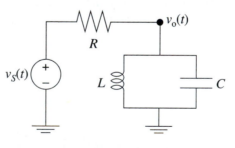

8.28 Determine the transfer function $\mathbf{H}(\omega) = \mathbf{V}_0/\mathbf{V}_S$ of the circuit below.

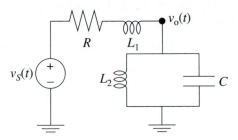

8.29 Determine the transfer function $\mathbf{H}(\omega) = \mathbf{V}_0/\mathbf{V}_S$ of the circuit below.

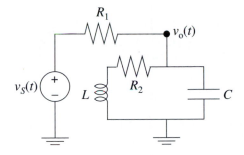

8.30 Consider several subsystems cascaded together, where the output of one feeds the input of the next:

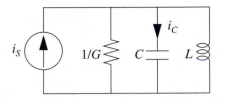

In this setup, the individual transfer functions $\mathbf{H}_1$, $\mathbf{H}_2$, etc. are defined as the ratio of the output to input phasors of each individual subsystem. Derive an expression for the overall transfer function $\mathbf{H}(\omega) = \mathbf{V}_0/\mathbf{V}_i$.

8.31 In Exercise 8.30, show that the magnitude $|\mathbf{H}(\omega)|$ of the overall transfer function, expressed in dB, is given by $|\mathbf{H}_1(\omega)|_{dB} + |\mathbf{H}_2(\omega)|_{dB} + \cdots$.

8.32 The circuit below is a parallel *GLC* circuit. Using the methods developed for the series *RLC* circuit, determine the current transfer function $\mathbf{I}_C/\mathbf{I}_S$; the frequency ω_m at which the transfer function is a maximum; the half-power bandwidth; and the quality factor.

8.33 Plot the transfer-function Bode diagrams for: (a) a parallel *GLC* circuit with $1/G = 1\ k\Omega$, $C = 1\ nF$, and $L = 1\ mH$ (with $\mathbf{H}(\omega) = \mathbf{I_C}/\mathbf{I_S}$); and (b) a series *RLC* circuit ($\mathbf{H}(\omega) = \mathbf{V_C}/\mathbf{V_S}$) with the same parameters. How do they compare?

8.34 Design a series *RLC* bandpass filter with the following parameters: $\Delta f \leq 0.5\ kHz, f_m = 2\ MHz$.

8.35 A certain communication-system circuit requires a lowpass filter with a half-power frequency of 1 MHz. It is possible to use an *RC* filter or an underdamped *RLC* filter to accomplish this. What, if anything, is gained by using the more complicated *RLC* filter?

8.36 Determine the transfer function in the bandstop filter circuit below.

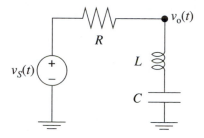

8.37 Plot the Bode diagrams for the circuit in Exercise 8.36, assuming $R = 1\ k\Omega, f_0 = 1\ MHz$, and $Q = 15$.

8.38 For the circuit below, determine the voltage transfer function $\mathbf{H}(\omega) = \mathbf{V_C}/\mathbf{V_S}$, the resonant frequency (if any), half-power bandwidth, and quality factor, and draw the magnitude and phase Bode plots.

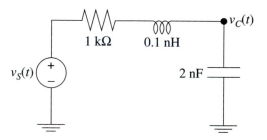

8.39 For the circuit below, determine the voltage transfer function $\mathbf{H}(\omega) = \mathbf{V_C}/\mathbf{V_S}$, the resonant frequency (if any), half-power bandwidth, and quality factor, and draw the magnitude and phase Bode plots.

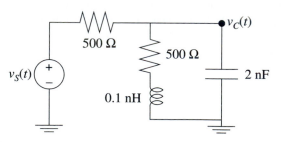

8.40 For the circuit below, determine the voltage transfer function $\mathbf{H}(\omega) = \mathbf{V}_L/\mathbf{V}_S$, the resonant frequency (if any), half-power bandwidth, and quality factor, and draw the magnitude and phase Bode plots.

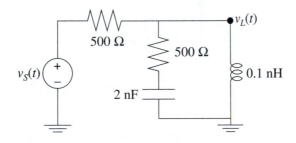

8.41 For the circuit below, determine the voltage transfer function $\mathbf{H}(\omega) = \mathbf{V}_C/\mathbf{V}_S$, Does this circuit have a resonant peak?

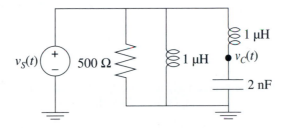

8.42 Derive the voltage transfer functions $\mathbf{V}_L/\mathbf{V}_S$ and $\mathbf{V}_C/\mathbf{V}_S$ for the circuit below.

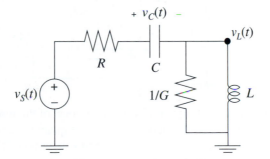

8.43 In the circuit below, the source $v_S(t)$ is a 100-MHz square wave operating between 0 V and 5 V. Determine and plot the voltage across the capacitor using the complex Fourier series representation of the square wave.

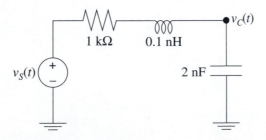

8.44 Repeat Exercise 8.43, assuming that the source $v_S(t)$ is the sawtooth wave below.

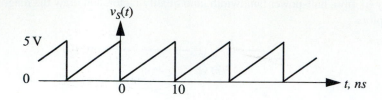

8.45 In the circuit below, the source $v_S(t)$ is a 400-MHz square wave operating between 0 V and 5 V. Determine and plot the voltage across the capacitor.

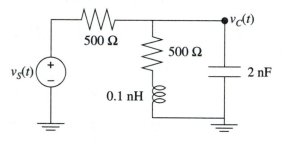

8.46 In the circuit below, the source $v_S(t)$ is a 400-MHz square wave operating between 0 V and 5 V. Determine and plot the voltage across the inductor.

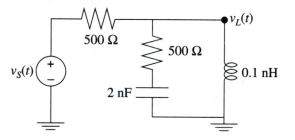

8.47 An elementary definition of "frequency" is the reciprocal of the fundamental period of a periodic function. This implies that frequency is a positive quantity; it is hard to imagine what the meaning of a negative frequency would be using this definition. However, we have seen that the complex Fourier representation of a signal requires both positive and negative frequencies. Can you think of a more sophisticated description of "frequency" which encompasses both positive and negative values?

Other Circuit-Analysis Concepts

9

9.1 Introduction

We have developed several circuit-analysis techniques which work well in approximating the switching speed of digital systems. These techniques have wide applicability over a range of problems. However, there are many other circuit ideas and techniques which we have not yet introduced.

In this chapter we discuss several of these important ideas. Because we wish to emphasize their generality, instead of showing how these methods can be applied to digital systems, we will use illustrations from across the spectrum of electrical and computer engineering applications.

9.2 Thevenin and Norton Equivalent Models

A complicated circuit containing many components is more difficult to analyze than a simpler circuit containing only a few components. We always try to simplify a circuit using simple concepts so as to generate fewer equations which must be solved. For example, we can combine several resistances in series into a single equivalent resistance; we ignore resistances in series with a current source or in parallel with a voltage source; we can often disregard the vanishingly small shunt conductance of a MOSFET gate. All of these actions help make a complicated circuit easier to solve.

It is possible to represent any connection of sources and components, as seen by two of the nodes, with just a single independent source and a single equivalent resistance (or impedance, in the case of ac steady state). Furthermore, the determination of the values of the source and equivalent resistance is straightforward. This result can be very helpful in simplifying circuits.

Consider the arbitrary circuit shown in Figure 9.1. The shaded region contains some connection of linear circuit elements. Two nodes of this circuit are accessible; we can connect a load device to these terminals and make current and voltage measurements. Suppose we bridge these two nodes with a resistor of value zero (i.e., a short circuit) and measure the resulting current. We call this current I_{sc}, the short-circuit current. With an infinite resistance connected (i.e., an open circuit), the voltage appearing across the terminals is V_{oc}, the open-circuit voltage. A load resistance between these two extremes results in a current I and voltage drop V across the two accessible nodes.

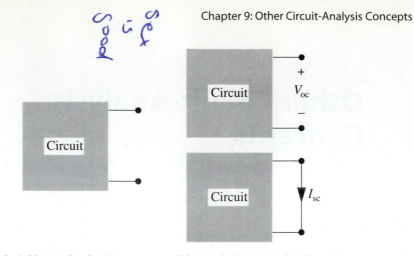

Figure 9.1 *Left:* Arbitrary circuit with two accessible terminals. *Top right:* The voltage across the two open-circuited terminals is called V_{oc}, the open-circuit voltage. *Bottom right:* The current through the two terminals when shorted is called I_{sc}, the short-circuit current.

We can describe the terminal characteristics of the circuit through either of the two equivalents shown on the right in Figure 9.2. The upper circuit, consisting of a series-connected voltage source V_{oc} and resistance R_T, is called a **Thevenin equivalent**. The lower circuit, a parallel combination of current source I_{sc} and resistance R_N, is known as a **Norton equivalent**. The equivalent circuit parameters are related through the Thevenin (Norton) resistance

$$R_T = R_N = \frac{V_{oc}}{I_{sc}} \tag{9.1}$$

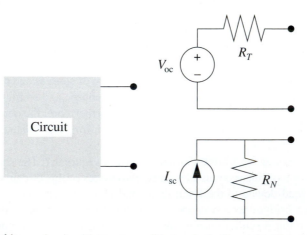

Figure 9.2 *Left:* Arbitrary circuit with two accessible terminals. *Top right:* Thevenin equivalent circuit. *Bottom right:* Norton equivalent circuit.

To use the equivalent models, we need determine only two of the three parameters. The open-circuit voltage and short-circuit current can be found by direct calculation; their ratio then determines the equivalent resistance. Alternatively, the Thevenin/Norton resistance can be found directly by determining the resistance across the two terminals when all independent sources in the circuit have been replaced by their dead (zero-valued) equivalents (short circuits for voltage sources, open circuits for current sources). We illustrate the technique with an example.

Example 9.1

Determine the Thevenin and Norton equivalents for the shaded part of the circuit in Figure 9.3(a).

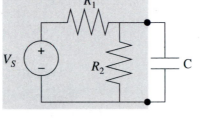

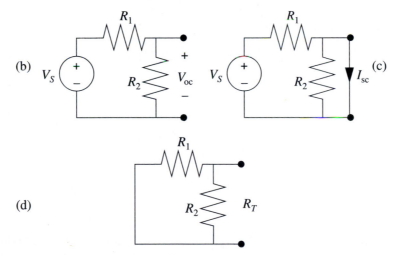

Figure 9.3 Steps in determining the equivalent-circuit parameters. (a) Equivalent resistance is found by replacing all independent sources with their dead equivalents. (b) Open-circuit voltage and (c) short-circuit current determination. (d) Finding the equivalent resistance by replacing independent sources with their dead equivalents.

Removing the capacitor (the **load** device) from the circuit results in the circuit Figure 9.3(b) with the two terminals open-circuited. The open-circuit voltage is found by applying the voltage divider rule:

$$V_{oc} = V_S \frac{R_2}{R_1 + R_2} \tag{9.2}$$

The equivalent resistance can be found by replacing the independent voltage source by its dead equivalent, a short circuit, as shown in Figure 9.3(d). The resistance seen at the two external terminals is the parallel combination

$$R_T = \frac{R_1 R_2}{R_1 + R_2} \tag{9.3}$$

The short-circuit current is found by determining the current flowing between the terminals when they are shorted, Figure 9.3(c):

$$I_{sc} = \frac{V_S}{R_1} \tag{9.4}$$

Alternatively, we can apply Eq. (9.1) by taking the ratio of Eqs. (9.2) and (9.3):

$$I_{sc} = \frac{V_{oc}}{R_T} = \frac{V_S \dfrac{R_2}{R_1 + R_2}}{\dfrac{R_1 R_2}{R_1 + R_2}} = \frac{V_S}{R_1}$$

which gives the same result.

The two equivalent circuits are shown in Figure 9.4.

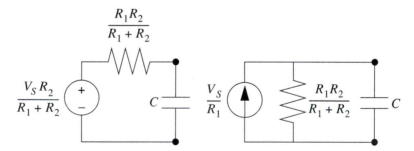

Figure 9.4 Thevenin equivalent (left) and Norton equivalent (right) representations of the circuit of Figure 9.3(a).

Note that if we were interested in the transient response of a circuit like that of Figure 9.3(a) (with a switch somewhere), we could use the equivalent circuit to easily calculate the time constant. From Figure 9.4, it is easy to see that the *RC* time constant is

$$\tau = \frac{R_1 R_2 C}{R_1 + R_2}$$

but this is not at all obvious from the original circuit.

Example 9.2

Determine the time constant governing the transient response of the circuit in Figure 9.5.

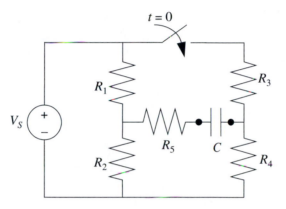

Figure 9.5 Circuit for Example 9.2. The time constant governing the transient response is easily found by determining the Thevenin equivalent resistance.

The two terminals of interest are the indicated nodes at either end of the capacitor. Replacing the independent voltage source with a short circuit, the circuit becomes (for $t > 0$)

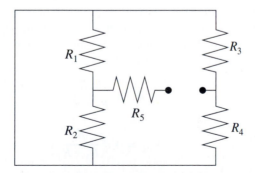

From this diagram we see that R_1 and R_2 are in parallel, as are R_3 and R_4. Hence, we can write the Thevenin resistance as

$$R_T = (R_1 \| R_2) + (R_3 \| R_4) + R_5$$

whereupon it follows that the time constant is

$$\tau = \left(\frac{R_1 R_2}{R_1 + R_2} + \frac{R_3 R_4}{R_3 + R_4} + R_5\right)C$$

Consider the circuit in Figure 9.6, where a load resistor of value R_L has been connected to the two terminals of interest. If R_L is either an open circuit ($\infty\,\Omega$) or a short circuit ($0\,\Omega$), no power will be dissipated in the load. For intermediate values, though, a finite amount of power will be dissipated. What value of R_L will result in maximum power transfer from the rest of the circuit into the load?

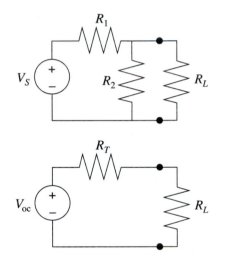

Figure 9.6 *Top:* Circuit for demonstrating maximum power transfer. *Bottom:* Thevenin equivalent circuit.

This problem, and the corresponding problem for more complicated circuits, is easy to handle by "Theveninizing" the rest of the circuit (i.e., the part connected to the two load terminals, absent the load), as shown in the bottom of Figure 9.6. The Thevenin resistance and open-circuit voltage are given by

$$R_T = \frac{\cdot R_1 R_2}{R_1 + R_2}$$

$$V_{oc} = V_S\left(\frac{R_2}{R_1 + R_2}\right)$$

The power dissipated in the load resistor is equal to R_L times the square of the current:

$$P_L = R_L\left(\frac{V_s}{R_T + R_L}\right)^2$$

To find the maximum power, we set $dP_L/dR_L = 0$:

$$\frac{d}{dR_L}\left[R_L\left(\frac{V_s}{R_T + R_L}\right)^2\right] = 0$$

$$(R_T + R_L)^2 - R_L 2(R_T + R_L) = 0$$

which is satisfied when

$$R_L = R_T$$

Maximum power transfer in a dc circuit occurs when the load resistance is equal to the equivalent resistance "seen" looking into the rest of the circuit. (The more general case of an ac steady-state circuit with complex impedances is considered in Exercise 9.12.)

9.3 Controlled Sources

In Chapter 1 we discussed ideal independent voltage and current sources. An ideal voltage source maintains the specified voltage across its terminals regardless of the current flowing through the source. "Independent" means that the voltage is either constant or a function of time in some specified manner. Similarly, the current through an ideal current source is not affected in any way by the voltage across the source. The current through an independent current source does not depend on voltages or currents elsewhere in the circuit.

Convenient abstractions useful for modeling active devices are **controlled sources** (also called **dependent** sources). The value of voltage or current associated with a controlled source is a function of a circuit parameter. The functional dependence must be specified explicitly in the circuit diagram. Symbols for controlled sources are shown in Figure 9.7.

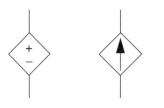

Figure 9.7 Symbols for controlled-voltage and controlled-current sources. The value associated with the dependent source is a function of a circuit parameter, usually a voltage or current.

Controlled sources are frequently encountered in models of semiconductor devices. For example, one model of a **bipolar junction transistor** useful over a limited range of operation is shown in Figure 9.8. The collector current, i_c, is linearly proportional to the base current i_b; the constant of proportionality is called the transistor beta.

For the most part, solving circuits with controlled sources is just like solving any other circuit. An exception is that controlled sources are not treated as "dead" when applying superposition or when finding the equivalent Thevenin/Norton resistance. Only independent sources are replaced by their dead equivalents when performing these operations.

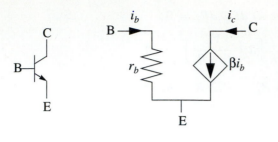

Figure 9.8 Schematic symbol (left) and one equivalent model (right) for a bipolar junction transistor. The collector current i_c is proportional to the base current i_b.

Example 9.3

Determine the Norton equivalent of the circuit in Figure 9.9.

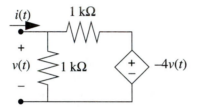

Figure 9.9 Circuit with controlled voltage for Example 9.3.

The equivalent resistance is found by replacing all independent sources in the circuit with their dead equivalents and determining the ratio v/i at the two terminals. Since there are no independent sources in this circuit, no replacements are made. The current $i(t)$ is found using KCL:

$$i(t) = \frac{v(t)}{1000} + \frac{v(t)-(-4v(t))}{1000} = \frac{6v(t)}{1000}$$

The Norton resistance is thus

$$R_N = \frac{v(t)}{i(t)} = \frac{1000}{6} = 167\ \Omega$$

The short-circuit current is the current flowing through the two terminals when they are bridged with a zero-value resistor. This causes the voltage $v(t)$ to be zero, which makes the value of the controlled-voltage source also zero. By inspection, the short-circuit current I_{sc} is therefore zero. The circuit in Figure 9.9 is equivalent to a single 167-Ω resistor.

9.4 Root-Mean-Square Values

A time-independent (dc) signal can be completely characterized by a single quantity, its amplitude. Time-varying signals, like the periodic signals in Figure 9.10, defy simple categorization. In this figure are shown sinusoidal, square, sawtooth, and triangle waveforms with identical frequencies and amplitude (2 V from peak to peak). Is it possible to define a quantity which somehow captures the "strength" or "effectiveness" of each waveform? For example, what should an ac voltmeter read when it is connected to each of these waveforms?

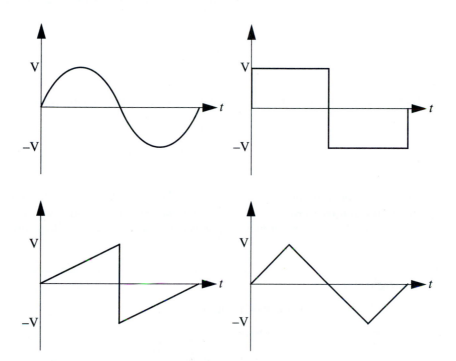

Figure 9.10 Representative ac voltage signals: sinusoid, square, sawtooth, and triangle waves. All have the same frequency, peak-to-peak amplitude (2V), and average value (zero).

The most straightforward ideas are to use either the peak-to-peak amplitude or the average value of the signal, but both of these have drawbacks which severely limit their usefulness. The peak-to-peak value does not distinguish at all among any of the waveforms shown, or others with the same peak-to-peak value. This measure depends only on the values of the signal at two instants of time and does not take into account what happens between those instants. Similarly, the average value is not useful, because symmetric signals like those shown in Figure 9.10 all have average values of zero, regardless of their amplitude.

Another approach is to define an effective voltage which relates how much power is being delivered to a resistive load. For example a dc source of value V_{DC} delivers power

$$P = \frac{V_{DC}^{2}}{R}$$

to load resistance R. A periodic signal $v(t)$, with period T, has an effective voltage V_{EFF} when this same power is dissipated in R:

$$\frac{V_{\text{EFF}}^2}{R} = \frac{V_{\text{DC}}^2}{R}$$

The power in the resistor is found by taking the average of the instantaneous power $v(t)i(t)$ over one period:

$$P = \frac{V_{\text{EFF}}^2}{R} = \frac{1}{T}\int_0^T v(t)i(t)\,dt$$

By Ohm's law, $i(t) = v(t)/R$, so the effective voltage, called the **rms voltage**, is defined as

$$V_{\text{rms}} = \sqrt{\frac{1}{T}\int_0^T v^2(t)\,dt} \tag{9.5}$$

Rms stands for "root-mean-square"; the rms value of a signal is found by taking the square root of the average (mean) squared value. Essentially, an ac voltage source with rms value V dissipates the same power in a resistor as a dc source of value V. Rms current is defined similarly.

Example 9.4

Calculate the rms value of a sinusoidal signal $v(t) = V\cos(\omega t)$.

Applying Eq. (9.5), we have

$$V_{\text{rms}} = \sqrt{\frac{1}{T}\int_0^T (V\cos\omega t)^2\,dt}$$

$$= \sqrt{V^2\frac{\omega}{2\pi}\int_0^{2\pi/\omega}\left(\frac{1}{2} + \frac{1}{2}\cos 2\omega t\right)dt}$$

Since the average value of a sinusoid over an integer number of periods is zero, the second term in the integral does not contribute. The rms value is thus

$$V_{\text{rms}} = \frac{1}{\sqrt{2}}V \qquad \text{(sinusoid)} \tag{9.6}$$

which is approximately 0.7071 times the amplitude. It must be emphasized that Eq. (9.6) is not a general definition of rms value; it applies only to sinusoidal signals.

The standard way of describing an ac voltage (or current) is understood to be its rms value, unless otherwise specified. For example, residential electricity in the United

States has a nominal value of 117 V at 60 Hz. The peak amplitude is thus $117 \times \sqrt{2} \approx 165$ V. An explicit expression for the voltage as a function of time is

$$v(t) = 117\sqrt{2}\cos 2\pi 60 t$$

A periodic signal $v(t)$ can be expressed as a Fourier series:

$$v(t) = A_0 + \sum_{n=1}^{\infty} A_n \cos(n\omega t + \theta_n)$$

The rms value is found from

$$V_{rms} = \sqrt{\frac{1}{T}\int_0^T v^2(t)\,dt}$$

$$= \sqrt{\frac{1}{T}\int_0^T \left[A_0 + \sum_{n=1}^{\infty} A_n \cos(n\omega t + \theta_n)\right]^2 dt}$$

The term in square brackets contains two kinds of terms. First, there are individual cosine terms multiplied by themselves, such as

$$[A_n \cos(n\omega t + \theta_n)]^2$$

which integrate to $A_n^2/2$. The other terms are cross products of sinusoids at two different frequencies, such as

$$2[A_m \cos(m\omega t + \theta_m)][A_k \cos(k\omega t + \theta_k)]$$

In deriving the Fourier-series representations, we saw that such terms integrate to zero. Thus, the rms value of the signal is given by

$$V_{rms} = \sqrt{A_0^2 + \sum_{n=1}^{\infty} \frac{A_n^2}{2}} = \sqrt{A_0^2 + \sum_{n=1}^{\infty}\left(\frac{A_n}{\sqrt{2}}\right)^2} \qquad (9.7)$$

Note that $A_n/\sqrt{2}$ is the rms value of the nth harmonic component.

9.5 Nodal Analysis

In this section we introduce a general analysis technique called **nodal analysis**, which can be applied to all circuits. This method involves systematically applying KCL at all of the nodes of a circuit to generate a set of differential equations (for time-domain analysis) or algebraic equations (for frequency-domain analysis).

9.5.1 Circuits with Grounded Voltage Sources

We begin our exploration of nodal analysis by considering a circuit where all of the voltage sources are grounded. The example circuit of Figure 9.11 is sufficiently general to illustrate many nodal-analysis principles. In this circuit, the rectangular boxes represent circuit elements and are labeled with their characteristic admittance Y, rather than resistance or impedance. Each element can be a resistance ($Y = 1/R$), capacitance ($Y = j\omega C$) or inductance ($Y = 1/j\omega L$), or some complicated combination of elements with equivalent admittance $Y = G + jB$.

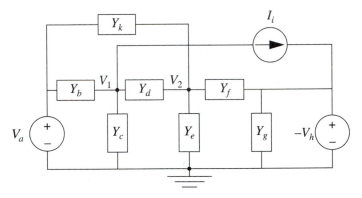

Figure 9.11 Example circuit for illustration of nodal analysis. The rectangular boxes represent admittances (**Y = I/V**).

The first step is to label the nodes. This circuit has five nodes. One of these is the ground node, which by definition is the voltage reference, so its voltage is zero. The extreme left and right nodes need not be labeled, as they are connected to grounded voltage sources; these nodes have known voltage values of V_a and $-V_h$. We label the remaining two nodes V_1 and V_2; we will call these the "unknown" node voltages, since their values have yet to be determined.

The next step is to use KCL to write equations describing the currents associated with each node. For consistency, we will associate a positive sign with currents leaving the node, and sum these positive currents to zero. Applying KCL at V_1:

$$Y_b(V_1 - V_a) + Y_c V_1 + Y_d(V_1 - V_2) + I_i = 0 \qquad (9.8)$$

and at node V_2:

$$Y_d(V_2 - V_1) + Y_e V_2 + Y_f(V_2 - (-V_h)) + Y_k(V_2 - V_a) = 0 \qquad (9.9)$$

Application of KCL at these nodes yields two circuit equations which can be solved directly for the two unknown voltages V_1 and V_2. Note that we did not write KCL equations for the nodes connected to the grounded voltage sources. These equations are unnecessary, since we already know the values of the voltages at these nodes, and the currents flowing through the sources automatically compensate for the other currents associated with such nodes. (It is, of course, possible to write KCL equations at such nodes, such as the far-left and far right nodes in Figure 9.11. However, this would entail introducing additional variables, complicating our system of equations, and these equations only serve to specify the currents through the voltage sources. It is simpler at this stage to consider only the unknown nodes.)

Note also that the admittance Y_g does not appear in either of the KCL equations Eq. (9.8) and Eq. (9.9). It does not appear because it is connected between two nodes with known voltages. It is trivial to determine the current through Y_g since it is in parallel with the voltage source $-V_h$:

$$I_g = -Y_g V_h$$

where I_g is the current through Y_g.

Now that we have applied KCL to the nodes with unknown voltages, we need to solve the resulting equations. First, we will rearrange Eqs. (9.8) and (9.9) so that all terms containing the unknown node voltages are on the left side, and all other terms are on the right. Collecting terms, we have

$$(Y_b + Y_c + Y_d)V_1 + (-Y_d)V_2 = Y_b V_a - I_i$$

$$(-Y_d)V_1 + (Y_d + Y_e + Y_f + Y_k)V_2 = Y_k V_a - Y_f V_h$$

These equations can be written in matrix form as

$$\begin{bmatrix} Y_b + Y_c + Y_d & -Y_d \\ -Y_d & Y_d + Y_e + Y_f + Y_k \end{bmatrix} \begin{bmatrix} V_1 \\ V_2 \end{bmatrix} = \begin{bmatrix} Y_b V_a - I_i \\ Y_k V_a - Y_f V_h \end{bmatrix} \tag{9.10}$$

Although we have considered a specific circuit in deriving Eq. (9.10), the node voltages in any linear circuit can be described by a set of linear equations in this form. In general, we can write

$$\mathbf{YV = I}$$

where $\mathbf{Y}$ is a matrix, called the **nodal admittance matrix**, $\mathbf{V}$ is a column vector of unknown node voltages, and $\mathbf{I}$ is a column vector of known branch currents. Once this system of equations has been generated from the circuit, they can be solved using traditional linear algebraic techniques. For small systems, with only a few rows, simple techniques like multiplying both sides by the inverse $\mathbf{Y}^{-1}$ may suffice, but for large circuits, which may have hundreds or thousands of rows, sophisticated numerical methods are usually needed. In either case, the objective is to determine the unknown node voltages, $\mathbf{V}$, from which any voltage or current in the circuit can be calculated. For example, the current I_a through the voltage source V_a in Figure 9.11 is given by

$$I_a = Y_b(V_1 - V_a) + Y_k(V_2 - V_a)$$

Similar expressions can be written for any current or voltage in this circuit.

As another example of nodal analysis, we will consider the three-lump ladder circuit shown in Figure 9.12. As before, the objective is to write a matrix equation $\mathbf{YV = I}$ which can be solved for the unknown node voltages $\mathbf{V}$. This circuit has five nodes; two are known (ground and V_S) and three are unknown (V_1, V_2, V_3). Applying KCL at the nodes with unknown voltages results in these three equations:

$$Y_1(V_1 - V_S) + Y_4 V_1 + Y_2(V_1 - V_2) = 0$$

$$Y_2(V_2 - V_1) + Y_5 V_2 + Y_3(V_2 - V_3) = 0$$

$$Y_3(V_3 - V_2) + Y_6 V_3 = 0$$

Expressed in matrix form:

$$\begin{bmatrix} Y_1 + Y_2 + Y_4 & -Y_2 & 0 \\ -Y_2 & Y_2 + Y_3 + Y_5 & -Y_3 \\ 0 & -Y_3 & Y_3 + Y_6 \end{bmatrix} \begin{bmatrix} V_1 \\ V_2 \\ V_3 \end{bmatrix} = \begin{bmatrix} Y_1 V_S \\ 0 \\ 0 \end{bmatrix} \quad (9.11)$$

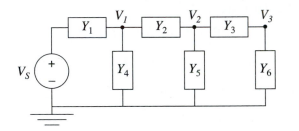

Figure 9.12 Example three-lump ladder circuit for illustration of nodal analysis. The rectangular boxes represent admittances (**Y = I/V**).

For a particular set of values, this system of equations can be solved by any convenient means. Algebraically, we can write an explicit solution for the unknown voltages by multiplying both sides by the inverse of the nodal admittance matrix:

$$\begin{bmatrix} V_1 \\ V_2 \\ V_3 \end{bmatrix} = \begin{bmatrix} Y_1 + Y_2 + Y_4 & -Y_2 & 0 \\ -Y_2 & Y_2 + Y_3 + Y_5 & -Y_3 \\ 0 & -Y_3 & Y_3 + Y_6 \end{bmatrix}^{-1} \begin{bmatrix} Y_1 V_S \\ 0 \\ 0 \end{bmatrix}$$

Comparing the matrix equations Eqs. (9.10) and (9.11) to their respective circuits, we observe some regular features:

- Each diagonal element Y_{mm} in the admittance matrix **Y** is the sum of all the admittances connected to node V_m.

- Each off-diagonal element Y_{mn} in the admittance matrix **Y** is the negative of the admittance connected between nodes V_m and V_n.

- Each element I_m in the current vector **I** corresponds to the sum of known currents flowing into node V_m.

This regularity can be understood by considering the currents flowing into a node, Figure 9.13. Shown in this figure is part of a circuit around a node with unknown voltage V_m. This node is connected to unknown nodes V_n and V_k through admittances Y_{mn} and Y_{mk}, and to a node with known voltage V' through admittance Y'. Also flowing into the node is known current I'. Applying KCL at this node,

$$Y_{mk}(V_m - V_k) + Y_{mn}(V_m - V_n) + Y'(V_m - V') - I' = 0$$

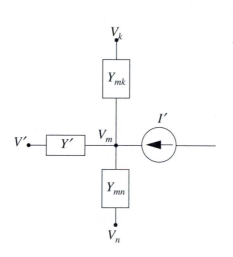

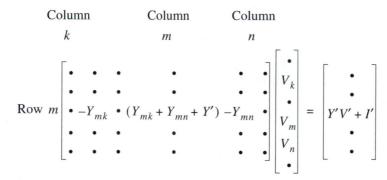

Figure 9.13 Application of KCL at node V_m. V' and I' are known values; V_m, V_k, and V_n are unknowns. Each of the currents contributes to the mth row in the nodal admittance formulation in a predictable way.

This KCL equation results in the following entries in the mth row of the nodal admittance matrix equation:

Each of the entries corresponds to one of the rules enumerated above. Instead of formally applying KCL, we can use these rules directly to generate the nodal admittance formulation. When applied to large circuits, use of these rules can generate an algebraic description more efficiently than straightforward application of KCL.

Consider the *RLGC* circuit in Figure 9.14. Transforming the circuit to the frequency domain, and specifying the circuit elements as admittances, results in the circuit shown in the bottom of this figure. The two unknown voltage phasors are V_1 and V_2. Using the rules above, we can write the nodal admittance equations as

$$\begin{bmatrix} \dfrac{1}{R} + \dfrac{1}{j\omega L} & -\dfrac{1}{j\omega L} \\[2ex] -\dfrac{1}{j\omega L} & \dfrac{1}{j\omega L} + j\omega C + G \end{bmatrix} \begin{bmatrix} V_1 \\[1ex] V_2 \end{bmatrix} = \begin{bmatrix} \dfrac{V_S}{R} \\[1ex] 0 \end{bmatrix} \qquad (9.12)$$

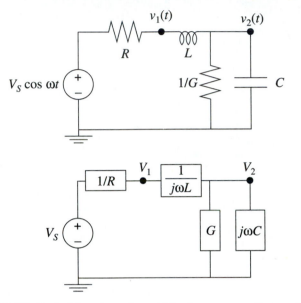

Figure 9.14 *Top: RLGC* circuit driven by a sinusoidal voltage source. *Bottom:* Frequency-domain representation, where the boxes represent the element admittances.

Suppose we wish to use this circuit as a filter. The information of interest is the transfer function $\mathbf{H}(\omega) = V_2/V_S$. This can be found from Eq. (9.12) by first multiplying both sides by the inverse of the nodal admittance matrix, which is given by

$$\begin{bmatrix} \dfrac{1}{R} + \dfrac{1}{j\omega L} & -\dfrac{1}{j\omega L} \\[2ex] -\dfrac{1}{j\omega L} & \dfrac{1}{j\omega L} + j\omega C + G \end{bmatrix}^{-1}$$

$$= \cfrac{1}{\left(\dfrac{1}{R} + \dfrac{1}{j\omega L}\right)\left(\dfrac{1}{j\omega L} + j\omega C + G\right) - \left(\dfrac{1}{j\omega L}\right)^2} \begin{bmatrix} \dfrac{1}{j\omega L} + j\omega C + G & \dfrac{1}{j\omega L} \\[2ex] \dfrac{1}{j\omega L} & \dfrac{1}{R} + \dfrac{1}{j\omega L} \end{bmatrix}$$

Applying the relationship $\mathbf{V} = \mathbf{Y}^{-1}\mathbf{I}$, we have

$$
\begin{bmatrix} V_1 \\ V_2 \end{bmatrix} = \cfrac{1}{\left(\dfrac{1}{R}+\dfrac{1}{j\omega L}\right)\left(\dfrac{1}{j\omega L}+j\omega C + G\right)-\left(\dfrac{1}{j\omega L}\right)^2}\begin{bmatrix} \dfrac{1}{j\omega L}+j\omega C + G & \dfrac{1}{j\omega L} \\ \dfrac{1}{j\omega L} & \dfrac{1}{R}+\dfrac{1}{j\omega L} \end{bmatrix}\begin{bmatrix} \dfrac{V_S}{R} \\ 0 \end{bmatrix}
$$

The transfer function is easily seen to be

$$
\mathbf{H}(\omega) = \frac{V_2}{V_S} = \cfrac{\dfrac{1}{j\omega L R}}{\left(\dfrac{1}{R}+\dfrac{1}{j\omega L}\right)\left(\dfrac{1}{j\omega L}+j\omega C + G\right)-\left(\dfrac{1}{j\omega L}\right)^2}
$$

9.5.2 Augmented Nodal Analysis

In some cases the application of nodal analysis to find the values of the circuit's unknown node voltages is not sufficient. We are often interested in knowing the values of the currents that flow through the grounded voltage sources. For example, it is necessary to know the drain on the battery in a portable computer so as to predict how long the computer can be used between charges. These currents can be written as a linear combination of the unknown node voltages, but it may be more convenient to include them in the overall problem formulation. Incorporation of these variables results in an extension of nodal analysis, known as **augmented nodal analysis**.

We will illustrate this extension by reexamining the circuit of Figure 9.11, which is redrawn in Figure 9.15. In this figure, we have explicitly labeled all of the nodes (other than ground) and the source currents I_a and I_h which we wish to solve for along with the unknown node voltages.

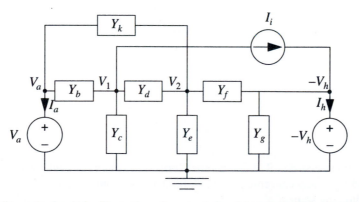

Figure 9.15 Example circuit for illustration of augmented nodal analysis. This circuit is identical to the circuit in Figure 9.11.

We proceed as before, except that we include in our voltage vector **V** all of the labeled nodes, including those with known voltages. We can apply the rules developed above to write the augmented nodal admittance equation. For example, in the row containing the V_a, the sum of admittances connected to V_a, $Y_b + Y_k$, should appear in the main diagonal, and so on. Application of this procedure results in the matrix equation

$$\begin{bmatrix} Y_b + Y_c + Y_d & -Y_d & -Y_b & 0 \\ -Y_d & Y_d + Y_e + Y_f + Y_k & -Y_k & -Y_f \\ -Y_b & -Y_k & Y_b + Y_k & 0 \\ 0 & -Y_f & 0 & Y_f + Y_g \end{bmatrix} \begin{bmatrix} V_1 \\ V_2 \\ V_a \\ -V_h \end{bmatrix} = \begin{bmatrix} -I_i \\ 0 \\ -I_a \\ I_i - I_h \end{bmatrix}$$

Unlike the matrix equations we have studied, this augmented nodal admittance form has some of the unknowns (V_1 and V_2) in the voltage vector, and other unknowns (I_a and I_h) in the current vector. We can deal with this mixture of variables by partitioning the nodal admittance matrix. We can think of this 4×4 nodal admittance matrix as a 2×2 matrix, where each element is itself a 2×2 matrix:

$$\left[\begin{array}{cc|cc} Y_b + Y_c + Y_d & -Y_d & -Y_b & 0 \\ -Y_d & Y_d + Y_e + Y_f + Y_k & -Y_k & -Y_f \\ \hline -Y_b & -Y_k & Y_b + Y_k & 0 \\ 0 & -Y_f & 0 & Y_f + Y_g \end{array} \right] = \begin{bmatrix} \mathbf{Y}_{11} & \mathbf{Y}_{12} \\ \mathbf{Y}_{21} & \mathbf{Y}_{22} \end{bmatrix}$$

Note that the upper left matrix, $\mathbf{Y}_{11}$, is the same matrix as in Eq. (9.10). We can write the augmented nodal admittance equations as

$$\begin{bmatrix} \mathbf{Y}_{11} & \mathbf{Y}_{12} \\ \mathbf{Y}_{21} & \mathbf{Y}_{22} \end{bmatrix} \begin{bmatrix} V_1 \\ V_2 \\ V_a \\ -V_h \end{bmatrix} = \begin{bmatrix} -I_i \\ 0 \\ -I_a \\ I_i - I_h \end{bmatrix} \tag{9.13}$$

The top row of this equation is

$$\mathbf{Y}_{11} \begin{bmatrix} V_1 \\ V_2 \end{bmatrix} + \mathbf{Y}_{12} \begin{bmatrix} V_a \\ -V_h \end{bmatrix} = \begin{bmatrix} -I_i \\ 0 \end{bmatrix}$$

which is just Eq. (9.10) in another form. This can be solved for the unknown voltages V_1 and V_2 by any convenient method. Formally, we can solve for the unknown voltages by the following steps:

$$\mathbf{Y}_{11} \begin{bmatrix} V_1 \\ V_2 \end{bmatrix} = \begin{bmatrix} -I_i \\ 0 \end{bmatrix} - \mathbf{Y}_{12} \begin{bmatrix} V_a \\ -V_h \end{bmatrix}$$

$$\begin{bmatrix} V_1 \\ V_2 \end{bmatrix} = \mathbf{Y}_{11}^{-1}\left(\begin{bmatrix} -I_i \\ 0 \end{bmatrix} - \mathbf{Y}_{12} \begin{bmatrix} V_a \\ -V_h \end{bmatrix} \right) \tag{9.14}$$

The second row of Eq. (9.13) is

$$\mathbf{Y}_{21} \begin{bmatrix} V_1 \\ V_2 \end{bmatrix} + \mathbf{Y}_{22} \begin{bmatrix} V_a \\ -V_h \end{bmatrix} = \begin{bmatrix} -I_a \\ I_i - I_h \end{bmatrix}$$

Solving for the unknown source currents I_a and I_h,

$$\begin{bmatrix} -I_a \\ I_i - I_h \end{bmatrix} = \begin{bmatrix} 0 \\ I_i \end{bmatrix} - \begin{bmatrix} I_a \\ I_h \end{bmatrix} = \mathbf{Y}_{21} \begin{bmatrix} V_1 \\ V_2 \end{bmatrix} + \mathbf{Y}_{22} \begin{bmatrix} V_a \\ -V_h \end{bmatrix}$$

$$\begin{bmatrix} I_a \\ I_h \end{bmatrix} = \begin{bmatrix} 0 \\ I_i \end{bmatrix} - \mathbf{Y}_{21} \begin{bmatrix} V_1 \\ V_2 \end{bmatrix} - \mathbf{Y}_{22} \begin{bmatrix} V_a \\ -V_h \end{bmatrix}$$

Substituting Eq. (9.14) for the "unknown" voltage vector yields an explicit expression for the source currents:

$$\begin{bmatrix} I_a \\ I_h \end{bmatrix} = \begin{bmatrix} 0 \\ I_i \end{bmatrix} - \mathbf{Y}_{21} \mathbf{Y}_{11}^{-1}\left(\begin{bmatrix} -I_i \\ 0 \end{bmatrix} - \mathbf{Y}_{12} \begin{bmatrix} V_a \\ -V_h \end{bmatrix} \right) - \mathbf{Y}_{22} \begin{bmatrix} V_a \\ -V_h \end{bmatrix}$$

9.5.3 Modified Nodal Analysis

The nodal analysis technique we have described is easily applied to circuits where the voltage sources are grounded. Circuits with **floating voltage sources** (i.e., sources which do not at either end connect to the circuit ground) require a modification to the nodal analysis method.

Consider the connection of elements shown in Figure 9.16, which could be part of a larger circuit. Assume that the unknown nodes are V_m and V_n, while the voltages V_a, V_b, and V_c and the current I_4 are known. As was the case with augmented nodal analysis, we include the (unknown) current I which flows through source V. Applying KCL at nodes V_m and V_n results in the following equations:

$$Y_3(V_m - V_c) + Y_5(V_m - V_n) + I - I_4 = 0 \tag{9.15}$$

$$Y_1(V_n - V_a) + Y_2(V_n - V_b) + Y_5(V_n - V_m) - I = 0 \tag{9.16}$$

These KCL equations result in the following entries in the rows corresponding to the V_m, V_n and I unknowns of the nodal admittance matrix equation:

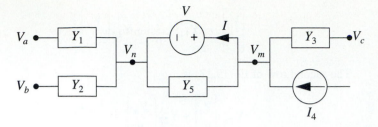

Figure 9.16 Part of a circuit which contains a floating voltage source. Neither end of the source V is connected to the ground node.

$$\begin{bmatrix} (Y_5 + Y_3) & -Y_5 & 1 \\ -Y_5 & (Y_1 + Y_2 + Y_5) & -1 \end{bmatrix} \begin{bmatrix} V_m \\ V_n \\ I \end{bmatrix} = \begin{bmatrix} I_4 + Y_3 V_c \\ Y_1 V_a + Y_2 V_b \end{bmatrix}$$

These two rows represent the two equations arising from applying KCL at the two nodes, but there are three unknowns: V_m, V_n, and I. The needed extra equation to complete the algebraic system is provided by applying KVL between the two unknown nodes, where the floating source is located:

$$V_m - V_n = V \tag{9.17}$$

Incorporating this equation in the matrix formulation results in the **modified nodal equations**:

$$\begin{bmatrix} Y_5 + Y_3 & -Y_5 & 1 \\ -Y_5 & Y_1 + Y_2 + Y_5 & -1 \\ 1 & -1 & 0 \end{bmatrix} \begin{bmatrix} V_m \\ V_n \\ I \end{bmatrix} = \begin{bmatrix} I_4 + Y_3 V_c \\ Y_1 V_a + Y_2 V_b \\ V \end{bmatrix} \tag{9.18}$$

which can be solved directly.

Some simplification to the system of equation (9.18) is possible if we are willing to dispense with finding the current I through the floating voltage source. Instead of the two rows which represent Eqs. (9.15) and (9.16), which arose from the application of KCL at nodes V_m, and V_n, we use a row which represents the sum of these two equations. Adding together Eqs. (9.15) and (9.16) eliminates the current I and yields

$$Y_3(V_m - V_c) + Y_1(V_n - V_a) + Y_2(V_n - V_b) - I_4 = 0 \tag{9.19}$$

which, when included with the KVL expression Eq. (9.17) results in the matrix formulation

$$\begin{bmatrix} Y_3 & (Y_1 + Y_2) \\ 1 & -1 \end{bmatrix} \begin{bmatrix} V_m \\ V_n \end{bmatrix} = \begin{bmatrix} I_4 + Y_1 V_a + Y_2 V_b + Y_3 V_c \\ V \end{bmatrix} \tag{9.20}$$

Notice that this system of equations does not contain the current I through the floating voltage source or, for that matter, the admittance Y_5 connected in parallel with the floating source.

Although it appears Eq. (9.19) was obtained by trickery, it actually has a fundamental circuit interpretation. Equation (9.19) is an example of a **supernode equation**, obtained by applying KCL to a region (supernode) of a circuit rather than a single node. KCL applies equally well to a region of a circuit, since the principle of conservation of charge applies both globally and locally. Considering the floating voltage source and its parallel admittance Y_5 to be a supernode, Figure 9.17, we see that there are four currents entering the supernode which must sum to zero:

$$Y_1(V_a - V_n) + Y_2(V_b - V_n) + Y_3(V_c - V_m) + I_4 = 0$$

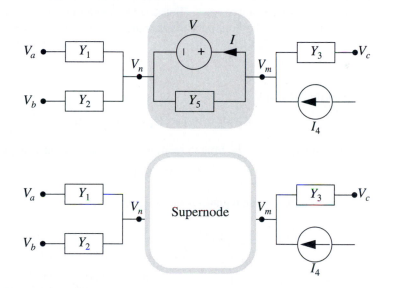

Figure 9.17 *Top:* Circuit of Figure 9.16 with supernode defined as the shaded region. *Bottom:* Application of KCL to the supernode results in Eq. (9.19), but with loss of information about the currents inside the supernode.

This equation is identical to Eq. (9.19), which was obtained by applying KCL individually to two different nodes and adding the resulting expressions. By treating the shaded region as a supernode, fewer equations result, which is a consequence of losing the information (i.e., the source current I) contained inside the region.

Example 9.5

Apply the concepts of modified nodal analysis to the circuit of Figure 9.18 to formulate a complete modified nodal admittance matrix system which describes the unknown node voltages of the circuit.

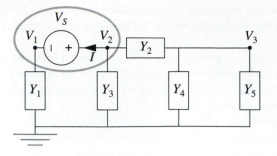

Figure 9.18 Example circuit with a floating voltage source.

There are three nonground node voltages in this circuit, therefore the system of equations will have three rows. (Because the problem statement does not ask for the current I through the floating voltage source, we will not include it in the formulation.) Writing a KCL expression at the supernode indicated by the shaded contour,

$$Y_1 V_1 + Y_3 V_2 + Y_2 (V_2 - V_3) = 0$$

Applying KCL at V_3:

$$Y_2 (V_3 - V_2) + (Y_4 + Y_5) V_3 = 0$$

The third equation is the branch relation across the floating voltage source:

$$V_2 - V_1 = V_S$$

Collecting these three equations into matrix form:

$$\begin{bmatrix} Y_1 & (Y_2 + Y_3) & (-Y)_2 \\ 0 & -Y_2 & (Y_2 + Y_4 + Y_5) \\ -1 & 1 & 0 \end{bmatrix} \begin{bmatrix} V_1 \\ V_2 \\ V_3 \end{bmatrix} = \begin{bmatrix} 0 \\ 0 \\ V_S \end{bmatrix}$$

9.5.4 Modified Nodal Analysis with Inductance

Nodal analysis is especially valuable in automating the process of finding the response of a circuit. By representing the circuit symbolically, the matrix equations can be generated in a systematic fashion. Solution of the resulting equations draws on the vast field of numerical analysis; thus circuits with hundreds or thousands of nodes can easily be solved with a computer.

A useful capability is to interpret the circuit response as a function of frequency. For example, in Section 9.5.1 we used nodal analysis to generate the transfer function of an *RLGC* circuit, reproduced in Figure 9.19. The nodal admittance formulation of this circuit is given by Eq. (9.12):

$$\begin{bmatrix} \left(\dfrac{1}{R} + \dfrac{1}{j\omega L}\right) & -\dfrac{1}{j\omega L} \\[2ex] -\dfrac{1}{j\omega L} & \left(\dfrac{1}{j\omega L} + j\omega C + G\right) \end{bmatrix} \begin{bmatrix} V_1 \\[1ex] V_2 \end{bmatrix} = \begin{bmatrix} \dfrac{V_S}{R} \\[1ex] 0 \end{bmatrix} \qquad (9.21)$$

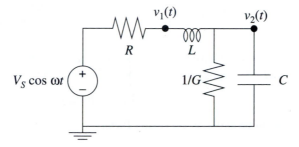

Figure 9.19 *RLGC* circuit. The admittance of the inductor is infinite at dc, which presents computational difficulties.

The voltage phasors V_1 and V_2 can be found for any nonzero frequency by a number of algebraic or numerical techniques.

Suppose, however, we want to determine the dc response of this circuit. Its is straightforward to find the dc response analytically; we simply replace inductors and capacitors with their dc equivalents, short circuits and open circuits. Alternatively, we can use the nodal admittance formulation and set the frequency $\omega = 0$. For circuits too large for hand analysis, this latter approach is necessary. However, the admittance of an inductor at dc is infinite, which leads to numerical difficulties when the system of equations is solved on a computer. For example, in Eq. (9.21), each entry in the nodal admittance matrix has a $1/j\omega L$ term. If we instruct the computer to evaluate the circuit at dc, it will result in divide-by-zero errors. Furthermore, this problem will occur in any circuit containing inductors.

A way to work around this difficulty is to write the equations so that the impedance, rather than the admittance, of the inductor is included. In Figure 9.20 we have represented the *RLGC* circuit in the frequency domain by using admittances for resistors and capacitors, and impedance for the inductor. Unlike the strict nodal admittance procedure, where element currents are specified whenever possible as admittances multiplied by a voltage difference, we will notate inductor currents with their own variables. Application of KCL at node V_1 is written as

$$\frac{1}{R}(V_1 - V_S) - I_L = 0$$

and at node V_2

$$I_L + (G + j\omega C)V_2 = 0$$

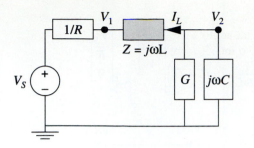

Figure 9.20 Modified frequency-domain representation of the *RLGC* circuit. Inductors are represented with impedances, while resistors and capacitors are specified by their admittances.

In this formulation, we have three unknowns, the node voltages V_1 and V_2, and the inductor current I_L. The third equation is the branch relation of the inductor, which we express as

$$(V_2 - V_1) = j\omega L I_L$$

Writing these three equations in matrix form,

$$
\begin{bmatrix}
\dfrac{1}{R} & 0 & -1 \\
0 & (G + j\omega C) & 1 \\
-1 & 1 & -j\omega L
\end{bmatrix}
\begin{bmatrix}
V_1 \\
V_2 \\
I_L
\end{bmatrix}
=
\begin{bmatrix}
\dfrac{V_S}{R} \\
0 \\
0
\end{bmatrix}
\tag{9.22}
$$

This representation does not have ω in the denominator of any term, and thus avoids the division by zero problem when calculating the dc response. Equation (9.22) is valid and useful at all frequencies.

The analogous situation with capacitors at high frequencies, where the admittance becomes large, is not a problem because the frequency (and capacitive admittance) are always represented as finite numbers in a computer.

As a final example which illustrates several of these principles, consider the circuit in Figure 9.21. We wish to construct the nodal admittance formulation describing the voltage at the four unknown nodes v_1 through v_4.

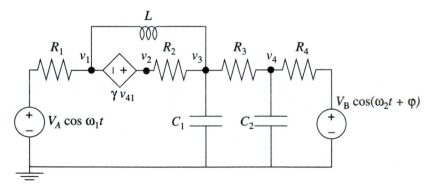

Figure 9.21 Multifrequency circuit containing an inductance and a floating controlled voltage source.

Redrawing the circuit in the frequency domain results in Figure 9.22. In this diagram, the open boxes are labeled with admittances, while the shaded box is labeled with impedance. We see that there are two independent voltage sources operating at different frequencies, so we will need to use superposition to find the total response.

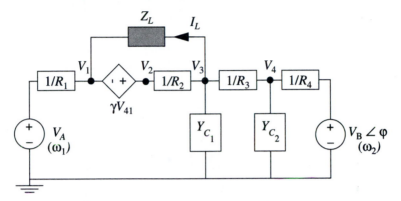

Figure 9.22 Frequency-domain representation of the circuit in Figure 9.21. Open boxes are admittances; the shaded box is labeled with the corresponding impedance.

The circuit is redrawn in Figure 9.23, first with the independent source on the right (operating at ω_2) replaced by its dead equivalent, and next with the other source (operating at ω_1) replaced by a short circuit. The variables in the top diagram have been superscripted with a single prime (i.e., V_1') to indicate they are the response resulting from the excitation at ω_1. Likewise, the variables in the bottom diagram are shown as V_1'' etc., where the double prime denotes a response at frequency ω_2.

We will first determine the response at ω_1. Applying KCL at the supernode formed by V_1', V_2', and the controlled voltage source:

$$\frac{1}{R_1}(V_1' - V_A) - I_L' + \frac{1}{R_2}(V_2' - V_3') = 0$$

KCL at V_3':

$$\frac{1}{R_2}(V_3' - V_2') + I_L' + j\omega_1 C_1 V_3' + \frac{1}{R_3}(V_3' - V_4') = 0$$

KCL at V_4':

$$\frac{1}{R_3}(V_4' - V_3') + j\omega_1 C_2 V_4' + \frac{1}{R_4}V_4' = 0$$

The branch relation of the controlled voltage source is

$$V_2' - V_1' = \gamma(V_4' - V_1')$$

The branch relation for the inductor is

$$j\omega_1 L I_L' = V_3' - V_1'$$

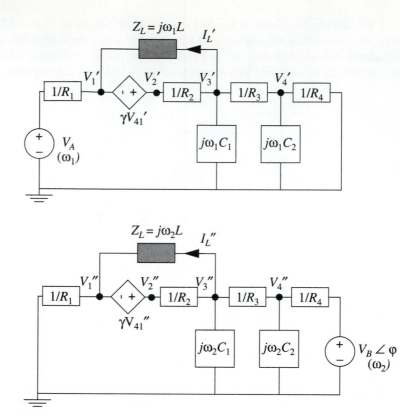

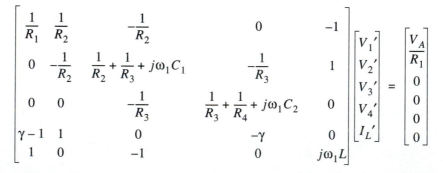

Figure 9.23 *Top:* Circuit with independent source at frequency ω_2 dead. *Bottom:* Circuit with independent source at frequency ω_1 dead.

Combining these five equations into a matrix representation,

$$
\begin{bmatrix}
\dfrac{1}{R_1}\dfrac{1}{R_2} & -\dfrac{1}{R_2} & 0 & -1 \\[2ex]
0 & -\dfrac{1}{R_2} & \dfrac{1}{R_2}+\dfrac{1}{R_3}+j\omega_1 C_1 & -\dfrac{1}{R_3} & 1 \\[2ex]
0 & 0 & -\dfrac{1}{R_3} & \dfrac{1}{R_3}+\dfrac{1}{R_4}+j\omega_1 C_2 & 0 \\[2ex]
\gamma-1 & 1 & 0 & -\gamma & 0 \\[1ex]
1 & 0 & -1 & 0 & j\omega_1 L
\end{bmatrix}
\begin{bmatrix}
V_1' \\[1ex] V_2' \\[1ex] V_3' \\[1ex] V_4' \\[1ex] I_L'
\end{bmatrix}
=
\begin{bmatrix}
\dfrac{V_A}{R_1} \\[2ex] 0 \\[1ex] 0 \\[1ex] 0 \\[1ex] 0
\end{bmatrix}
$$

We next determine the response arising from the voltage source at frequency ω_2. Applying KCL at the supernode formed by V_1'', V_2'', and the controlled voltage source:

$$
\frac{1}{R_1}V_1'' - I_L'' + \frac{1}{R_2}(V_2'' - V_3'') = 0
$$

KCL at V_3'':

$$\frac{1}{R_2}(V_3'' - V_2'') + I_L'' + j\omega_2 C_1 V_3'' + \frac{1}{R_3}(V_3'' - V_4'') = 0$$

KCL at V_4'':

$$\frac{1}{R_3}(V_4'' - V_3'') + j\omega_2 C_2 V_4'' + \frac{1}{R_4}(V_4'' - (V_B \angle \varphi)) = 0$$

The branch relation of the controlled voltage source is

$$V_2'' - V_1'' = \gamma(V_4'' - V_1'')$$

The branch relation for the inductor is

$$j\omega_2 L I_L'' = V_3'' - V_1''$$

Combining these five equations into a matrix representation,

$$
\begin{bmatrix}
\dfrac{1}{R_1} \dfrac{1}{R_2} & -\dfrac{1}{R_2} & 0 & -1 \\[2ex]
0 \;\; -\dfrac{1}{R_2} & \dfrac{1}{R_2} + \dfrac{1}{R_3} + j\omega_2 C_1 & -\dfrac{1}{R_3} & 1 \\[2ex]
0 \quad 0 & -\dfrac{1}{R_3} & \dfrac{1}{R_3} + \dfrac{1}{R_4} + j\omega_2 C_2 & 0 \\[2ex]
\gamma - 1 \quad 1 & 0 & -\gamma & 0 \\[1ex]
1 \quad 0 & -1 & 0 & j\omega_2 L
\end{bmatrix}
\begin{bmatrix}
V_1'' \\ V_2'' \\ V_3'' \\ V_4'' \\ I_L''
\end{bmatrix}
=
\begin{bmatrix}
0 \\ 0 \\ \dfrac{V_B \angle \varphi}{R_4} \\ 0 \\ 0
\end{bmatrix}
$$

These two matrix equations can be solved for the variables V_1', ..., I_L'' and the results superposed. For example, the phasors corresponding to node voltage $v_3(t)$ will have magnitude and phase

$$V_3' = |V_3'| \angle \theta_3'$$
$$V_3'' = |V_3''| \angle \theta_3''$$

The time-domain representation of the voltage at this node is

$$v_3(t) = |V_3'| \cos(\omega_1 t + \theta_3') + |V_3''| \cos(\omega_2 t + \theta_3'')$$

and similarly for the other voltages (and inductor current).

9.6 Summary

The terminal characteristics of a circuit can be expressed by a series combination of a voltage source (value V_{oc}) and a resistance R_T, known as a **Thevenin equivalent** circuit. Alternately, the same terminal characteristics are given by a parallel connection of a current source (value I_{sc}) and a resistance R_N, which is called a **Norton equivalent** circuit. The relationship between these parameters are

$$R_T = R_N = \frac{V_{oc}}{I_{sc}}$$

The voltage V_{oc} is the open-circuit voltage at the two terminals of interest, while the current I_{sc} is the current flowing through a short circuit connecting the two terminals. The equivalent resistance is determined by replacing all independent voltage and current sources in the circuit with their dead equivalents.

The **root-mean-square** or **rms** value of a voltage $v(t)$ is defined as

$$V_{rms} = \sqrt{\frac{1}{T}\int_0^T v^2(t)\,dt}$$

An ac source with value V_{rms} will dissipate the same power in a resistive load as a dc source of identical value. Ac voltmeters are calibrated to read rms values. The rms value of a sinusoid with peak amplitude V is $V/\sqrt{2}$.

A **controlled source** is a voltage or current source whose value depends on a parameter other than time. Often this parameter is a voltage or current elsewhere in the circuit. In analyzing circuits with controlled sources, we do not replace them with dead equivalents when applying superposition, or when finding the Thevenin/Norton equivalent circuit.

Nodal analysis is a systematic method of finding a symbolic representation of the voltages and currents in a circuit. Application of KCL at the unknown nodes (i.e., those whose value is to be determined) of a circuit results in a system of linearly independent nodal equations, which can be expressed in matrix form as

$$\mathbf{YV = I}$$

In this equation, $\mathbf{Y}$ is a matrix called the **nodal admittance matrix**, $\mathbf{V}$ is a vector of unknown node voltages, and $\mathbf{I}$ is a vector of known current relations. More generally, the vector $\mathbf{V}$ can also include branch currents flowing through voltage sources; this results in additional equations and variables and is known as **augmented nodal analysis**. Floating voltage sources are easily handled by applying KCL at a **supernode**, which is a region of a circuit enclosed by a continuous curve. Alternatively, the branch currents through the floating sources can be included in the formulation. Either of these **modified nodal analysis** methods will result in a system which can be solved for the desired unknowns. Potential divide-by-zero problems, which can occur in calculating the dc response of circuits containing inductors, can be avoided by using an impedance formulation of the inductor current, rather than the usual admittance expression.

Problems

9.1   Find the Norton equivalent parameters for the circuit below.

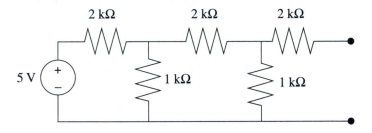

9.2   Determine the Thevenin equivalent parameters for the circuit below.

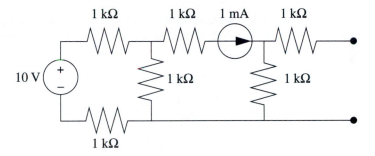

9.3   The switch in the circuit below closes at $t = 0$. Determine $v_C(t)$ for $t > 0$, assuming $v_C(0) = 0$ V.

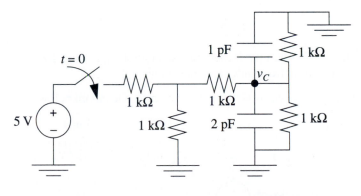

9.4   The switch in the circuit below closes at $t = 0$. Determine $v_C(t)$ for $t > 0$, assuming that the
switch has been open for a long time.

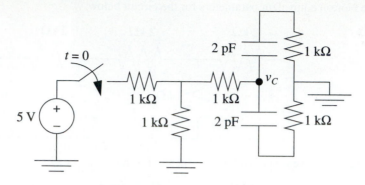

9.5   Show that an ideal independent voltage source cannot be represented as a Norton equivalent
circuit; similarly, an ideal independent current source cannot be represented as a Thevenin
equivalent.

9.6   Determine the Thevenin equivalent representation for the circuit below.

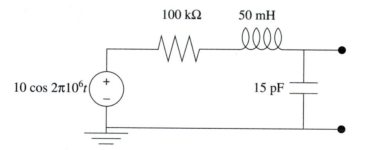

9.7   Recalculate the Thevenin parameters in Exercise 9.6, assuming a source frequency equal to
the undamped natural frequency ω_0 of the circuit.

9.8   Determine the Norton equivalent representation for the circuit below.

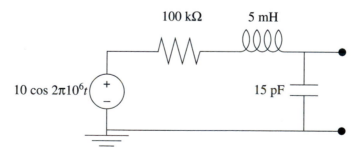

9.9   Determine the Thevenin equivalent representation for the circuit below.

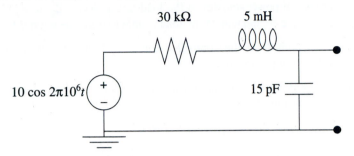

9.10 Determine the Thevenin equivalent representation for the circuit below.

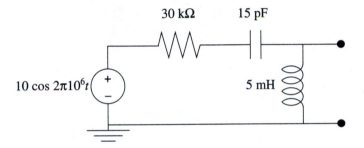

9.11 Determine the Thevenin equivalent representation for the circuit below.

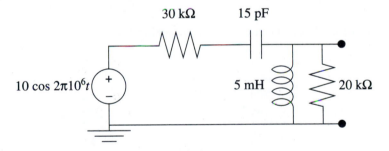

9.12 Consider an ac steady-state circuit, with a Thevenin equivalent representation as shown below:

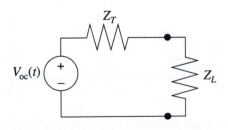

In this circuit, $v_{oc}(t)$ is a sinusoidal voltage and Z_T and Z_L represent complex impedances. Show that maximum power transfer to the load occurs when $Z_T = Z_L^*$. (*Hint:* Remember that only the resistive component of the load is able to dissipate energy.)

9.13 Show that if the polarity of the controlled source in Figure 9.9 is reversed, the equivalent resistance of the circuit is negative.

9.14 The circuit below is a model for the biasing network of a bipolar junction transistor. Determine I_B and V_{CE}, assuming $\beta = 100$.

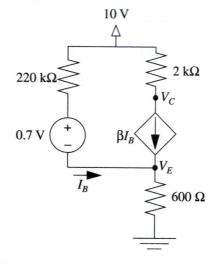

9.15 Determine I_B and V_{CE} in the circuit below, assuming $\beta = 100$.

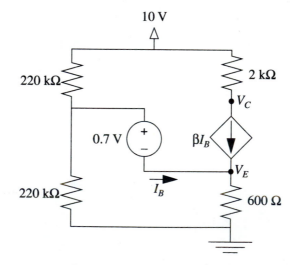

9.16 The circuit below is a simplified model of an amplifier incorporating a bipolar junction transistor. The parameter h_{fe} is known as the small-signal current gain and typically has a value between 80 and 200. v_{in} and v_o are ac input and output voltages. Derive an expression for the small-signal voltage gain, defined by $A_V = v_o/v_{in}$.

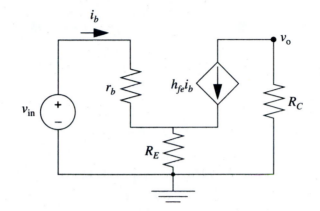

9.17 In Exercise 9.16, show that the small-signal voltage gain A_V is nearly independent of h_{fe} for sufficiently large values of R_E. What determines the gain in this case?

9.18 In Exercise 9.16, show that the magnitude of the small-signal voltage gain $|A_V|$ can be increased by placing a capacitor in parallel with R_E.

9.19 The circuit below, containing two identical transistors, is known as a **differential amplifier**. Using the model for a bipolar junction transistor, Figure 9.8, determine the relationship between the output v_o and the two input signals v_{in1} and v_{in2}.

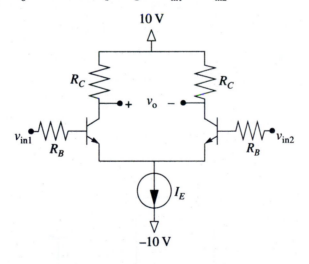

9.20 An **ideal transformer** is governed by the relations $v_2 = Nv_1$ and $i_2 = i_1/N$, where N is the
turns ratio and the voltages and currents (which must be ac) are defined as shown:

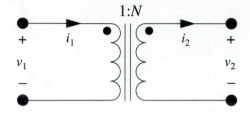

Two ideal transformers, each with $N = 2$, are used in the circuit below. Determine the rms
value of v_0, assuming $V_m = 10$ V.

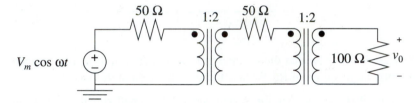

9.21 (a) Determine the Norton equivalent for the circuit below.

(b) Suppose a 100-Ω resistor is connected across the two terminals. Determine the power dissipated by the 200-Ω resistor.

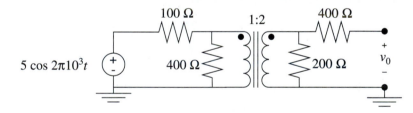

9.22 Determine the rms values of the periodic square-wave voltage signal below.

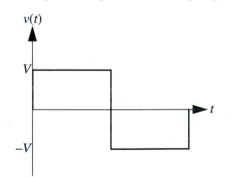

9.23 Determine the rms values of the periodic asymmetric square-wave voltage signal below.

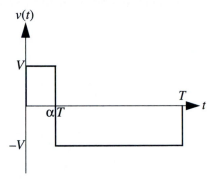

9.24 Determine the rms values of the periodic sawtooth-wave voltage signal below.

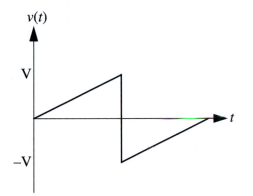

9.25 Determine the rms values of the periodic triangle-wave voltage signal below.

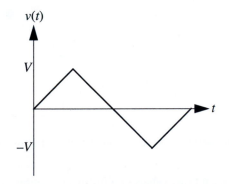

9.26 Consider the periodic triangle-wave voltage signal below, which has a nonzero average value of V_{DC} and a peak-to-peak amplitude of 2 V.

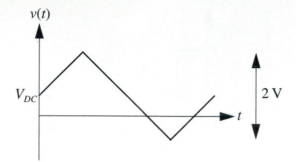

Use the definition of rms value, Eq. (9.5), to determine the effective value of this waveform.

9.27 Show that the rms value derived in Exercise 9.26 can alternatively be obtained by using the effective value of the waveform with a zero dc value (from Exercise 9.25) in conjunction with Eq. (9.7).

9.28 Determine the rms voltage of the series combination below, for the following pairs of source frequencies and phases:

(a) $f_1 = 10^6$ Hz, $\phi_1 = 0$ rad, $f_2 = 10^3$ Hz, $\phi_2 = 0$ rad

(b) $f_1 = 10^6$ Hz, $\phi_1 = 0$ rad, $f_2 = 0$ Hz, $\phi_2 = 0$ rad

(c) $f_1 = 10^6$ Hz, $\phi_1 = 0$ rad, $f_2 = 10^6$ Hz, $\phi_2 = 0$ rad

(d) $f_1 = 10^6$ Hz, $\phi_1 = 0$ rad, $f_2 = 10^6$ Hz, $\phi_2 = \pi$ rad

(e) $f_1 = 10^6$ Hz, $\phi_1 = 0$ rad, $f_2 = 10^6$ Hz, $\phi_2 = \pi/2$ rad

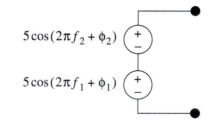

9.29 Determine the Fourier series representation for $v(t) = 10 \sin^5 (\omega t)$. What is the rms value of this voltage?

9.30 Determine the average and rms values of a full-wave rectified sine wave, $v(t) = |V \sin (\omega t)|$.

9.31 A periodic signal is expressed using a complex exponential Fourier series:

$$v(t) = \sum_{n=-\infty}^{\infty} f_n e^{jn\omega t}$$

where the coefficients are given by

$$f_n = \frac{1}{T}\int_0^T v(t)e^{-jn\omega t}\,dt$$

Derive an expression, analogous to Eq. (9.7), which gives the rms value of the signal in terms of the complex Fourier coefficients $f_0, f_{\pm 1}, f_{\pm 2}, \ldots, f_{\pm n}$.

9.32 Modern digital voltmeters are known as *true-rms-responding* instruments, meaning that they respond to and read the actual rms voltage value, regardless of the waveshape. However, older instruments generally respond to the average of a rectified version of an ac voltage signal and then multiply that average value by a constant to display the rms value. These *average-responding* instruments will give correct readings only for sinusoidal input signals.

Suppose an instrument is constructed such that the following steps are taken in reading the voltage:

- The signal $v(t)$ is passed through a single ideal diode, to produce a half-wave rectified version of the input.

- The resulting signal is averaged over one period; call this result X.

- The value of X is multiplied by an internal calibration factor, C, which is selected such that $X*C$ is the rms value of the original input signal, IF that signal is a pure sinusoid with no dc component.

- $X*C$ is displayed on the instrument readout.

(a) Determine the value of C.

(b) Suppose you need to determine the rms voltage of the signal below. What is the true rms value, and what value does the average-responding instrument display?

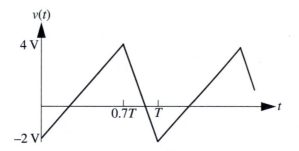

(c) Repeat part (b) for a symmetric square wave at 60 Hz which alternates between -3 V and 9 V.

(d) What will the instrument display if the input signal is a 16 V dc?

9.33 In the ac steady-state RC circuit below, determine the following:

(a) $v_C(t)$, $v_R(t)$, and $i(t)$

(b) the phase difference, in radians, between $v_C(t)$ and $v_R(t)$

(c) the rms values of v_C, v_R, and i

(d) the energy dissipated per second in the resistor

(e) the energy supplied per second by the dc source

(f) the energy supplied per second by the ac source

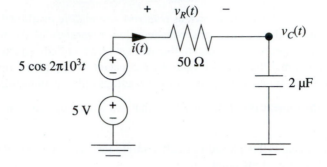

9.34 Write the nodal admittance equation $\mathbf{YV} = \mathbf{I}$ for the circuit below.

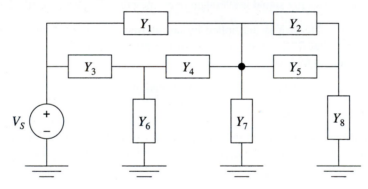

9.35 Write the nodal admittance equation $\mathbf{YV} = \mathbf{I}$ for the circuit below.

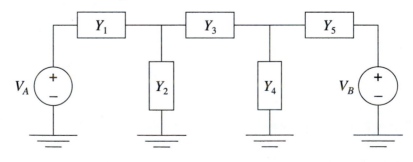

9.36 Write the nodal admittance equation **YV = I** for the circuit below.

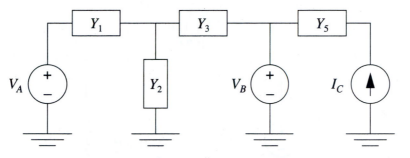

9.37 Write the nodal admittance equation **YV = I** for the circuit below.

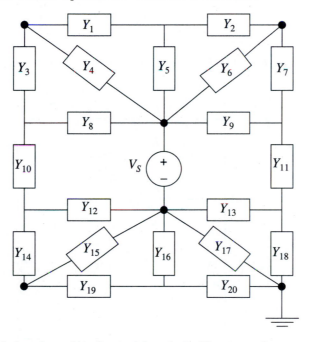

9.38 Solve the circuit below for $v_C(t)$ using nodal analysis. The source frequency is 455 kHz.

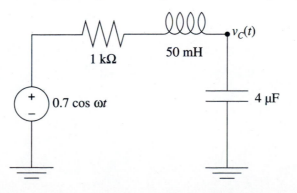

9.39 Solve the circuit below for $v_L(t)$ using nodal analysis. The source frequency is 45 kHz.

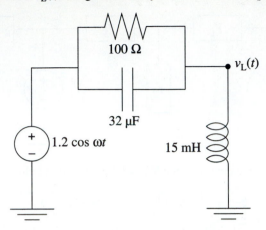

9.40 In the circuit below, $R = 50\ \Omega$, $L = 12$ mH, and $C = 10\ \mu$F. Determine all of the node voltages.

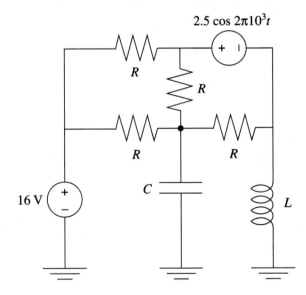

Selected Answers

Chapter 1

1.3 2 kΩ, 40 nF, 48 μs

1.5 70 nF

1.9b 3.16 mA

1.13 3.4 mJ

1.24a 3.05 kΩ

1.26   absorbing

1.28 4.6 kΩ

1.34 $715 \cos 2\pi 10^3$ mA

Chapter 2

2.12 $8e^{-\frac{t}{12\times10^{-6}}} - 3$ V

2.15 69.4 ns

2.16d $v_C(t) = 6e^{-\frac{t}{30\times10^{-9}}}$ V, $i_C(t) = -20e^{-\frac{t}{30\times10^{-9}}}$ mA

2.20b 3.22 ns

2.22 867 mV

2.28c 12.5τ

2.31c 0 A

2.32c 12 C

Chapter 3

3.3 28 Ω, 3.37 pF

3.18 2.5 V

3.21 $v_1(t) = -e^{-250 \times 10^6 t} - 5e^{-1 \times 10^6 t} + 3$ V,

   $v_2(t) = -2e^{-250 \times 10^6 t} + 5e^{-1 \times 10^6 t} + 3$ V

3.26   -3.7%

Chapter 4

4.2   12.1 MHz

4.6   4.14 MHz

4.12 d) N = 6

4.20   118 ns

4.26   117 ns, -264 mV

4.30

$$
\begin{bmatrix}
\dfrac{C_1 + C_C}{2} & 0 & -\dfrac{C_C}{2} & 0 \\[2ex]
0 & C_{in} + \dfrac{C_1 + C_C}{2} & 0 & -\dfrac{C_C}{2} \\[2ex]
-\dfrac{C_C}{2} & 0 & \dfrac{C_2 + C_C}{2} & 0 \\[2ex]
0 & -\dfrac{C_C}{2} & 0 & \dfrac{C_2 + C_C}{2}
\end{bmatrix}
\begin{bmatrix}
\dfrac{dv_1}{dt} \\[2ex]
\dfrac{dv_2}{dt} \\[2ex]
\dfrac{dv_3}{dt} \\[2ex]
\dfrac{dv_4}{dt}
\end{bmatrix}
$$

$$
= \left(
\begin{bmatrix}
-\left(\dfrac{2}{2R_{out} + R_1} + \dfrac{2}{R_1}\right) & \dfrac{2}{R_1} & 0 & 0 \\[2ex]
\dfrac{2}{R_1} & -\dfrac{2}{R_1} & 0 & 0 \\[2ex]
0 & 0 & -\dfrac{2}{R_2} & \dfrac{2}{R_2} \\[2ex]
0 & 0 & \dfrac{2}{R_2} & -\dfrac{2}{R_2}
\end{bmatrix}
\begin{bmatrix}
v_1 \\ v_2 \\ v_3 \\ v_4
\end{bmatrix}
+
\begin{bmatrix}
\dfrac{2V_S}{2R_{out} + R_1} \\[2ex] 0 \\[1ex] 0 \\[1ex] 0
\end{bmatrix}
\right)
$$

4.38

$$
\begin{bmatrix}
C_1 + C_3 & -C_1 & -C_3 \\
-C_1 & C_1 + C_2 & 0 \\
-C_3 & 0 & C_3 + C_4
\end{bmatrix}
\begin{bmatrix}
\dfrac{dv_0}{dt} \\[2ex]
\dfrac{dv_1}{dt} \\[2ex]
\dfrac{dv_2}{dt}
\end{bmatrix}
=
\begin{bmatrix}
-\dfrac{1}{R_1} & 0 & 0 \\[2ex]
0 & -\dfrac{1}{R_2} & \dfrac{1}{R_2} \\[2ex]
0 & \dfrac{1}{R_2} & -\dfrac{1}{R_2}
\end{bmatrix}
\begin{bmatrix}
v_0 \\ v_1 \\ v_2
\end{bmatrix}
+
\begin{bmatrix}
\dfrac{V_S}{R_1} \\[2ex] 0 \\[1ex] 0
\end{bmatrix}
$$

Chapter 5

5.1 $-15e^{-\frac{t}{417\times10^{-9}}}$ V

5.4b   668 V

5.9 $v_C(t) = e^{-4.17\times10^9 t}(-5\cos 8.12\times10^9 t - 2.56\sin 8.12\times10^9 t) + 5$ V

        $i_L(t) = e^{-4.17\times10^9 t}(0.0513\sin 8.12\times10^9 t)$ A

5.15 $v_C(t) = 112\times10^{-6}\sin 224\times10^9 t$ V

        $i_L(t) = (-5.00\times10^{-6}\cos 224\times10^9 t) + 5.00\times10^{-6}$ A

5.30   50 mA

5.32 $C = 1$ pF

Chapter 6

6.5 $v(0, t) = \begin{cases} 2.5\text{ V} & 0 < t < 2T_D \\ 0\text{ V} & t > 2T_D \end{cases}$

        $v(l, t) = 0$ V

6.12 $v(0, t) = 0.4$ V

        $v(l, t) = \begin{cases} 0\text{ V} & 0 < t < T_D \\ 0.4\text{ V} & t > T_D \end{cases}$

6.15 $v(0, t) = \begin{cases} 0.25\text{ V} & 0 < t < 2T_D \\ 0.5\text{ V} & t > 2T_D \end{cases}$

        $v(l, t) = \begin{cases} 0\text{ V} & 0 < t < T_D \\ 0.5\text{ V} & t > T_D \end{cases}$

Chapter 7

7.2   45.1 MHz

7.11 $\dfrac{V}{2} + \displaystyle\sum_{n=1}^{\infty} \dfrac{V}{n\pi}\sin n\omega t$

7.21 $\dfrac{2V}{\pi} + \displaystyle\sum_{\substack{n=2 \\ n\ \text{even}}}^{\infty} \dfrac{4V}{\pi}\dfrac{1}{1-n^2}\cos n\omega t$

7.29 $(-1)^n f_n$

7.32 $8.76\left(\cos 2\pi 60t + \cos\left(2\pi 60t + \dfrac{\pi}{4}\right)\right) + 0.377\left(\sin 2\pi 60t + \sin\left(2\pi 60t + \dfrac{\pi}{4}\right)\right)$ V

7.38a   amplitude modulation

Chapter 8

8.3 $3\angle\dfrac{\pi}{4}$, $2.13\angle 0.278$, $V\angle\phi$, $-\dfrac{V}{2}\angle\left(-\phi - \dfrac{\pi}{2}\right)$

8.15   23.8 MHz

8.41 $\dfrac{1}{1-\omega^2 LC}$

Chapter 9

9.20   2.57 V

9.24 $\dfrac{V}{\sqrt{3}}$

9.29 $V_{rms} = 4.96$ V

9.32c   6.71 V, 9.99 V

9.33d   70.8 mW

Index

A

ac steady-state 334, 339
active line 189
admittance 397, 464
amplitude 340, 383
angular frequency 241, 351, 383
approximate analysis 260
argument 386
augmented nodal analysis 469
average value 340, 341, 347, 461
average values 246

B

band
  rock 1
  rubber 427
bandpass filter 431
bandstop filter 450
bandwidth 418
battery 39, 55
binomial expansion 262
Bode plot 416
bonding wire 214
branch 14
buffer 163

C

capacitance 22
  dc model 100, 215, 247
  input 66, 171, 311
  matrix 196
capacitive coupling 179, 189
characteristic equation 225, 241,
    259, 419
characteristic impedance 284
characteristic velocity 284
charge 5, 256
circuit 12
  Norton equivalent 454
  open 29, 297
  resistive 42
  short 29, 297

  Thevenin equivalent 454
clock 83, 87, 327, 344, 346, 433
clock skew 328
  frequency-domain analysis
    410
  time-domain analysis 334
CMOS 63
coaxial cable 256
combinational logic 327
complex Fourier series 366
conductance 256, 402, 446
  gate 256
controlled sources 459
corner frequency 416
critical damping 225, 231, 245,
    272
crosstalk 180, 189, 196, 438
current 6
  continuity of inductor 48, 215,
    255
  direct 8
  short-circuit 453
current discontinuity 260
current divider 41
current source 20

D

damped natural frequency 239,
    251, 420
damping factor 233, 239, 245,
    419
dB 416
dc component 340, 347, 361
dc steady-state 76, 136, 193, 217,
    221, 228, 300
  as average values 246
dead equivalent 364, 455, 459
decade 417
decibel 416
delay line 279
delay time 284, 296, 314
dependent sources 459

determinant 141
dielectric constant 23, 180
differential equation 73, 134, 154,
    190, 223, 337, 357, 427,
    463
dimensionless damping ratio 421,
    430
DRAM (dynamic random access
    memory) 256
duals 219
duty cycle 83

E

effective voltage 461
eigenvalues 142, 225, 259
electromagnetic wave 279
energy 45
  dissipated during logic
    transition 94
  stored by capacitor 46, 97,
    245, 428
  stored by inductor 48, 245,
    428
  stored in RLC circuit 429
energy storage elements 221, 257
envelope 245
Euler's identity 240, 385
even symmetry 346
excitation 337
exponential sawtooth 333, 351,
    362

F

fanout 171
Faraday's law 26
filters 418
flip–flops 327
floating line 185
floating voltage source 471
forbidden combinations of
    elements 38
forced response 78

Fourier analysis 339
Fourier coefficients
    derivation 341
    of exponential sawtooth 356
    of square wave 345, 351, 369
Fourier series 339, 343, 403, 433
    complex form 366, 436
    rms value of 463
frequency
    corner 416
    fundamental 330, 339
frequency-domain analysis 327
    of interconnects 406
    the big picture 337
fringing fields 129, 198
fundamental frequency 330, 339

G
gate conductance 256
gate delay 61, 79
general solutions 146, 193, 216,
    224, 226, 233, 240
GLC circuits 273, 449
ground node 14
grounded line 182

H
half-power bandwidth 418
harmonics 330, 340, 405, 434
hertz 330

I
IC 1, 196, 330, 440
ideal transformer 486
identity matrix 140, 147
impedance 390
    characteristic 284
    mismatch 295, 311
    of capacitors 391
    of inductors 392
    of resistors 391
    parallel combination 396
    reactive component 400
    resistive component 400
    series combination 394
independent capacitors 154
independent inductors 215
inductance 25
    dc model 215, 247
    interconnect 213

package 214, 403
    speed-up 277
initial condition 74, 137, 151,
    153, 194, 242
integrated circuit 1, 185, 196, 327
interconnect delay 122, 156
interconnects 115
    active adjacent line 189, 439
    and fanout 175
    capacitance 116, 180
    floating adjacent line 185
    frequency-domain analysis
        406
    grounded adjacent line 182
    inductance 213
    models of 119
    multiple 175
    n-LC-lump model 280
    n-RC-lump model 119
    pi-connection model 180
    resistance 116
    scaling 129
    single RC lump model 119,
        153, 175, 220
    three-RC-lump model 159
    two-RC-lump model 130, 133,
        154, 175
inverter model 64, 66
inverter pull-down 76, 85, 92,
    124, 174, 251
inverter pull-up 78, 84, 91, 124,
    174, 254

K
KCL 19, 222, 258, 460, 463, 473
Kirchhoff's current law 19
Kirchhoff's voltage law 16
KVL 16, 215, 222, 237, 258, 359

L
LC line 279
leakage conductance 256, 259
left traveling wave 287, 296, 302
load 455
logic threshold 64
loop 17
lossless transmission line 279,
    297
lossy waveguides 256
lowpass filter 418

M
magnitude 415
matched load 296, 300, 303
matched source 300, 311
matrix determinant 141
matrix formulation
    of coupled interconnects 190
    of dc circuits 44
    of differential equations 135,
        146
    of RLGC circuit 258
    of series RLC circuit 223
matrix inverse 193, 465
maximum power transfer 458
MCM 279
model limitations 40, 47, 118,
    129, 130, 159, 199, 247,
    256, 260
modified nodal analysis 471
multi–chip modules 279

N
natural frequency 139, 142, 147,
    153, 193, 241, 419
    damped 239, 420
    dominant 155, 158
    of series RLC circuit 224
    undamped 239, 246, 420
nodal admittance matrix 465
nodal analysis 463
node 13
    common 14
noise immunity 93
nonsingular matrix 141
nontrivial solutions 141, 225
Norton equivalent 454
Norton resistance 454

O
odd symmetry 346
Ohm's law 22, 295, 391, 462
open-circuit voltage 453
overdamped response 225

P
package inductance 214, 403
pad driver 204
parallel connection 33, 171
    virtually parallel capacitors
        34, 154

partial derivatives 283
peak-to-peak amplitude 461
period 330
periodic signals 330
permeability 27, 284
permittivity 23, 180, 285
phase 340, 383, 415
phasors 366, 385
    CCW and CW rotating 436
pi-connection model 180
polar notation 389
polysilicon 126
power 10, 45
    dissipation by computer 98
    maximum power transfer 458
    of resistor 45
    of switch 28
    reference convention 11

Q

Q (see quality factor)
quality factor 430

R

RC circuits 71, 99, 171, 314, 352
reactance 400
    asymptotic behavior 402
reactive component 400
reciprocity 199
rectangular notation 389
reference directions
    associated 10, 45
    unassociated 11
reflection coefficient 296
reflections 294, 303, 307
resistance 21, 400
    Norton 454
    Thevenin 454
resistive component 400
resistivity 21, 116, 126, 128
resonance 422, 424, 434
    in mechanical systems 427
right traveling wave 287, 296, 300
ring oscillator 88
ringing 245, 435
RL circuits 215
RLC circuits 221

RLGC circuits 256, 467
rms 460
root locus 274
root-mean-square 460
rust 337

S

selectivity 432
sequential logic 327
series connection 29
series RLC circuit 221
short-circuit current 453
siemens 256, 397
signal
    symmetry 346
signals
    periodic 330
    periodic lagging 384
    rectified sinusoid 376
    rms value of sinusoidal 462
    sawtooth 461
    sinusoidal 337, 383
    triangle 461
solenoid 25, 284
source
    controlled 459
    current 20
    dead equivalent 364, 455
    dependent 459
    floating voltage 471
    voltage 19
speed-up inductor 277
square wave 83, 327, 330, 344,
    346
standard unit 3
start-up transient 334
steady state 76, 136, 217, 297
step function 47
supernode 473
superposition 339, 357, 477
susceptance 402, 446
switch 27
switching speed 83, 171, 185,
    251, 328, 413
synchronous 328, 330

T

Taylor expansion 237
Telegrapher's equations 283
Thevenin equivalent 454
Thevenin resistance 454
threshold voltage 79, 138, 158,
    196
time constant 75, 80, 124, 153,
    154, 173, 184, 217, 335,
    359, 457
time-domain analysis 327, 358
toast 8, 28, 93, 98, 245, 264
total response 75, 136, 153
transfer function 414, 420, 434
transformer 486
transient response 75
transmission line 256, 279
    capacitively loaded 311
    dc model 297
    load end model 299
    reflections 294
    source end model 298
traveling waves 287, 294

U

undamped natural frequency 239,
    246, 420, 428
undamped response 226, 245
underdamped response 226, 239,
    424
unmatched load 296, 307
unmatched source 303, 307, 316

V

vacuum tube 1
virtually parallel capacitors 34,
    154
voltage 8
    branch 15
    continuity of capacitor 46, 137
    drop 17
    node 15
    open-circuit 453
    rise 17
voltage divider 40
voltage source 19
voltmeter 461